T0145220

Lecture Notes in Networks and Systems 1026

Series Editor

The series "Lecture Notes in Networks and Systems" publishes the latest developments in Networks and Systems—quickly, informally and with high quality. Original research reported in proceedings and post-proceedings represents the core of LNNS.

Volumes published in LNNS embrace all aspects and subfields of, as well as new challenges in, Networks and Systems.

The series contains proceedings and edited volumes in systems and networks, spanning the areas of Cyber-Physical Systems, Autonomous Systems, Sensor Networks, Control Systems, Energy Systems, Automotive Systems, Biological Systems, Vehicular Networking and Connected Vehicles, Aerospace Systems, Automation, Manufacturing, Smart Grids, Nonlinear Systems, Power Systems, Robotics, Social Systems, Economic Systems and other. Of particular value to both the contributors and the readership are the short publication timeframe and the worldwide distribution and exposure which enable both a wide and rapid dissemination of research output.

The series covers the theory, applications, and perspectives on the state of the art and future developments relevant to systems and networks, decision making, control, complex processes and related areas, as embedded in the fields of interdisciplinary and applied sciences, engineering, computer science, physics, economics, social, and life sciences, as well as the paradigms and methodologies behind them.

Indexed by SCOPUS, INSPEC, WTI Frankfurt eG, zbMATH, SCImago.

All books published in the series are submitted for consideration in Web of Science.

For proposals from Asia please contact Aninda Bose (aninda.bose@springer.com).

Wojciech Zamojski · Jacek Mazurkiewicz ·
Jarosław Sugier · Tomasz Walkowiak ·
Janusz Kacprzyk
Editors

System Dependability - Theory and Applications

Proceedings of the Nineteenth International
Conference on Dependability of Computer
Systems DepCoS-RELCOMEX.
July 1–5, 2024, Brunów, Poland

 Springer

Editors
Wojciech Zamojski
Department of Computer Engineering
Wrocław University of Science
and Technology
Wrocław, Poland

Jacek Mazurkiewicz
Department of Computer Engineering
Wrocław University of Science
and Technology
Wrocław, Poland

Jarosław Sugier
Department of Computer Engineering
Wrocław University of Science
and Technology
Wrocław, Poland

Tomasz Walkowiak
Department of Computer Engineering
Wrocław University of Science
and Technology
Wrocław, Poland

Janusz Kacprzyk 🄳
Polish Academy of Sciences
Systems Research Institute
Warsaw, Poland

ISSN 2367-3370   ISSN 2367-3389 (electronic)
Lecture Notes in Networks and Systems
ISBN 978-3-031-61856-7   ISBN 978-3-031-61857-4 (eBook)
https://doi.org/10.1007/978-3-031-61857-4

Preface

In this volume we would like to present proceedings of the 19 International Conference on Dependability of Computer Systems DepCoS-RELCOMEX which is scheduled to be held stationary in Brunów Palace, Poland, from 1 to 5 July 2024. It is the second time, after a three-year interruption caused by the COVID-19 pandemic, when the Conference will be organized in a regular manner in a beautiful Brunów Palace—our traditional venue—close to many palaces and castles located in the charming Lower Silesia's Valley of Palaces and Gardens.

DepCoS-RELCOMEX scope has always been focused on diverse issues which are constantly arising in performability and dependability analysis of contemporary computer systems and networks. Dependability of computer processing means obtaining reliable (true and timely) results in the conditions of processing both quantitative and qualitative data, using precise and "fuzzy/imitating" models and algorithms. It should be emphasized that artificial intelligence algorithms and tools are increasingly used in modern information technology and computer engineering, and therefore we are expanding our view on the dependability of systems that are progressively using methods based on cognitive systems and deep learning tool. In our opinion, this approach (**dependability** as the credibility of systems, **AI** tools and computer applications) meets the challenges which the computer science presently faces, both in its theoretical studies and in engineering. Ever-growing number of research methods being continuously developed for such analyses apply the newest results of artificial intelligence (AI) and computational intelligence (CI). Topical diversity of papers in these proceedings illustrate broad variety of multi-disciplinary subjects which should be considered in contemporary dependability explorations.

The Conference is organized annually since 2006 by the Department of Computer Engineering at the Faculty of Information and Communication Technology, Wrocław University of Science and Technology, but its roots go back to the heritage of two much older events: RELCOMEX (1977–1989) and Microcomputer School (1985–1995) which were organized by the Institute of Engineering Cybernetics (predecessor of the Department) under the leadership of prof. Wojciech Zamojski, now also the DepCoS chairman. Since 2006 the proceedings were printed, chronologically, first by the IEEE Computer Society (till 2009), then by Wrocław University of Science and Technology Publishing House (2010–2012), and recently by Springer Nature in the "Advances in Intelligent Systems and Computing" volumes no. 97 (2011), 170 (2012), 224 (2013), 286 (2014), 365 (2015), 479 (2016), 582 (2017), 761 (2018), 987 (2019), 1173 (2020), and 1389 (2021). Since 2022 DepCoS proceedings are a part of the "Lecture Notes in Networks and Systems" series, and the two previous editions were included in vols. 484 and 737. Springer Nature is one of the largest and most prestigious scientific publishers, with the LNNS titles being submitted for indexing in CORE Computing Research & Education database, Web of Science, SCOPUS, INSPEC, DBLP, and other indexing services.

It is our pleasant and honorable obligation now to thank everyone who participated in organization of the Conference and in preparation of this volume: the authors, members of the Program and the Organizing Committees, and all other individuals who assisted in creation of this book. Special recognition should be given to the reviewers whose opinions and comments helped immensely to select and enhance the submissions. Alphabetically, our sincerest thanks this year go to: Ilona Bluemke, Frank Coolen, Łukasz Jeleń, Alexander Grakovski, Alexey Lastovetsky, Urszula Kużelewska, Jecek Mazurkiewicz, Jan Magott, Marek Młyńczak, Yiannis Papadopoulos, Czesław Smutnicki, Janusz Sosnowski, Robert Sobolewski, Jarosław Sugier, Kamil Szyc, Tomasz Walkowiak, Marek Woda, and Wojciech Zamojski. Their work, not mentioned anywhere else in this volume, deserves to be highlighted in this introduction.

At the end of this preface, we would like to thank all authors who selected DepCoS-RELCOMEX as *the* platform to publish and discuss their research results. We believe that included papers will contribute to advances in design, analysis, and engineering of dependable computer systems and networks, offering an interesting source material for scientists, researchers, engineers, and students working in these areas.

Wojciech Zamojski
Jacek Mazurkiewicz
Jarosław Sugier
Tomasz Walkowiak
Janusz Kacprzyk

Organization

Nineteenth International Conference on Dependability
of Computer Systems DepCoS-RELCOMEX
Brunów, Poland, July 1–5, 2024

Program Committee

Wojciech Zamojski (Chairman)	Wrocław University of Science and Technology, Poland
Michael Affenzeller	Upper Austria University of Applied Sciences, Austria
Ali Al-Dahoud	Al-Zaytoonah University, Amman, Jordan
Andrzej Białas	Research Network ŁUKASIEWICZ - Institute of Innovative Technologies EMAG, Katowice, Poland
Ilona Bluemke	Warsaw University of Technology, Poland
Magdalena Bogalecka	Gdynia Maritime University, Poland
Wojciech Bożejko	Wrocław University of Science and Technology, Poland
Eugene Brezhniev	National Aerospace University "KhAI", Kharkov, Ukraine
Dariusz Caban	Wrocław University of Science and Technology, Poland
De-Jiu Chen	KTH Royal Institute of Technology, Stockholm, Sweden
Frank Coolen	Durham University, UK
Denny B. Czejdo	Fayetteville State University, USA
Wiktor B. Daszczuk	Warsaw University of Science and Technology, Poland
Mieczysław Drabowski	Cracow University of Technology, Poland
Francesco Flammini	University of Linnaeus, Sweden
Peter Galambos	Óbuda University, Hungary
Manuel Gill Perez	University of Murcia, Spain
Aleksander Grakowskis	Transport and Telecommunication Institute, Riga, Latvia
Laszlo Gulacsi	Óbuda University, Hungary
Atsushi Ito	Chuo University, Tokyo, Japan
Dariusz Jagielski	4th Military Hospital, Wrocław, Poland

Łukasz Jeleń	Wrocław University of Science and Technology, Poland
Ireneusz Jóźwiak	Wrocław University of Science and Technology, Poland
Igor Kabashkin	Transport and Telecommunication Institute, Riga, Latvia
Janusz Kacprzyk	Polish Academy of Sciences, Warsaw, Poland
Vyacheslav S. Kharchenko	National Aerospace University "KhAI", Kharkov, Ukraine
Ryszard Klempous	Wrocław University of Science and Technology, Poland
Krzysztof Kołowrocki	Gdynia Maritime University, Poland
Leszek Kotulski	AGH University of Science and Technology, Krakow, Poland
Vasilis P. Koutras	University of the Aegean, Chios, Greece
Levente Kovacs	Óbuda University, Hungary
Henryk Krawczyk	Gdansk University of Technology, Poland
Dariusz Król	Wrocław University of Science and Technology, Poland
Andrzej Kucharski	Wrocław University of Science and Technology, Poland
Marek Kulbacki	Polish – Japanese Academy of Information Technology, Warsaw, Poland
Urszula Kużelewska	Bialystok University of Technology, Białystok, Poland
Alexey Lastovetsky	University College Dublin, Ireland
Henryk Maciejewski	Wrocław University of Science and Technology, Poland
Jan Magott	Wrocław University of Science and Technology, Poland
Jacek Mazurkiewicz	Wrocław University of Science and Technology, Poland
Daniel Medyński	Collegium Witelona, Legnica, Poland
Marek Młyńczak	Wrocław University of Science and Technology, Poland
Yiannis Papadopoulos	Hull University, UK
Andrzej Pawłowski	University of Brescia, Italy
Ewaryst Rafajłowicz	Wrocław University of Science and Technology, Poland
Przemysław Rodwald	Polish Naval Academy, Gdynia, Poland
Jerzy Rozenblit	Arizona University, Tucson, USA
Imre Rudas	Óbuda University, Hungary
Mirosław Siergiejczyk	Warsaw University of Technology, Poland

Czesław Smutnicki	Wrocław University of Science and Technology, Poland
Robert Sobolewski	Bialystok University of Technology, Poland
Janusz Sosnowski	Warsaw University of Technology, Poland
Carmen Paz Suarez-Araujo	Universidad de Las Palmas de Gran Canaria, Spain
Jarosław Sugier	Wrocław University of Science and Technology, Poland
Laszlo Szilagyi	Sapientia Hungarian University of Transylvania, Romania
Tomasz Walkowiak	Wrocław University of Science and Technology, Poland
Max Walter	Siemens, Germany
Tadeusz Więckowski	Wrocław University of Science and Technology, Poland
Bernd E. Wolfinger	University of Hamburg, Germany
Min Xie	City University of Hong Kong, Hong Kong SAR, China
Irina Yatskiv	Transport and Telecommunication Institute, Riga, Latvia

Organizing Committee

Chair

Wojciech Zamojski	Wrocław University of Science and Technology, Poland

Members

Jacek Mazurkiewicz	Wrocław University of Science and Technology, Poland
Jarosław Sugier	Wrocław University of Science and Technology, Poland
Tomasz Walkowiak	Wrocław University of Science and Technology, Poland
Tomasz Zamojski	Wrocław University of Science and Technology, Poland
Mirosława Nurek	Wrocław University of Science and Technology, Poland

Contents

Large Language Models for Data Extraction in Slot-Filling Tasks

Marek Bazan[1,2(✉)] ⬥, Tomasz Gniazdowski[2] ⬥, Dawid Wolkiewicz[2] ⬥,
Juliusz Sarna[2] ⬥, and Maciej E. Marchwiany[2] ⬥

[1] Department of Computer Engineering, Wroclaw University of Science
and Technology, ul. Janiszewskiego 11/17, 50-370 Wrocław, Poland
marek.bazan@pwr.edu.pl

[2] JT Weston sp. z o.o, Atrium Plaza, al. Jana Pawła II 29, 00-867 Warszawa, Poland

Abstract. Large language models (LLMs) have turned out recently to
be a powerful tool for solving natural language processing and understanding tasks. In this paper, we investigate the usage of three open-
source large language models in a slot-filling task, which is a crucial task
in chatbot development. Apart from testing the method on an in-house
created dataset, we checked the methodology on two main benchmarks in
this field. The obtained results for models with 7B parameters are comparable with those achieved by closed-source chatGPT family models,
which are more than 20 times bigger.

Keywords: Open source LLMs · Slot Filling · LLaMA · Orca · Mistral

1 Introduction

Large language models (LLMs) are commonly used for data extraction and
understanding tasks such as: named entity recognition [38], slot-filling tasks
[7,28,34], event extraction [9], relation extraction [18,37] as well as information
extraction, in general [8,11,31].

A general, effective solution of a slot-filling problem is the main component
of modern chatbots. A slot-filling task is encountered e.g. while using chatbots
to guide a customer in filling forms for e-commerce web pages. It can also be
found in analyses of long detailed case studies that are not driven by a single
question chatbot.

The overall pipeline of using an LLM in the above scenario requires:

1. an automatic prompt and context generation,
2. sending the prompt and context to the LLM,
3. parsing the LLM's response and sending back results.

In this paper, we study the performance of LLMs: Orca [22,25], LLaMA [36]
and Mistral [13]. Our investigations cover an in-house built dataset (that we

ⓒ The Author(s), under exclusive license to Springer Nature Switzerland AG 2024
W. Zamojski et al. (Eds.): DepCoS-RELCOMEX 2024, LNNS 1026, pp. 1–18, 2024.
https://doi.org/10.1007/978-3-031-61857-4_1

publish) in Polish and two benchmarks for a slot-filling task in English named SNIPS [5] and ATIS [19]. To our knowledge, the presented results are the first attempt at the evaluation of Open Source LLMs on a slot-filling task that can be compared with closed models of the ChatGPT family [27].

The few-shot learning technique for LLMs was discovered at the beginning of 2020s and presented by Braun et al. [4]. Based on this research we focus on three types of prompt creation methods: zero-shot learning, few-shot learning, and chain of thought. Moreover, following [15] we investigated much bigger models.

Benchmarks that we used are in English. However, the application of our priority is message understanding in Polish. The in-house datasets coming from the business processes of *JT Weston sp. z o.o.* are in Polish. Since LLMs exploited in this paper have been trained on Polish language corpuses, our method can be used for both languages.

The remainder of the paper is organized in the following way. First, we review related research literature in the area of slot-filling tasks in the deep learning domain as well as LLMs. We then describe usage of three prompting methods. In Sect. 4 we present the results on the most popular publicly available benchmarks as well as on our in-house datasets. Results on benchmarks are compared with ChatGPT family models. Finally, we comment on the current limitations of open source LLMs and conclude the paper.

2   Related Work

A comprehensive survey on applications of LLMs can be found in [10]. A wide validation of closed LLMs (the ChatGPT family) on many natural language processing tasks can be found in [14].

Surveys on slot-filling tasks with the intent classification and the named entity recognition tasks can be found in [40] and [12], respectively.

A slot-filling task is also encountered in chatbots for business processes in the corporate world [1,33]. The Neula Assistant is one of the examples of such a platform where business processes are tracked by executing BPMN diagrams.

The most commonly used benchmarks for slot-filling tasks are SNIPS, which contains 7 intents and 39 slot types, and ATIS, which contains 21 intents and 128 slot types. The splits for train, validation and test subsets of the above dataset may be found in [40].

The first attempt at a slot-filling task was based on Maximum Entropy models and Hidden Markov models [23]. Conditional random fields [30] had been common approaches used to solve sequence tagging problems. The history of the application of deep neural networks (DNN) for this task dates back to deep convex networks [6] and with kernel learning based on n-grams. The application of the LSTMs for this purpose was first shown in [41]. The first application of BERT to solve this task, shown in [17], included a comparison with other models such as BiGRU and a convolutional neural network. Apart from deep neural architectures, conditional random fields also achieve good results and have been investigated in parallel [42] to DNNs. The current state-of-the-art results for the ATIS and SNIPS datasets are achieved with BERT architecture enhanced with convolutional layers [24]. Recently, LLMs were used for information extraction in a few-shot scenarios in [20]. The results a slot-filling task achieved for closed-source ChatGPT models are presented in [28].

Another approach to retrieve general information from text using LLMs is in-context learning [3]. It differs from zero-shot, few-shot and chain of thought. However, it also uses samples for feeding prompt fields.

3   Methods

In this section, we describe three prompting methods to command LLMs that we have investigated as well as the approach to process the response from models.

3.1   Prompting

Zero-Shot. Zero-shot prompting is a method in which LLM receives a natural language description of a given task [4]. An example of a zero-shot prompt in English and Polish can be found in Appendices A.1 and B.1, respectively.

Few-Shot. Few-shot prompting is a method that incorporates example pairs of context and completion, followed by the actual user question [4]. In our solution, five examples (general tasks of slot-filling) were provided each time during testing. Examples were not related to the actual task of the following user question. An example of few-shot prompts in English and Polish can be found in Appendices A.2 and B.2, respectively.

Zero-Shot Chain of Thought. The chain of thought (CoT) prompting (few-shot-CoT prompting) method [39] was proposed as a simple improvement over the original Few-shot prompting method. The main goal was to simplify an example tasks by showing a model of step-by-step solutions. The performance boost was impressive, especially with very sizable LLMs [27,36]. The main disadvantage of this method was the need to prepare task-specific examples with explanations for the model. The zero-shot-CoT prompting method [16] solves this problem with a simple solution, by adding a *"Let's think step by step"* phrase at the end of the zero-shot prompt. An example of zero-shot-CoT prompts in English and Polish can be found in Appendices A.3 and B.3, respectively.

3.2   Receiving the Output

Key Matching Algorithm. During the experiments, it was noticed that the LLMs tend to change the name of slots (entities) to be filled. Most often, models do this by using synonyms or by changing the language of the slot name (most often from Polish to English). To solve this problem we implemented a greedy-matching algorithm. First, the slot names were automatically translated to the proper language (English or Polish). Then, sentence-level embeddings of slot names from the original prompt and the LLM were calculated with a Fasttext algorithm [2]. The original and the LLM's names were matched by their maximum cosine similarity value (see Algorithm 1). One iteration of the algorithm is presented. After one iteration, the matching threshold value was decreased. The algorithm terminated when all slot names were matched.

Algorithm 1. One iteration of the greedy slot names matching algorithm

1: **for** $i \leftarrow 1$ to N **do**   ▷ N is the number of slot names
2:     $slotNameLLMVec \leftarrow slotNamesLLMVec[i]$
3:     **for** $j \leftarrow 1$ to $length(slotNamesOrgVec)$ **do**
4:         $slotNameOrgVec \leftarrow slotNamesOrgVec[j]$
5:         $cosSimilarities[j] \leftarrow cosSimilarity(slotNameLLMVec, slotNameOrgVec)$
6:     **end for**
7:     **if** $max(cosSimilarities) \geq 0.9 - \eta$ **then**
8:         $matchedSlotNames[i] \leftarrow slotNamesOrgVec[argMax(cosSimilarities)]$
9:         $slotNamesOrgVec.pop(argMax(cosSimilarities))$
10:    **end if**
11: **end for**   ▷ η lowers the threshold

---

4 Experiments

In this section, we describe the datasets on which we tested the LLMs, the evaluation process with four different metrics as well as discuss some limitations of one of these metrics. Finally, we share some comparison observations about the behavior of the models we tested on a lot-filling task.

4.1 Datasets

The statistical experiments we carried out were on test partitions of SNIPS [5] and ATIS [32] datasets.

Moreover, we created two slot-filling Polish datasets within *JT Weston*: (i) *leaves* based on a process of taking days off and (ii) *delegations* based on a process of accounting for a business trip. The main assumption of datasets prepared with the business team was to make the data reflect real business situations and problems. Datasets were prepared independently by people from *JT Weston* and verified by the business department. Many field names have to be inferred based on other values and various date formats were used. Statistics of all datasets used in experiments are presented in Appendix C.

4.2   Evaluation

LLMs' hallucinations are more challenging in Polish than in English. Polish is sensitive to not only word order but also on extremely complicated inflection (which can change the meaning). Moreover, any given entity may have multiple values in the input text. This is why there was no one-to-one correspondence between the predicted and ground truth entities (as in the *IOB* format). It was not possible to determine the number of all false negative predictions (and $F1$ metric) in in-house datasets. For this reason, only precision was calculated for in-house datasets. Moreover, from the business perspective precision is the most important metric. For the SNIPS and ATIS datasets it was possible to calculate the $F1$ score, so for these datasets, we calculated a slot $F1$ metric [35].

Algorithm 2. Precision source metric

1: **for** $i \leftarrow 1$ to N **do**   ▷ N is the number of entity predictions
2:     $entityPredClass \leftarrow entitiesPreds[i].class$
3:     $entityPred \leftarrow entitiesPreds[i].value$
4:     **if** $entityPred \neq$ "n/a" **then**
5:         $notNAPredCounter \leftarrow notNAPredCounter + 1$
6:         $entityPredTokenLemma \leftarrow lemmatizeTokens(tokenizeText(entityPred))$
7:         $inputTextTokenLemma \leftarrow lemmatizeTokens(tokenizeText(inputText))$
8:         **for** $j \leftarrow 1$ to k **do**   ▷ k is a parameter of a evaluation algorithm
9:             $entityPredNGrams \leftarrow nGrams(entityPredTokenLemma, j)$
10:            $inputTextNGrams \leftarrow nGrams(inputTextTokenLemma, j)$
11:            $commonPartPredText \leftarrow commonPart(entityPredNGrams, inputTextNGrams)$
12:            **if** $commonPartPredText \neq \emptyset$ **then**
13:                **if** $j \geq 2 \lor commonPartPredText \notin stopWordsSet$ **then**
14:                    $notHallucinations \leftarrow notHallucinations + 1$
15:                **end if**
16:            **end if**
17:        **end for**
18:    **end if**
19: **end for**
20: $precisionSource \leftarrow notHallucinations/notNAPredCounter$ ▷ precision source metric value
   is calculated for all entities predictions for all input texts

---

To calculate the *precision source* metric [26], we began with checking the entity value hallucination. Several steps were necessary. First the input text and predicted entity value were tokenized and lemmatized (lines 6–7 in Algorithm 2). The lemmatization step is our development of the original method, because LLMs often change the form of words, especially in Polish. Then all the words were changed to lowercase. Finally, a common part of any n-grams from the prediction of the entity value and the input text was calculated (line 12 in Algorithm 2) (for $n = 1$, we additionally check if the common part was not a stop word). If a common part exists and it was not a stop word, then the prediction was not recognized as hallucination. The precision source is considered as a *percentage of named entities in the summary that can be found in the source* [26]. In our case, it was a percentage of entity predictions that were present in the input text (line 20 in Algorithm 2).

We calculated precision for the *leaves* and *delegations JT Weston* datasets. Due to the frequent changes in word forms in Polish, we considered the predictions to be correct when the cosine similarity between sentence-level embedding vectors of ground truth and entity value prediction calculated with the *Fasttext* (c.f. [2]) algorithm was above a given threshold (0.8 and 0.9). We mark the metrics as $precision_{0.8}$ and $precision_{0.9}$. The results for $precision_{0.9}$ are presented in Fig. 1. All numerical values can be found in Appendix D.

Similarly, for the SNIPS and ATIS datasets, the $F1$ metric with thresholds 0.8 and 0.9 was used, which enabled us to compute $F1_{0.8}$ and $F1_{0.9}$. The slot $F1$ score with threshold is calculated for a slot with an assumption that the answer is correct if cosine similarity between the sentence-level embedding vectors of the predicted value and ground truth is above a given threshold. For the slot $F1$ score, a perfect match between the predicted value and ground truth was necessary to recognize the prediction as a true positive. Numerical values are presented in Table 1. For the *Orca mini 3* we used the few-shot method with examples in the system prompt. For the *Mistral 02 instruct* the few-shot prompting method was used, with examples in the user prompt. For the *LLaMA 2 chat* we used the zero-shot chain of thought method.

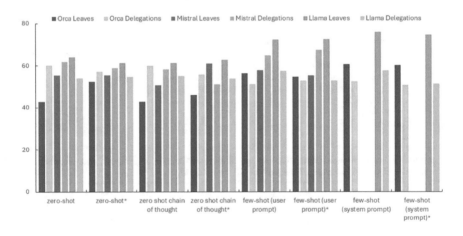

Fig. 1. Comparison of $F1_{0.9}$ for different models and prompting techniques. Prompting methods marked with a star symbol (*) are the prompts, in which an additional dataset name was passed as the sentence - *"Text is about leave"* or *"Text is about delegation"* in order to to emphasize the context of the message.

Table 1. Slot $F1$ metric calculated for *Orca mini 3, Mistral 02 instruct* and *LLaMA 2 chat* for the SNIPS and ATIS datasets. Results of our experiments were compared with the results of *Codex, Chat GPT, GPT-3.5* [28] and State-of-The-Art results (SOTA).

model	SNIPS		ATIS	
	slot $F1$	[%] of SOTA	slot $F1$	[%] of SOTA
Orca mini 3 (**our experiments**)	56.48	57.46	8.85	8.99
Mistral 02 instruct (**our experiments**)	55.56	56.52	51.93	52.74
LLaMA 2 chat (**our experiments**)	35.13	35.74	1.71	1.74
Codex [21]	**68.90**	**70.09**	**57.29**	**58.19**
GPT-3.5	**68.90**	**70.09**	55.72	56.59
Chat GPT	58.24	59.25	15.71	15.96
finetuned SOTA (*CTRAN*) [29]	98.30	–	98.46	–

4.3 Reasoning Problem and Hallucinations with the Precision Source Algorithm

We require models to make simple reasoning in the prompts. However we observed varied hallucinations. The precision source algorithm cannot generalize synonymous phrases, for example if there is "sick leave", then the reason for the leave is "sickness". However, this leads to the prediction being considered a hallucination by a precision source metric (if the word "sickness", or its other form does not appear elsewhere in the input text). (The lemma of the word "sick" is "sick" and for "sickness" is "sickness" - these are different parts of speech). Models also changed the format of data or use synonyms (e.g., "n/a" may be changed into "no information", or "1.12.2023" into "the first of December" by a model). This creates a much more difficult problem of hallucination detection. The detector has to differentiate incorrect values from synonyms or format changes. Moreover, we noticed that if the embedding vectors of these two phrases (e.g., "1.12.2024" and "tomorrow") are close enough, the hallucinations will not be detected by commonly used algorithms (e.g., word embeddings). This seems as a very complex task and the problem is left open.

4.4 Output Generated by Models

During the research, several conclusions were drawn about the responses created by the *LLMs*.

The *Orca 3 mini* model rarely added hallucinations or additional, unnecessary information to the generated output. For both zero-shot and few-shot prompts, it usually returned only the extracted data. The format of the JSON returned by *Orca 3 mini* was also usually correct.

The *Mistral 02 instruct* changed the date format very often (e.g., May 5 to 5/5). This model often invents non-existent words in Polish. In most examples, it returned JSON in the valid format with some additional data. The *Mistral 02 instruct* sometimes hallucinated - it returned its own examples in a few-shot prompting method that were not present in the input prompt.

The *LLaMA 2 chat* changed the format of the returned data a lot and it converted JSON into formats that cannot be parsed. It added a lot of redundant text explaining its decisions (instead of returning just JSON with the extracted data). The most difficult task was to extract the returned data from the model response.

To generate the correct output (valid JSON format), the LLM had to process the same input data many times. All investigated models were prompted for each message (up to 5 times) until they returned the output that could have been parsed. The performance of models measured by requested number of unconnected output format is presented in Appendix D. We observed that prompting techniques can change performance of models, it is specialty visible for no-English texts. In general LLMs has better performance for English. For English prompts and Polish texts, LLMs most often automatically translated the extracted data and entity names into English. This made the approach of prompting models in English with Polish data less effective. We also tested single value extraction by prompt (with multiple prompts for single test) , what usually produced an output in an incorrect format and was much slower. Models that extracted a single entity value instead of multiple ones produced more hallucinations. Therefore this approach was also abandoned.

5   Conclusions

In this paper we have presented the results of the application of chosen small (up to 7B parameters), open-source LLMs to a slot-filling tasks for complex texts. This task is a key ingredient of natural language understanding of dialogs for chatbot development. We compare our methodology with two benchmarks with the chatGPT 3.5 family models. The achieved results show that open-source LLMs can be as good, and in some cases even better than closed-source models.

For model evaluation, four metrics were used: (i) precision based on cosine similarity with thresholds 0.8 and 0.9, (ii) $F1$ based on cosine similarity with thresholds 0.8 and 0.9, (iii) slot-$F1$ and (iv) precision source. The slot-$F1$ was calculated for the SNIPS and ATIS benchmarks. *LLaMA 2* models such as *LLaMA chat* and *Orca* achieve very low performance on this metric due to many false positives, though the results of *Mistral* outperform even ChatGPT.

The best performance (measured by precision source) on in-house datasets was achieved with the *LLaMA 2 Chat*, for independent benchmarks, the *Mistral* model was the best, since it limits the output of the information that is not supported by the input message, that is visible on the ATIS benchmark. Based on our results, the best-performing prompt techniques have been chosen: Few-shot for the *Orca 3 mini* and *Mistral*, and the zero-shot chain of thought for the *LLaMA chat*. Detailed information can be found in Appendix D.

In our investigations, we experimented with different LLMs in the slot-filling task. Our problem was complicated. The texts were written in Polish and were written in a tricky way. Moreover, we used various date types, and many slot values had to be inferred by other values or from the input text. Obtaining the valid JSON format from the model also turned out to be a difficult task. During the research, numerous indirect problems, such as changing the name of the keys in the returned JSON were solved. Due to the high level of complexity, various LLMs turned out to be too weak to be used in our problem. The *MPT, Falcon, Bloomz* or even base *LLaMA 2* models were rejected for this reason.

Acknowledgement. This work was partially financed from the grant POIR.01.01.01-00-0930/19 entitled "Development of self-configuring personal assistant module on the BPM platform using natural language and artificial intelligence processing tools and algorithms".

A   Examples of Prompts in English

A.1   Example of a Zero-Shot Prompt

You are an AI assistant who is very good at extracting data from short texts written in English. Extract the given values: ENITY_NAMES_TO_EXTRACT. Return the extracted data in valid json format. Return key names and values in English. If a given value is not in the text or you are not sure of the answer - return "n/a". If a given value is not found in the text, try to infer it from other values. Return extracted values in double quotes:"". Do not return other information. Help as much you can.
TEXT_FROM_WHICH_DATA_IS_EXTRACTED

A.2   Example of a Few-Shot Prompt

You are an AI assistant who is very good at extracting data from short texts written in English. Return only extracted data in valid json format. Examples are presented below.
Text: I noticed a fault in the elevator on the fourth floor and in the elevator on the fifth floor (the doors get stuck when opening). Best regards, Anna Nowak (555-123-456)
Entities: ["Name", "Surname", "Contact telephone number", "Description of the fault", "Location of the fault", "Date of the fault noticed"]
Output:
{
"Name": ["Anna"],
"Surname": ["Nowak"],
"Contact telephone number": ["555-123-456"],
"Description of the fault": ["door stuck when opening"],
"Location of the fault": ["elevators on the fourth floor", "elevators on the fifth floor"],
"Date of the fault noticed": "n/a"
}
Text: As part of the new "Business Process Automation" project, we are determining the details. The project will last for 12 months. The total budget of the project is PLN 1,000,000. We will implement the project in Warsaw. Regards

Mariusz.
Entities: ["Project name", "Project budget", "Project team name", "Project duration", "City of project implementation", "Country of project implementation"]
Output:
{
"Project name": ["Business Process Automation"],
"Project budget": ["PLN 1,000,000"],
"Project team name": "n/a",
"Project duration": ["12 months"],
"City of project implementation": ["Warsaw"],
"Country of project implementation": ["Poland"]
}
Text: I am an employee of ABC Electronics, my order number is 20231001. My shopping cart is deleting (emptying) itself. The second error is that it logs me out of the site at random moments. Please help.
Entities: ["Customer name", "Order number", "Department", "Importance", "Report date", "Error description", "Country", "Supervisor"]
Output:
{
"Customer name": ["ABC Electronics"],
"Order number": ["20231001"],
"Department": "n/a",
"Importance": "n/a",
"Report date": "n/a",
"Error description": ["deleting the shopping cart", "logging out of the website at random moments"],
"Country": "n/a",
"Supervisor": "n/a"
}
Text: I am reporting a problem with the printer. Marek
Entities: ["Device name", "Name"]
Output:
{
"Device name": ["printer"],
"Name": ["Marek"]
}
Text: This is my candidacy for the position of data analyst.
Entities: ["Position"]
Output:
{
"Position": ["data analyst"]
}
Text: TEXT_FROM_WHICH_DATA_IS_EXTRACTED
Entities: ENTITY_NAMES_TO_EXTRACT
Output:

A.3   Example of a Zero-Shot-CoT Prompt

You are an AI assistant who is very good at extracting data from short texts written in English. Extract the given values from the text: ENTITY_NAMES_TO_EXTRACT. Return the extracted data in valid json format. Return key names and values in English. If a given value is not in the text or you are not sure of the answer - return "n/a". If a given value is not found in the text, try to infer it from other values. Return extracted values in double quotes: "". Do not return other information. Let's think step by step.
TEXT_FROM_WHICH_DATA_IS_EXTRACTED

B   Examples of Prompts in Polish

B.1   Example of a Zero-Shot Prompt

Jesteś asystentem AI, który bardzo dobrze wydobywa dane z krótkich tekstów napisanych po polsku. Wyciągnij z tekstu podane wartości: ENITY_NAMES_TO_EXTRACT. Zwróć wydobyte dane w prawidłowym formacie json. Nazwy kluczy oraz wartości zwróć w języku polskim. Jeśli w tekście nie ma danej wartości lub nie jesteś pewny odpowiedzi - zwróć "n/a". Jeśli w tekście nie ma danej wartości, spróbuj ją wywnioskować na podstawie innych wartości. Wyciągnięte wartości zwróć w apostrofach: "". Nie zwracaj innych informacji. Pomóż jak najlepiej potrafisz.
TEXT_FROM_WHICH_DATA_IS_EXTRACTED

B.2   Example of a Few-Shot Prompt

Jesteś asystentem AI, który bardzo dobrze wydobywa dane z krótkich tekstów napisanych po polsku. Zwróć jedynie wyciągnięte dane w prawidłowym formacie json. Przykłady zaprezentowano poniżej.
Text: Zauważyłem usterkę w windzie na czwartym piętrze oraz w windzie na piątym piętrze (drzwi się zacinają przy otwieraniu). Pozdrawiam, Anna Nowak (555-123-456)
Entities: ["Imię", "Nazwisko", "Telefon kontaktowy", "Opis usterki", "Miejsce usterki", "Data zauważenia usterki"]
Output:
{
"Imię": ["Anna"],
"Nazwisko": ["Nowak"],
"Telefon kontaktowy": ["555-123-456"],
"Opis usterki": ["zacinanie się drzwi przy otwieraniu"],
"Miejsce usterki": ["windy na czwartym piętrze", "windy na piątym piętrze"],
"Data zauważenia usterki": "n/a"
}
Text: W ramach nowego projektu "Automatyzacja Procesów Biznesowych" ustalamy szczegóły. Projekt będzie trwać przez 12 miesięcy. Całkowity budżet projektu

to 1,000,000 PLN. Projekt będziemy realizować w Warszawie. Pozdrawiam, Mariusz.

Entities: ["Nazwa projektu", "Budżet projektu", "Nazwa zespołu projektu", "Czas trwania projektu", "Miasto realizacji projektu", "Państwo realizacji projektu"]

Output:

{
"Nazwa projektu": ["Automatyzacja Procesów Biznesowych"],
"Budżet projektu": ["1,000,000 PLN"],
"Nazwa zespołu projektu": "n/a",
"Czas trwania projektu": ["12 miesięcy"],
"Miasto realizacji projektu": ["Warszawa"],
"Państwo realizacji projektu": ["Polska"]
}

Text: Jestem pracownikem ABC Electronics, mój numer zamówienia to 20231001. Mój koszyk zakupowy sam się usuwa (opróżnia). Drugi błąd polega na wylogowywaniu mnie ze strony w losowych momentach. Proszę o pomoc.

Entities: ["Nazwa klienta", "Numer zamówienia", "Oddział", "Waga błędu", "Data zgłoszenia", "Opis błędów", "Kraj", "Przełożony"]

Output:

{
"Nazwa klienta": ["ABC Electronics"],
"Numer zamówienia": ["20231001"],
"Oddział": "n/a",
"Waga błędu": "n/a",
"Data zgłoszenia": "n/a",
"Opis błędów": ["usuwanie koszyka zakupowego", "wylogowywanie ze strony w losowych momentach"],
"Kraj": "n/a",
"Przełożony": "n/a"
}

Text: Zgłaszam problem z drukarką. Marek

Entities: ["Nazwa urządzenia", "Imię"]

Output:

{
"Nazwa urządzenia": ["drukarka"],
"Imię": ["Marek"]
}

Text: To moja kandydatura na stanowisko analityka danych.

Entities: ["Stanowisko"]

Output:

{
"Stanowisko": ["analityk danych"]
}

Text: TEXT_FROM_WHICH_DATA_IS_EXTRACTED

Entities: ENITY_NAMES_TO_EXTRACT
Output:

B.3   Example of a Zero-Shot-CoT Prompt

Jesteś asystentem AI, który bardzo dobrze wydobywa dane z krótkich tekstów napisanych po polsku. Wyciągnij z tekstu podane wartości: ENITY_NAMES_TO_ EXTRACT. Zwróć wydobyte dane w prawidłowym formacie json. Nazwy kluczy oraz wartości zwróć w języku polskim. Jeśli w tekście nie ma danej wartości lub nie jesteś pewny odpowiedzi - zwróć "n/a". Jeśli w tekście nie ma danej wartości, spróbuj ją wywnioskować na podstawie innych wartości. Wyciągnięte wartości zwróć w apostrofach: "". Nie zwracaj innych informacji. Myśl krok po kroku.
TEXT_FROM_WHICH_DATA_IS_EXTRACTED

C   Dataset Details

The details of the datasets are presented in Table 2.

Table 2. Statistics of the datasets that were used in our experiments. Texts count refers to the total number of texts, slots to fill refers to the total number of slots to fill, slots sets refer to the number of slots sets, unique slots refer to the number of unique slots in the given dataset. Some slots were present in multiple slot sets.

dataset name	text count	slots to fill	slots sets	unique slots
JT Weston leaves	44	169	1	5
JT Weston delegations	30	314	1	12
SNIPS [5]	700	1790	7	39
ATIS [32]	893	2837	20	69

D   Results

(See Tables 3, 4, 5 and 6).

Table 3. Metric values for the *Orca 3 mini*, *Mistral 02 instruct* and *LLaMA 2 chat* models received for the *JT Weston leaves* dataset with metrics: $precision_{0.8}$, $precision_{0.9}$ and precision source (see Sect. 4.2). Prompting methods marked with a star symbol (*) had an additional dataset name passed as a sentence - *Text is about leave.*

prompting method	$precision_{0.8}$	$precision_{0.9}$	precision source
Orca 3 mini			
zero-shot	51.81	42.77	81.33
zero-shot*	59.75	52.2	82.39
zero-shot chain of thought	53.01	42.77	78.92
zero-shot chain of thought*	52.35	45.88	78.24
few-shot (user prompt)	66.25	56.25	88.13
few-shot (user prompt)*	66.06	54.55	89.7
few-shot (system prompt)	72.22	**60.49**	**93.21**
few-shot (system prompt)*	**73.01**	60.12	92.64
Mistral 02 instruct			
zero-shot	62.5	55.36	82.14
zero-shot*	63.16	55.56	81.29
zero-shot chain of thought	62.07	50.58	80.46
zero-shot chain of thought*	**69.77**	**61.05**	80.23
few-shot (user prompt)	69.13	57.72	81.21
few-shot (user prompt)*	67.11	55.26	**84.21**
few-shot (system prompt)	–	–	–
few-shot (system prompt)*	–	–	–
LLaMA 2 chat			
zero-shot	71.43	64.00	82.86
zero-shot*	68.89	61.11	85.00
zero-shot chain of thought	68.33	61.11	83.89
zero-shot chain of thought*	70.62	62.71	85.88
few-shot (user prompt)	77.37	72.26	89.78
few-shot (user prompt)*	80.44	72.46	90.58
few-shot (system prompt)	**80.88**	**75.74**	**94.85**
few-shot (system prompt)*	80.85	74.47	91.49

Table 4. Metric values for the *Orca 3 mini, Mistral 02 instruct* and *LLaMA 2 chat* models received for the *JT Weston delegations* dataset. Metrics: $precision_{0.8}$, $precision_{0.9}$ and precision source (see Sect. 4.2). Prompting methods marked with a star symbol (*) had an additional dataset name passed as a sentence - *Text is about delegation.*

prompting method	$precision_{0.8}$	$precision_{0.9}$	precision source
Orca 3 mini			
zero-shot	**68.05**	**60.06**	73.67
zero-shot*	63.91	57.10	**74.56**
zero-shot chain of thought	67.81	59.83	73.79
zero-shot chain of thought*	64.27	55.62	72.62
few-shot (user prompt)	56.40	50.99	71.68
few-shot (user prompt)*	57.71	52.74	70.15
few-shot (system prompt)	57.28	52.35	72.35
few-shot (system prompt)*	55.93	50.61	70.94
Mistral 02 instruct			
zero-shot	66.23	61.91	69.26
zero-shot*	63.74	58.79	73.08
zero-shot chain of thought	60.66	58.29	69.67
zero-shot chain of thought*	55.09	50.94	69.43
few-shot (user prompt)	66.67	64.87	**78.98**
few-shot (user prompt)*	**67.98**	**67.37**	77.04
few-shot (system prompt)	–	–	–
few-shot (system prompt)*	–	–	–
LLaMA 2 chat			
zero-shot	**62.11**	54.04	72.98
zero-shot*	61.88	54.69	74.38
zero-shot chain of thought	61.42	54.94	72.84
zero-shot chain of thought*	61.34	53.78	72.38
few-shot (user prompt)	61.48	57.41	79.26
few-shot (user prompt)*	56.36	52.73	65.00
few-shot (system prompt)	60.79	**57.55**	**83.45**
few-shot (system prompt)*	53.98	51.14	63.07

Table 5. Metric values for the *Orca mini 3*, *Mistral 02 instruct* and *LLaMA 2 chat* models received for the SNIPS and ATIS datasets. For the *Orca mini 3* we used the few-shot method with the system prompt. For *Mistral 02 instruct* the few-shot prompting method was used, with the user prompt. For *LLaMA 2 chat* we used the zero-shot-chain of thought method.

model	slot $F1_{0.8}$	slot $F1_{0.9}$	precision source
SNIPS			
Orca mini 3	**67.52**	**64.48**	67.68
Mistral 02 instruct	67.08	63.89	**72.03**
LLaMA 2 chat	56.82	53.26	61.77
ATIS			
Orca mini 3	23.82	22.93	25.69
Mistral 02 instruct	**66.99**	**64.56**	**77.98**
LLaMA 2 chat	10.83	9.96	20.03

Table 6. The number of times for a given model for the *leaves* and *delegations* datasets when the data returned by the model was not parsable. Prompting methods marked with a star symbol (*) had an additional dataset name passed as a sentence - *Text is about leave* (for *leaves* dataset) and *Text is about delegation* (for *delegations* dataset).

	leaves			delegations		
	Orca 3 mini	LLaMA 2 chat	Mistral 02 instruct	Orca 3 mini	LLaMA 2 chat	Mistral 02 instruct
zero-shot	0	0	0	0	4	9
zero-shot*	0	0	0	1	1	12
zero-shot chain of thought	0	0	0	0	2	11
zero-shot chain of thought*	0	0	0	0	0	7
few-shot (user prompt)	0	0	0	0	3	0
few-shot (user prompt)*	0	2	0	0	5	0
few-shot (system prompt)	0	0	–	0	4	–
few-shot (system prompt)*	1	0	–	0	8	–

References

1. Barros, A., Sindhgatta, R., Nili, A.: Scaling up chatbots for corporate service delivery systems. Commun. ACM **64**(8), 88–97 (2021)
2. Bojanowski, P., Grave, E., Joulin, A., Mikolov, T.: Enriching word vectors with subword information. Trans. Assoc. Comput. Linguist. **5**, 135–146 (2017)
3. Bölücü, N., Rybinski, M., Wan, S.: Impact of sample selection on in-context learning for entity extraction from scientific writing. In: Findings of the Association for Computational Linguistics: EMNLP 2023, pp. 5090–5107 (2023)
4. Brown, T.E.A.: Language models are few-shot learners. Adv. Neural. Inf. Process. Syst. **33**, 1877–1901 (2020)

5. Coucke, A., et al.: Snips voice platform: an embedded spoken language understanding system for private-by-design voice interfaces. arXiv preprint arXiv:1805.10190 (2018)
6. Deng, L., Tur, G., He, X., Hakkani-Tur, D.: Use of kernel deep convex networks and end-to-end learning for spoken language understanding. In: 2012 IEEE Spoken Language Technology Workshop (SLT), pp. 210–215. IEEE (2012)
7. Dong, G., et al.: Revisit input perturbation problems for llms: a unified robustness evaluation framework for noisy slot filling task. In: Liu, F., Duan, N., Xu, Q., Hong, Y. (eds.) NLPCC 2023. LNCS, vol. 14302, pp. 682–694. Springer, Heidelberg (2023)
8. Eisenstein, J.: Natural language processing, **14**, 2022 (2018). verfügbar unter https://princeton-nlpgithub.io/cos484/readings/eisenstein-nlp-notes.pdf. zuletzt geprüft am
9. Gao, J., Zhao, H., Yu, C., Xu, R.: Exploring the feasibility of chatgpt for event extraction. arXiv preprint arXiv:2303.03836 (2023)
10. Hadi, M.U., et al.: Large language models: a comprehensive survey of its applications, challenges, limitations, and future prospects. Authorea Preprints (2023)
11. Han, R., Peng, T., Yang, C., Wang, B., Liu, L., Wan, X.: Is information extraction solved by chatgpt? an analysis of performance, evaluation criteria, robustness and errors. arXiv preprint arXiv:2305.14450 (2023)
12. Jehangir, B., Radhakrishnan, S., Agarwal, R.: A survey on named entity recognition-datasets, tools, and methodologies. Nat. Lang. Process. J. **3**, 100017 (2023)
13. Jiang, A.Q., et al.: Mistral 7b (2023)
14. Kocoń, J., et al.: Chatgpt: jack of all trades, master of none. Inf. Fusion. 101861 (2023)
15. Kojima, T., Gu, S.S., Reid, M., Matsuo, Y., Iwasawa, Y.: Large language models are zero-shot reasoners. Adv. Neural. Inf. Process. Syst. **35**, 22199–22213 (2022)
16. Kojima, T., Gu, S.S., Reid, M., Matsuo, Y., Iwasawa, Y.: Large language models are zero-shot reasoners (2023)
17. Korpusik, M., Liu, Z., Glass, J.R.: A comparison of deep learning methods for language understanding. In: Interspeech, pp. 849–853 (2019)
18. Kotnis, B., et al.: MILIE: modular & iterative multilingual open information extraction, pp. 6939–6950 (2022). https://doi.org/10.18653/v1/2022.acl-long.478. https://aclanthology.org/2022.acl-long.478
19. Liu, X., Eshghi, A., Swietojanski, P., Rieser, V.: Benchmarking natural language understanding services for building conversational agents. In: Marchi, E., Siniscalchi, S.M., Cumani, S., Salerno, V.M., Li, H. (eds.) Increasing Naturalness and Flexibility in Spoken Dialogue Interaction. LNEE, vol. 714, pp. 165–183. Springer, Singapore (2021). https://doi.org/10.1007/978-981-15-9323-9_15
20. Ma, Y., Cao, Y., Hong, Y., Sun, A.: Large language model is not a good few-shot information extractor, but a good reranker for hard samples!, pp. 10572–10601 (2023). https://doi.org/10.18653/v1/2023.findings-emnlp.710. https://aclanthology.org/2023.findings-emnlp.710
21. Mark, C., et al.: Evaluating large language models trained on code (2021)
22. Mathur, P.: orca_mini_v3_7b: An explain tuned llama2-7b model (2023)
23. McCallum, A., Freitag, D., Pereira, F.C., et al.: Maximum entropy markov models for information extraction and segmentation. In: ICML, vol. 17, pp. 591–598 (2000)
24. Mehrish, A., Majumder, N., Bharadwaj, R., Mihalcea, R., Poria, S.: A review of deep learning techniques for speech processing. Inf. Fusion, 101869 (2023)
25. Mukherjee, S., Mitra, A., Jawahar, G., Agarwal, S., Palangi, H., Awadallah, A.: Orca: progressive learning from complex explanation traces of gpt-4 (2023)

26. Nan, F., et al.: Entity-level factual consistency of abstractive text summarization (2021)
27. Achiam, J., et al.: OpenAI: GPT-4 technical report (2023)
28. Pan, W., Chen, Q., Xu, X., Che, W., Qin, L.: A preliminary evaluation of chatgpt for zero-shot dialogue understanding. arXiv preprint arXiv:2304.04256 (2023)
29. Rafiepour, M., Sartakhti, J.S.: CTRAN: CNN-transformer-based network for natural language understanding. Eng. Appl. Artif. Intell. **126**, 107013 (2023). https://doi.org/10.1016/j.engappai.2023.107013
30. Raymond, C., Riccardi, G.: Generative and discriminative algorithms for spoken language understanding. In: Interspeech 2007-8th Annual Conference of the International Speech Communication Association (2007)
31. Schacht, S., Kamath Barkur, S., Lanquillon, C.: Promptie-information extraction with prompt-engineering and large language models. In: Stephanidis, C., Antona, M., Ntoa, S., Salvendy, G. (eds.) HCII 2023, vol. 1836, pp. 507–514. Springer, Heidelberg (2023). https://doi.org/10.1007/978-3-031-36004-6_69
32. Shivakumar, P.G., Yang, M., Georgiou, P.: Spoken language intent detection using confusion2vec. arXiv preprint arXiv:1904.03576 (2019)
33. Simić, S.D., Starčić, T., Ferlatti, A., Etinger, D., Tankovic, N.: A business process model driven chatbot architecture. In: Intelligent Human Systems Integration (IHSI 2022): Integrating People and Intelligent Systems, vol. 22, no. 22 (2022)
34. Sun, G., Feng, S., Jiang, D., Zhang, C., Gašić, M., Woodland, P.C.: Speech-based slot filling using large language models. arXiv preprint arXiv:2311.07418 (2023)
35. Tjong Kim Sang, E.F., De Meulder, F.: Introduction to the CoNLL-2003 shared task: language-independent named entity recognition. In: Proceedings of the Seventh Conference on Natural Language Learning at HLT-NAACL 2003, pp. 142–147 (2003)
36. Touvron, H., et al.: Llama 2: open foundation and fine-tuned chat models (2023)
37. Wadhwa, S., Amir, S., Wallace, B.: Revisiting relation extraction in the era of large language models, pp. 15566–15589 (2023). https://doi.org/10.18653/v1/2023.acl-long.868. https://aclanthology.org/2023.acl-long.868
38. Wang, S., et al.: GPT-NER: named entity recognition via large language models (2023)
39. Wei, J., et al.: Chain-of-thought prompting elicits reasoning in large language models (2023)
40. Weld, H., Huang, X., Long, S., Poon, J., Han, S.C.: A survey of joint intent detection and slot filling models in natural language understanding. ACM Comput. Surv. **55**(8), 1–38 (2022)
41. Yao, K., Peng, B., Zhang, Y., Yu, D., Zweig, G., Shi, Y.: Spoken language understanding using long short-term memory neural networks. In: 2014 IEEE Spoken Language Technology Workshop (SLT), pp. 189–194. IEEE (2014)
42. Zhu, S., Zhao, Z., Ma, R., Yu, K.: Prior knowledge driven label embedding for slot filling in natural language understanding. IEEE/ACM Trans. Audio Speech Lang. Process. **28**, 1440–1451 (2020)

Anonymization of Bids in Blockchain Auctions Using Zero-Knowledge Proof

Marlena Broniszewska[1], Wiktor B. Daszczuk[1](✉) ⓘ, and Denny B. Czejdo[2] ⓘ

[1] Department of Electronics and Information Technology, Warsaw University of Technology,
Nowowiejska Street 15/19, 00-665 Warsaw, Poland
`wiktor.daszczuk@pw.edu.pl`

[2] Department of Mathematics and Computer Science, Fayetteville State University, Fayetteville,
NC 28301, USA
`bczejdo@uncfsu.edu`

Abstract. The recent COVID-19 pandemic accelerated the world's digitalization including transformation of traditional auctions. Although buyers cannot examine an item in online auctions, they become a more popular option. They are more convenient since buyers can participate from all around the world. Additional advantage is that the possibility to make auctions more anonymous. Specifically, it may be crucial for the user to keep their interest in the commodity hidden from other buyers and even the seller. This article explores the problem of anonymization in web services by using the example of anonymous online auctions. A system for anonymous, blockchain-based online auctions using zero-knowledge proof is proposed. The system enables a party to attend an auction anonymously digitally isolated from other competitors and the sellers. The system consists of a smart contract on a blockchain and a decentralized web application that facilitates zero-knowledge proof calculation and interactions with the smart contract.

Keywords: blockchain · zero knowledge proof · auction · anonymity · Zk-SNARK

1 Introduction

Evolution of technology [1] and COVID-19 accelerated digitalization. Online shopping and traditional auctions moved to the "e-world" for convenience and global access, despite the inability to examine items beforehand.

Digitalization has advantages but also risks, such as personal data disclosure. Collected data can be used for malicious purposes. Anonymity during auctions may be necessary for privacy.

Bitcoin Whitepaper [2] accelerated blockchain technology development. It allows decentralized, transparent systems without centralized control. Blockchain-based systems can prevent malicious auction bidding.

Zero-knowledge proof is a cryptographic method that provides privacy and security. It enables one party to prove possession of information without revealing it. It can be used in blockchain-based systems for anonymity.

W. Zamojski et al. (Eds.): DepCoS-RELCOMEX 2024, LNNS 1026, pp. 19–28, 2024.
https://doi.org/10.1007/978-3-031-61857-4_2

This article investigates how to create an anonymous online auction system using blockchain and zero-knowledge proof. The contribution is hiding the fact of participation in the auction without third-party involvement. Compared to other concepts described in the next section, our approach goes the furthest in hiding.

2   Related Work

The surge in online selling, buying, and bidding has made electronic auctions a pivotal part of e-commerce [3]. While they offer efficiency benefits, security concerns persist, especially in comparison to traditional auctions [4]. Centralized systems demand trust in third-party intermediaries, raising transparency, integrity, and traceability issues. Cryptography addresses central auctioneer problems [5], but alliances among auctioneers pose risks [6]. Blockchain emerges as a privacy-focused solution [7], but challenges like collusion among authorities remain [8]. Anonymous auctions using time-release encryption are proposed [9], yet reliance on third-party auctioneers persists. Privacy-centric blockchain solutions extend to payments [10] and smart contracts replacing central auctioneer [15]. The authors of [11] use the zero-knowledge proof for the secrecy and validity of the transaction. Still, bid privacy challenges persist, raising the need for innovative schemes.

In the broader blockchain context, zero-knowledge proof technology assures privacy and anonymity [12]. Several projects integrate zero-knowledge proof, such as confidentiality assurance in transactions [13] and fund origin concealment in mixers [14]. Platforms like Polygon Hermez leverage zero-knowledge protocols to trim data storage on blockchains [15].

3   Blockchain

Blockchain is a decentralized digital ledger that stores data in blocks shared across nodes, making it nearly impossible to modify or remove data once stored. It has three key features: immutability - transactions are unchangeable, decentralization - execution without a central authority, and consensus ensuring security and validating transactions [16]. Various blockchains use different consensus mechanisms (e.g., Proof of Work or Proof of Stake).

Many blockchain networks are used worldwide, such as the oldest Bitcoin 2010, private IBM blockchain, or one of the most established - Ethereum. The Ethereum blockchain is used in this article, so any further references to the blockchain network apply to the Ethereum network.

The blockchain stores data in blocks, each with a header and a batch of transactions. Blocks reference previous blocks through cryptographic hashes for a continuous order. Nodes keep their own copy, updating when new blocks are added. Miners add transactions, solve puzzles to add blocks and receive rewards. Ethereum switched from Proof of Work (PoW) to Proof of Stake (PoS) in September 2022.

Proof of Work entails miners solving complex puzzles to add blocks, consuming significant energy. [17]. To address this, Ethereum transitioned to Proof of Stake, where validators stake ETH coins as collateral, eliminating the need for energy-intensive puzzles. Dis-honest behavior risks the destruction of staked coins as a penalty in PoS, which does not involve puzzle-solving.

3.1 Ethereum Accounts

There are two types of accounts in the Ethereum blockchain, both used in the solution:

Externally Owned Accounts. EOA is an account that can hold, receive and send funds controlled by a public and private key pair. It has a unique 42-character hexadecimal address and can only store balance information [18].

Smart Contract Accounts. Smart Contract Accounts are a type of Ethereum blockchain account with a unique address and balance, as well as associated code. They can store any type of data and perform actions through transactions initiated by EOA or SC Accounts. To create a Smart Contract Account, a Smart Contract must be deployed to the Ethereum network.

A wallet manages public and private keys for EOA, allowing users to send and receive cryptocurrencies and interact with their accounts.

3.2 Transactions

A transaction is a signed message that can change the state of the blockchain network. It has a specified structure and contains data such as sender and recipient addresses, the amount of cryptocurrency being sent, and a signature created using the sender's private key to ensure its validity. Each transaction must contain, among others:

- Recipient - the recipient address
- Value - the amount of Ether coin sent to the recipient
- Data - data sent to a Smart Contract
- A ECDSA digital signature - signature created using a private key associated with the account that is sending a transaction

If a party wants to send a transaction, they must pay for a gas (computation resources) used to process such transaction by a node. The currency used in Ethereum is Ether coin. Ether is a cryptocurrency used for paying for executing transactions such as smart contract deployment. Each blockchain can have its own cryptocurrency that is used to pay for sending transactions, e.g., for Polygon blockchain, the Matic cryptocurrency is used. There are two types of transactions:

- Smart contract creation (deployment) transaction
- Message call transaction

  - Ether transfer transaction from one account to another
  - Smart Contract interaction transaction

The Ethereum network is a public network, hence anyone can read and write data to the blockchain. It means that every transaction sent to the Ethereum blockchain is publicly available, and anyone can access it. To send a transaction to the Ethereum blockchain, a transaction fee must be paid. In this particular blockchain, Ether (ETH) cryptocurrency is used for such payment.

A smart contract is blockchain code with its state, changed only by transactions. Ethereum, using Solidity language, employs the EVM (Ethereum Virtual Machine) to execute and update after each block. Smart contract interactions are irreversible, stressing the importance of careful code writing and verification. Transactions on Ethereum are atomic; if an issue arises, the entire transaction rolls back, preventing state changes in the smart contract.

4 Zero-Knowledge Proof

ZKP [19] is a cryptography method where a party can prove to another party the validity of a statement without disclosing the knowledge itself. In a verification process, a prover proves to a verifier that they possess specific knowledge without revealing it. An example use case is A proving to B that their income is in the required range for a mortgage without revealing the exact income value.

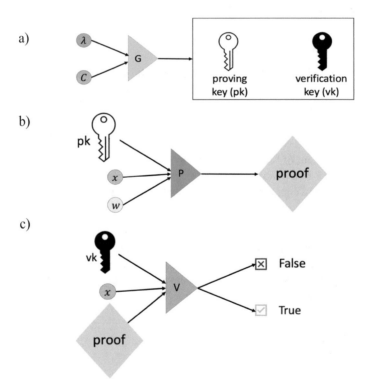

Fig. 1. The ZKP functions: a) keys generation G, b) prover P, c) verifier V

A simple example of how zero-knowledge proof works can be Alice proving to Bob, that she knows the value of x, such that $f(x)$: $g^x = h \pmod{p}$, where g and h are publicly known values and p is a large prime number. It is simple to know x and compute $f(x)$, but it is difficult to do it conversely. Alice can prove to Bob that she knows x without revealing anything about the x. She can do it by choosing a random value r and sending g^r (mod p) and $h \times g^{-r}$ (mod p) values to Bob. Then, Bob can verify that $g^r \times (h \times g^{-r})$ (mod p) $= g^x \times g^{-r} = h$ (mod p), without knowing the value of x.

Two types of ZKP are used: interactive and non-interactive. As the proof must be made by a smart contract procedure, in blockchain non-interactive proofs are applied.

Zk-SNARK is an abbreviation for zero-knowledge succinct non-interactive argument of knowledge, and it is a ZKP-based protocol increasingly used in blockchain applications. Zk-SNARK protocol has additional features [20] in comparison to zero-knowledge protocol, which are:

- Succinct—The proof must be neat enough to be verified in milliseconds.
- Non-interactive—A single message from the prover to the verifier is required to execute the proof transcript.
- Argument of knowledge—It is nearly impossible to create a valid proof without access to the witness (the secret information), and the constructed proof fulfills the'soundness' requirement.

There are three main functions in Zk-SNARK protocol to conduct the proof computation and verification: G, P, V (Fig. 1).

The key generation function G (Fig. 1a) takes two parameters as input, which are a randomly generated, secret λ parameter and a public, specific program C. As an output, function G creates 2 keys publicly accessible: proving key pk and the verification key vk. For a given program C, these keys need to be generated only once. These keys are not related to EOA keys.

The prover function P (Fig. 1b) takes the proving key pk, created by the G function, public input x and private witness w (the secret information) as input. The x parameter is a SHA256 hash of w. This means the x does not have to be private since it is impossible to recreate the original information (the witness w) back from the x. As an output, the P function creates proof that the prover knows the secret information (witness) and the witness is valid.

The verifier function V (Fig. 1c) takes the verification key vk generated by the G function, the proof received from the prover P, and the public input x. The x value is the same as for generating the proof (input of P function). Function V verifies the proof, and if the proof received from the prover is valid the V function returns *true*, if the proof is invalid the function returns *false*.

The Zk-SNARK protocol allows for proving information without revealing it, but generating public parameters for the proving and verification keys (also known as the Common Reference String (CRS) [21]) can be a weak point. Access to the secret λ parameter can lead to the creation of fake proofs. Multi-party computation of public parameters can increase security, but requires trust in each party [22]. The Zk-STARK protocol eliminates the need for trust by using a different approach, but it is not discussed in this article.

5 Proposed Architecture of Zero-Knowledge Proof Based Blockchain Auctions

5.1 Proposed Solution

This article proposes a system for anonymous blockchain auctions using zero-knowledge proof. Bidders lodge their public keys with the auction administrator to create the auction system while maintaining privacy.

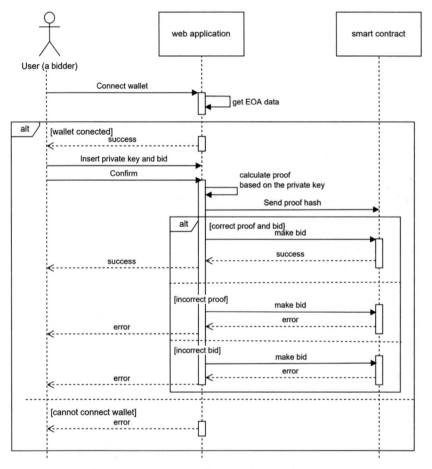

Fig. 2. UML sequence diagram of the auction system

Users can join the auction by creating a zero-knowledge proof of private key owner-ship associated with a public key lodged with the auction administrator. The proof and bid are sent to the auction smart contract, which verifies their correctness. The bidder must offer a higher bid than the current highest one to have the transaction accepted.

5.2 Architecture

The system uses a smart contract on a blockchain. The seller creates the contract with auction details and authorized users. Users must register with their public keys to participate. Bidders can make bids by sending blockchain transactions. Users must provide evidence of eligibility to participate.

To keep the participant's identity private, the system uses zero-knowledge proof. The smart contract stores public keys instead of personal information. Bidders must send a proof of ownership of their private key associated with the public key in the smart contract. The zero-knowledge proof only confirms if a user is eligible to participate.

Using zero-knowledge proof in blockchain auctions helps keep the bidders interest in attending the auction privately. Bidders can place their bid once, and the proof of private key ownership ensures authenticity. The proposed application is a decentralized application (dApp) that runs autonomously on a blockchain network, providing several advantages such as being resistant to centralized control.

The system has two main elements: a blockchain smart contract and a web application dApp. Users interact with the dApp, which communicates with the auction smart contract. Web3 is a new version of the web that hosts decentralized apps on blockchain technology, as opposed to the read-only and read-write webs of the past [23].

The web app dApp has two functions (Fig. 2): prove private key ownership and send proof to auction contract via EOA. To send a transaction, the user needs an Ethereum account with enough cryptocurrency. There are two ways to pay for the commodity: prepay and get reimbursed if not winning, or charge only the winner after the auction ends.

This approach does not guarantee the seller gets paid and avoids multiple charging transactions. Users can participate in multiple auctions with a lower EOA balance. To use the account, the user connects a wallet and authorizes transactions to prevent misuse. After a successful connection, the user inserts their private key and bid amount.

Inserting bid amount and private key allows the app to send two transactions to the smart contract: one with the proof hash and EOA address, the other with proof and bid. The app waits for the SC response and the bidder becomes temporary winner if the bid is correct and highest.

To join an auction, user sends proof SHA256 hash once, stored with their EOA address in the contract. Only the original sender can use this proof to bid. Each proof hash can only be saved once and cannot be modified. After sending proof and bid, contract takes action.

Smart contract validates proof first. If it's valid, it proceeds to validate the bid's correctness (higher than current highest bid). Correct bids are saved with the proof, making the sender the winner. Public/private Ethereum keys differ from keys to create auction proof. Figure 2 displays the cooperation of dApp with Auction smart contract.

The user connects the wallet, enters the keys and bid. dApp sends the transaction to Auction smart contract for verification. Correct bids are saved, and incorrect ones are rejected.

6 Implementation of Zero-Knowledge Proof Based Blockchain Auction

6.1 Tools

The implementation of the designed auction system was carried out using the tools: ZoKrates for creating zero-knowledge proof procedures and generating Solidity smart contracts based on them, React.js for building web UI, Ethers.js to connect and interact with the Polygon network and other Ethereum-based blockchains.

6.2 Security Tests

Security is crucial in smart contract development as the code cannot be modified or removed once deployed. There are various predefined security vulnerabilities to consider.

Front running attack [24] is a known vulnerability where dishonest miners can steal a user's transaction by submitting their own transaction with the same proof before the original transaction is included in a new block. To prevent this, users must first send a transaction with a proof hash, and then with the proof and bid. The Auction smart contract stores proof hashes and sender addresses to distinguish authorized users. If a miner steals the proof hash, they cannot participate in the auction as they are unable to generate the actual proof from its hash.

The Auction smart contract is not susceptible to the reentrancy attack as it does not use external contract functions or loops [24].

Function default visibility vulnerability arises when public functions can be executed by anyone. The Auction smart contract has two public functions for saving proof hashes and placing bids. A private proof verification function, exclusively callable by the Auction contract, is unnecessary for user participation in the auction.

6.3 Financial Analysis

Gas is used to determine the transaction cost in Ethereum blockchain. Gas is the amount of computational effort required to execute a transaction. The total cost of a transaction depends on the amount of gas used and the current gas cost. The analysis shows that each activity: deployment of an auction, saving a proof hash and making a bid are about 22 thousand times cheaper in Polygon than in Ethereum. These costs in Etehereum are about 93USD, 1.70USD and 11.6USD, respectively. Polygon is cheaper due to lower gas price and higher processing capability.

7 Summary

This article proposes a decentralized system for anonymous blockchain auction, consisting of an Auction smart contract and a web dApp. The smart contract stores a list of public keys and a ZK proof validation mechanism, while the dApp allows users to create a ZK proof and bid without revealing their participation.

The system maintains users' privacy by using zero-knowledge protocol. Users must create a ZK proof to participate in the auction, which reveals nothing about their identity. The web application simplifies the process of creating the proof and sending a transaction to the blockchain.

Blockchain prevents bids and data manipulations by recording each transaction and making it verifiable by anyone. It also makes it difficult to send malicious transactions and manipulate or remove data sent to the contract. The Auction smart contract sends a SHA256 hash of the proof to the contract before the actual transaction.

Financial analysis was conducted to compare the total cost of using the proposed auction system on Polygon and Ethereum blockchains. The results showed that using the system on Polygon is cheaper, which can be a significant factor for users in deciding whether to use the system or not.

To make the system profitable, entrance and service fees could be charged. An entrance fee could be charged to any user participating in the auction after placing their first bid. A service fee could be charged to the seller and can be a percentage of the final commodity price or a predetermined value.

The auction system is open to updates and adjustments to incorporate new features in the zero-knowledge protocol and blockchain technology.

References

1. McCain, A.: How fast is technology advancing? growing, evolving, and accelerating at exponential rates (2023). https://www.zippia.com/advice/how-fast-is-technology-advancing/
2. Nakamoto, S.: Bitcoin: a peer-to-peer electronic cash system (2008). https://bitcoinwhitepaper.co/
3. Bajari, P., Hortaçsu, A.: Economic Insights from internet auctions. J. Econ. Lit. **42**, 457–486 (2004). https://doi.org/10.1257/0022051041409075
4. Trevathan, J.: Privacy and security in online auctions (2007). https://researchonline.jcu.edu.au/1788/2/02whole.pdf
5. Franklin, M.K., Reiter, M.K.: The design and implementation of a secure auction service. IEEE Trans. Softw. Eng. **22**, 302–312 (1996). https://doi.org/10.1109/32.502223
6. Montenegro, J.A., Fischer, M.J., Lopez, J., Peralta, R.: Secure sealed-bid online auctions using discreet cryptographic proofs. Math. Comput. Model. **57**, 2583–2595 (2013). https://doi.org/10.1016/j.mcm.2011.07.027
7. Sharma, G., Verstraeten, D., Saraswat, V., Dricot, J.-M., Markowitch, O.: Anonymous sealed-bid auction on ethereum. Electronics **10**, 2340 (2021). https://doi.org/10.3390/electronics10192340
8. Sonnino, A., Król, M., Tasiopoulos, A.G., Psaras, I.: AStERISK: auction-based shared economy resolution markets for blockchain platforms. In: 2019 Workshop on Decentralized IoT Systems and Security. Internet Society, Reston (2019). https://doi.org/10.14722/diss.2019.230001
9. Xiong, J., Wang, Q.: Anonymous auction protocol based on time-released encryption atop consortium blockchain. Int. J. Adv. Inf. Technol. **09**, 01–16 (2019). https://doi.org/10.5121/ijait.2019.9101
10. Yang, X., Li, W.: A zero-knowledge-proof-based digital identity management scheme in blockchain. Comput. Secur. **99**, 102050 (2020). https://doi.org/10.1016/j.cose.2020.102050

11. Galal, H.S., Youssef, A.M.: Verifiable sealed-bid auction on the ethereum blockchain. In: Zohar, A. (ed.) FC 2018. LNCS, vol. 10958, pp. 265–278. Springer, Heidelberg (2019). https://doi.org/10.1007/978-3-662-58820-8_18

12. Wang, Z., et al.: On how zero-knowledge proof blockchain mixers improve, and worsen user privacy. In: ACM Web Conference 2023, Austin, TX, 30 April–4 May 2023, pp. 2022–2032. ACM, New York (2023). https://doi.org/10.1145/3543507.3583217

13. Hopwood, D.E., Bowe, S., Hornby, T., Wilcox, N.: Zcash protocol specification (2023). https://raw.githubusercontent.com/zcash/zips/master/protocol/protocol.pdf

14. Pertsev, A., Semenov, R., Storm, R.: Tornado cash privacy solution (2019). https://crebaco.com/planner/admin/uploads/whitepapers/2982941Tornado.cash_whitepaper_v1.4.pdf

15. Sutopo, A.H.: Blockchain programming smart contract on polygon (2023). ISBN: 9786239285241

16. Wright, C.S.: Bitcoin: a peer-to-peer electronic cash system. SSRN Electron. J. (2008). https://doi.org/10.2139/ssrn.3440802

17. Ethereum's energy expenditure. https://ethereum.org/en/energy-consumption/

18. What is an ethereum address? https://info.etherscan.com/what-is-an-ethereum-address/

19. Goldwasser, S., Micali, S., Rackoff, C.: The knowledge complexity of interactive proof-systems. In: 17th Annual ACM Symposium on Theory of Computing - STOC 1985, December 1985, pp. 291–304. ACM Press, New York (1985). https://doi.org/10.1145/22145.22178

20. Petkus, M.: Why and how zk-SNARK works: definitive explanation. http://infosec.pusan.ac.kr/wp-content/uploads/2019/06/WhyAndHowZkSnarkWorks.pdf

21. Baumann, P., Campalani, P., Yu, J., Misev, D.: Finding my CRS: a systematic way of identifying CRSs. In: 20th International Conference on Advances in Geographic Information Systems, Redondo Beach, CA, 6–9 November 2012, pp. 71–78. ACM, New York (2012). https://doi.org/10.1145/2424321.2424332

22. Park, C., Chung, M., Ryu, D.: A Blockchain-based protocol of trusted setup ceremony for zero-knowledge proof. In: 2023 5th Blockchain and Internet of Things Conference, Osaka, Japan, 21–23 July 2023, pp. 35–40. ACM, New York (2023). https://doi.org/10.1145/3625078.3625083

23. Murray, A., Kim, D., Combs, J.: The promise of a decentralized internet: what is Web3 and how can firms prepare? Bus. Horiz. **66**, 191–202 (2023). https://doi.org/10.1016/j.bushor.2022.06.002

24. Samreen, N.F., Alalfi, M.H.: A survey of security vulnerabilities in ethereum smart contracts. In: CASCON 2020: the 30th Annual International Conference on Computer Science and Software Engineering, 10–13 November 2020, pp. 73–82 (2020). https://doi.org/10.5555/3432601.3432611

Survival Signature for Reliability Quantification of Large Systems and Networks

Frank P. A. Coolen$^{(\boxtimes)}$ and Tahani Coolen-Maturi

Durham University, Durham DH1 3LE, UK
{frank.coolen,tahani.maturi}@durham.ac.uk

Abstract. The survival signature is a useful tool for quantification of reliability of large systems and networks with relatively few types of components. This paper provides an introductory overview of the survival signature, with emphasis on recent developments and challenges to enable its use for practical applications. Topics discussed include different survival signatures for specific scenarios, the level of detail in reliability modelling, and computational aspects.

In the literature, system reliability quantification is mostly focused on binary state systems with typically only few components in relatively straightforward configurations and with single functions. Real-world systems, on the other hand, often have multiple levels of functioning and consist of many components in a variety of configurations while they may need to perform multiple functions, leading to substantial challenges for reliability quantification.

In the twelve years since its introduction, the survival signature has gained much attention in the literature, and progress has been made on the challenges indicated above. However, many challenges remain, including some theoretical questions about the very nature of system reliability in real-world situations.

Keywords: Large systems · Survival signature · System reliability · Networks

1 Survival Signature

Coolen and Coolen-Maturi [6] introduced the survival signature for quantification of reliability of binary state systems with binary state components. Consider a system with $K \geq 1$ types of components, with n_k components of type $k \in \{1, 2, \ldots, K\}$ and $\sum_{k=1}^{K} n_k = n$. The essential assumption is that the random failure times of components of the same type are exchangeable [9,13]. The state vector $\underline{x} \in \{0, 1\}^n$ of the system describes the states of its components, with 1 representing that a component functions and 0 that it does not function. The system structure function $\phi(\underline{x}) \in \{0, 1\}$ describes the functioning of the system given the component states $\underline{x}$, where 1 represents that the system

© The Author(s), under exclusive license to Springer Nature Switzerland AG 2024
W. Zamojski et al. (Eds.): DepCoS-RELCOMEX 2024, LNNS 1026, pp. 29–37, 2024.
https://doi.org/10.1007/978-3-031-61857-4_3

functions and 0 that it does not function. Due to the arbitrary ordering of the components in the state vector, components of the same type can be grouped together, leading to a state vector that can be written as $\underline{x} = (\underline{x}^1, \underline{x}^2, \ldots, \underline{x}^K)$, with $\underline{x}^k = (x_1^k, x_2^k, \ldots, x_{n_k}^k)$ the sub-vector representing the states of the components of type k.

The *survival signature*, denoted by $\Phi(l_1, l_2, \ldots, l_K)$, with $l_k = 0, 1, \ldots, n_k$ for $k = 1, \ldots, K$, is defined as the probability that the system functions given that *precisely* l_k of its n_k components of type k function, for each $k \in \{1, 2, \ldots, K\}$. There are $\binom{n_k}{l_k}$ state vectors $\underline{x}^k$ with $\sum_{i=1}^{n_k} x_i^k = l_k$; let S_l^k denote the set of these state vectors for components of type k and let $S_{l_1, \ldots, l_K}$ denote the set of all state vectors for the whole system for which $\sum_{i=1}^{n_k} x_i^k = l_k$, $k = 1, 2, \ldots, K$. Due to the exchangeability assumption for the failure times of the n_k components of type k, all the state vectors $\underline{x}^k \in S_l^k$ are equally likely to occur, hence

$$\Phi(l_1, \ldots, l_K) = \left[\prod_{k=1}^{K} \binom{n_k}{l_k}^{-1} \right] \times \sum_{\underline{x} \in S_{l_1, \ldots, l_K}} \phi(\underline{x}) \tag{1}$$

The survival signature is useful for deriving the probability for the event that the system functions at time $t > 0$, so for $T_S > t$, where T_S is the random system failure time. Let $C_k(t) \in \{0, 1, \ldots, n_k\}$ denote the number of components of type k in the system which function at time $t > 0$, then

$$P(T_S > t) = \sum_{l_1=0}^{n_1} \cdots \sum_{l_K=0}^{n_K} \left\{ \Phi(l_1, \ldots, l_K) P\left(\bigcap_{k=1}^{K} \{C_k(t) = l_k\} \right) \right\} \tag{2}$$

Equation (2) is the central result for the use of survival signatures to quantify system reliability through the system survival function, and, with that function, important metrics such as the system's expected failure time or the remaining time until system failure at any moment during its functioning. The key aspect here, as seen from Equation (2), is that the information about the system structure is contained in one factor of the formula, namely the survival signature, which is completely separated from information about the component failure times, which is included through the probabilities for the numbers of functioning components at time t. This has several major advantages, including the opportunity to easily investigate the effect of changes to the system structure on its reliability and the possibility to easily compare different system structures [28].

The survival signature requires specification at $\prod_{k=1}^{K} (n_k + 1)$ inputs while the structure function must be specified at 2^n inputs; in particular for large systems (large values of n) with relatively few component types (small values of K), this can lead to a substantial difference which also simplifies storage of the function. It should be noted that the survival signature is entirely equivalent to the structure function in case all components are of different types (so $K = n$). It generalizes the system signature, presented by Samaniego [26, 27] to multiple types of components, as Samaniego's signature is only defined when all components in the system are of a single type. Whilst Samaniego's signature has had a substantial impact on the literature on system reliability, its practical value was limited because very few large practical systems and networks consist of only a single type of components.

Equation (2), and hence the whole survival signature approach to system reliability, requires only the assumption that failure times of components of the same type are exchangeable. Further simplifications to the derivation of the system survival function occur if one adds additional assumptions. In particular, assuming that the failure times of components of different types are independent leads to

$$P(T_S > t) = \sum_{l_1=0}^{n_1} \cdots \sum_{l_K=0}^{n_K} \left\{ \Phi(l_1, \ldots, l_K) \prod_{k=1}^{K} P(C_k(t) = l_k) \right\} \tag{3}$$

The further assumption that the failure times of components of the same type are independent and identically distributed (*iid*) with cumulative distribution function (CDF) $F_k(t)$ for type k, leads to

$$P(T_S > t) = \sum_{l_1=0}^{n_1} \cdots \sum_{l_K=0}^{n_K} \left\{ \Phi(l_1, \ldots, l_K) \prod_{k=1}^{K} \binom{n_k}{l_k} [F_k(t)]^{n_k-l_k} [1 - F_k(t)]^{l_k} \right\} \tag{4}$$

One can also assume a parametric CDF to enable learning about the parameter based on data, e.g. using Bayesian statistics [2], or use a frequentist statistical method, for example Nonparametric Predictive Inference [10,11]. The general formula for the system survival function, Eq. (2), can also be applied if components' failure times are dependent, for example there may be common-cause failure modes, a risk of cascading failures, load sharing between components and so on. Initial studies into several of such possibilities have been published [7,15,16] and there are many related research challenges, in particular on modelling actual dependencies in real-world scenarios.

In the remainder of this paper, a brief overview of the theoretical developments of the survival signature concept is presented, together with a discussion of some key challenges for practical implementation of the concept to challenging real-world scenarios. Section 2 discusses several generalizations of the survival signature while computational aspects are considered in Sect. 3. Some possibly less obvious issues for practical modelling to quantify reliability of large-scale systems are discussed in Sect. 4. The paper ends with brief mentioning of some further results and challenges in Sect. 5.

2   Generalized Survival Signatures

The survival signature concept has been generalized in several ways. A crucial generalization is for multi-state systems, where both the system and components can have multiple states ranging from perfect functioning to failure. Qin and Coolen [23] presented this generalization, enabling a wide variety of applications to be developed, including support for inspection and replacement decisions. Of course, the processes of state changes for components must be modelled and mapping the components' states to the system state can be a complex task, but this is anyhow required if one aims at reliability quantification of multi-state systems.

Coolen-Maturi et al. [12] generalised the concept of the survival signature for multiple systems with multiple types of components and with some components shared between systems. A particularly important feature is that the functioning of these systems can be considered at different times, enabling computation of relevant conditional probabilities with regard to a system's functioning conditional on the status of another system with which it shares components. This is important for many practical systems and networks, for example computer networks, but the theory has a wider relevance as it can directly be applied to a system which performs multiple functions, which is the case for many practical systems. This has led to a substantial area of research, typically considering specific reliability scenarios or restricted system structures, e.g. Yi et al. [31] consider systems with a monotone structure function.

Huang et al. [19] presented survival signatures for general phased-missions scenarios, which have several additional challenges such as the possibility that not all components of one type function in the same phases. Whilst such modelling can become rather complex, it does provide a framework for decision support, for example on optimal ordering of phases or re-ordering in case some components are known to have failed during the process.

3   Computational Aspects

For reliability of small systems and networks one can simply derive the system structure function and use Eq. (1) to compute the survival signature. This approach has been implemented in the statistical software R [1], and can be used for small to medium-sized systems and networks. Reed [24] presented a substantial improvement on the required computation time by using binary decision diagrams, which can also be used for reliability of multi-terminal networks [25]. Using basic combinatorics, one can compute the survival signature of a system consisting of two subsystems in either series or parallel configuration, if the survival signatures of those subsystems are available; this enables quick computation of the survival signatures of series-parellel systems of any size [11]. As a generalization of this combinatorial method, the survival signature for a multi-state system can be easily derived from the survival signatures of its subsystems if the state of the system is a function of the states of the subsystems [23].

The main reason for the introduction of the survival signature is to enable quantification of system reliability, and related statistical inferences, for large real-world systems and networks, for which one normally would not have the full structure function available. We can think here, for example, about large industrial systems or transportation networks with thousands of components. For such cases, one may need to approximate the survival signature. To do so, it is particularly useful that the survival signature of a coherent system is an increasing function. Approximating the survival signature has received much attention. For example, Behrensdorf et al. [4] use percolation theory to exclude areas of the input space of the survival signature where its value does not increase, followed by approximation of the survival signature in the other parts of the input space by Monte Carlo (MC) methods. They illustrate their method on a model of the Great Britain (GB) electricity transmission network, consisting of 29 nodes of two types, and on a model of the Berlin metro network, consisting of 306 nodes and 350 edges, with the nodes divided into two types based on their degree. Also using MC, Di Maio et al. [14] use entropy to direct the sampling towards non-trivial areas of the input space, and they illustrate their method on the same GB electricity transmission network. Recently, Lopes da Silva and Sullivan [29] have presented a powerful method to approximate the survival signature for two-terminal networks with two types of components. They show that each MC replication to estimate the survival signature entails solving a multi-objective maximum capacity path problem, and adapt a Dijkstra-like bi-objective shortest path algorithm to solve this problem. They show the efficiency of their algorithm compared to other approaches, which increases with the size of the network, by application to several networks including a power system, which has 4,000 nodes and 29,336 arcs and includes cycles and self-loops.

Once the survival signature of a system or network has been derived, or approximated, it is a useful tool for a range of objectives. To investigate system reliability one often has to rely on simulation methods, and Patelli et al. [22] have shown that the use of the survival signature, instead of the system structure function, can lead to extremely efficient simulations. The key idea here is that one simulates n_k components failure times for each component type k and orders all these times. Then moving from time 0 on, one uses the survival signature in the natural way to derive what can be considered as a kind of empirical survival function, instead of just a single system failure time which such a simulation would provide when based on the structure function. So, in a single simulation run, when gets a simulated value for the probability that the system functions for every time point t, which enables good estimation of the system survival function after relatively few simulation runs, with the accuracy based on the standard estimation theory for binomial proportions and hence with pointwise confidence intervals quantifying the accuracy of the estimates. This simulation approach was extended by George-Williams et al. [18] for the case of dependence between component failure times.

The survival signature approach is also helpful for statistical inference on system reliability, which is important due to the need to learn from data, and possibly also including expert judgements, in practical applications. Of course, if failure data are available for the individual component types, for example from previous use or experimental settings, then this is quite straightforward to take into account in Eq. (2), leading to the corresponding inferences for the system reliability. A frequentist and fully nonparametric approach for such inferences has been presented [3], which uses Nonparametric Predictive Inference [10], a statistical approach based on few modelling assumptions made possible by the use of imprecise probabilities. This appraoch can also be used to derive bounds for the system survival function if one only has partial information about either the component failure data or the system structure [11].

In many practical applications one may have limited data on component failures, in which case Bayesian methods may be useful, in particular as these enable the inclusion of expert judgements. The combination of Bayesian methods for inference on the component failure times and the survival signature has been presented [2]. Walter et al. [30] presented a generalized Bayesian approach combined with the survival signature by using sets of priors, as typically done in theory of robust Bayesian methods. An interesting feature of their work is that, through a specific way of defining the sets of prior distributions, conflict between data and prior judgements can be highlighted during the Bayesian updating process. This may be particularly important in applications where it would turn out the prior judgements may have been too optimistic, where such prior-data conflict would indicate that the system reliability could be substantially lower than was originally thought.

4   Modelling for System Reliability

The main challenges for reliability quantification for large systems and networks result from the size and complexity of real-world systems and networks. This includes many factors, such as functional requirements and environmental circumstances, and it may well be that no two components are believed to have exchangeable failure times. Also, one may wish to distinguish many different functioning states for components and systems. However, the key issue in developing a mathematical model for the reliability of a practical system is the target aim of the model, which is typically to support some decision processes. Crucially, there tend to be limited time and resources for the modelling, which will imply that one does not need to, or even can, include all factors that could distinguish between the random failure types of components in the model. Crucially, exchangeability assumptions for the failure times of different components should not be interpreted as strong judgements of their failure time distributions being identical, but instead they are choices with regard to the level of modelling of the system [9]. Similarly, the number of functioning states is typically determined by the corresponding decision problems for which the modelling is undertaken. The main consideration here, from practical perspective, is to choose levels of modelling which are suitable for the tasks whilst being achievable given practical constraints, and not to aim at models which are more detailed representations of reality than is necessary.

An aspect of quantification of system reliability which has received relatively little attention in the literature, but is of great practical importance, is the choice of which components to include in the study. Intuitively, one would consider it necessary to provide a complete model, but for large systems the definition of what components actually are, and which are relevant for describing the reliability of the system for a specific application, has received little attention. From this perspective, Coolen and Coolen-Maturi [8] argued in favour of a change of the nature of the system structure function, from deterministic to stochastic, meaning that the system structure function value for given states of the components is a probability distribution over possible states, rather than a single state. This would enable modelling of system reliability based on only the states of a subset of its components, whilst statistical inference would remain possible based on data from the process. It could also reflect uncertain influences on the system reliability which may not be taken into account explicitly, such as variable environmental circumstances or variations in the use of the system. This is an area where substantial research progress would be needed, which would best be based on practical applications.

5   Concluding Remarks

This paper has provided an introductory overview of the survival signature with some discussion of recent developments and main challenges. Many examples of powerful methodology for system reliability quantification enabled by the use of survival signatures have not been discussed, these include new component reliability importance measures [17], resilience achieved by swapping components within a system [21], reliability-redundancy allocation [20], stochastic comparison of different systems [28] and stochastic processes to describe the system reliability over time with varying assumptions on loads or failure processes [5]. It should be noticed that quite some well-known decision processes, in particular management of systems which requires planning of inspections and maintenance activities, fit very well with the level of modelling corresponding to the survival signature. For example, on determining stocks of spare components it is typically not relevant, if there are multiple components of the same type in a system, to consider which specific component may fail at some future time, only its type may be relevant. There may also be opportunities to apply the survival signature concept to scenarios away from traditional engineering, for example the resilience of big organisational structures could be modelled with survival signatures if an organisation has a workforce which mainly consists of groups of workers with exchangeable skills.

Since its introduction, just over a decade ago, the survival signature has received increasing attention from academic researchers and has been acknowledged to be a powerful tool for quantification of reliability of systems and networks. The main intention has always been to provide a practical tool that enables upscaling of reliability quantification, and related statistical inference and decision support, to large scale practical systems and networks, consisting of thousands of components. The route to achieve this still has several major challenges, including further development of computational methods and new ways to model the practically relevant aspects of systems. The achievements of the survival signature, as reported in the literature thus far, are substantial and indicate that a step-change in reliability quantification for large-scale practical systems and networks is feasible.

References

1. Aslett, L.J.M.: Reliability theory: tools for structural reliability analysis. R package (2012)
2. Aslett, L.J.M., Coolen, F.P.A., Wilson, S.P.: Bayesian inference for reliability of systems and networks using the survival signature. Risk Anal. **35**, 1640–1651 (2015)
3. Augustin, T., Coolen, F.P.A., de Cooman, G., Troffaes, M.C.M.: Introduction to Imprecise Probabilities. Wiley, Chichester (2014)
4. Behrensdorf, J., Regenhardt, T.E., Broggi, M., Beer, M.: Numerically efficient computation of the survival signature for the reliability analysis of large networks. Reliabil. Eng. Syst. Saf. **216**, 107935 (2021)
5. Chang, M., Huang, X., Coolen, F.P.A., Coolen-Maturi, T.: New reliability model for complex systems based on stochastic processes and survival signature. Eur. J. Oper. Res. **309**, 1349–1364 (2023)
6. Coolen, F.P.A., Coolen-Maturi, T.: Generalizing the signature to systems with multiple types of components. In: Zamojski, W., et al. (eds.) Complex Syst. Dependabil., vol. 170, pp. 115–130. Springer, Heidelberg (2012). https://doi.org/10.1007/978-3-642-30662-4_8
7. Coolen, F.P.A., Coolen-Maturi, T.: Predictive inference for system reliability after common-cause component failures. Reliabil. Eng. Syst. Saf. **135**, 27–33 (2015)
8. Coolen, F.P.A., Coolen-Maturi, T.: The structure function for system reliability as predictive (imprecise) probability. Reliabil. Eng. Syst. Saf. **154**, 180–187 (2016)
9. Coolen, F.P.A., Coolen-Maturi, T.: The survival signature for quantifying system reliability: an introductory overview from practical perspective. In: van Gulijk, C., Zaitseva, E. (eds.) Reliability Engineering and Computational Intelligence. SCI, vol. 976, pp. 23–37. Springer, Cham (2021). https://doi.org/10.1007/978-3-030-74556-1_2
10. Coolen, F.P.A., Coolen-Maturi, T.: Nonparametric predictive inference. In: Lovric, M. (ed.) International Encyclopedia of Statistical Science, 2nd edn. Springer, Heidelberg (2024)
11. Coolen, F.P.A., Coolen-Maturi, T., Al-nefaiee, A.H.: Nonparametric predictive inference for system reliability using the survival signature. Proc. Inst. Mech. Eng. Part O: J. Risk Reliabil. **228**, 437–448 (2014)
12. Coolen-Maturi, T., Coolen, F.P.A., Balakrishnan, N.: The joint survival signature of coherent systems with shared components. Reliabil. Eng. Syst. Saf. **207**, 107350 (2021)
13. De Finetti, B.: Theory of Probability. Wiley, London (1974)
14. Di Maio, F., Pettorossi, C., Zio, E.: Entropy-driven monte carlo simulation method for approximating the survival signature of complex infrastructures. Reliabil. Eng. Syst. Saf. **231**, 108982 (2023)
15. Eryilmaz, S., Coolen, F.P.A., Coolen-Maturi, T.: Marginal and joint reliability importance based on survival signature. Reliabil. Eng. Syst. Saf. **172**, 118–128 (2018)
16. Eryilmaz, S., Coolen, F.P.A., Coolen-Maturi, T.: Mean residual life of coherent systems consisting of multiple types of dependent components. Nav. Res. Logist. **65**, 86–97 (2018)
17. Feng, G., Patelli, E., Beer, M., Coolen, F.P.A.: Imprecise system reliability and component importance based on survival signature. Reliabil. Eng. Syst. Saf. **150**, 116–125 (2016)

18. George-Williams, H., Feng, G., Coolen, F.P.A., Beer, M., Patelli, E.: Extending the survival signature paradigm to complex systems with non-repairable dependent failures. Proc. Inst. Mech. Eng. Part O: J. Risk Reliabil. **233**, 505–519 (2019)
19. Huang, X., Aslett, L.J.M., Coolen, F.P.A.: Reliability analysis of general phased mission systems with a new survival signature. Reliabil. Eng. Syst. Saf. **189**, 416–422 (2019)
20. Huang, X., Coolen, F.P.A., Coolen-Maturi, T.: A heuristic survival signature based approach for reliability-redundancy allocation. Reliabil. Eng. Syst. Saf. **185**, 511–517 (2019)
21. Najem, A., Coolen, F.P.A.: System reliability and component importance when components can be swapped upon failure. Appl. Stoch. Model. Bus. Ind. **35**, 399–413 (2019)
22. Patelli, E., Feng, G., Coolen, F.P.A., Coolen-Maturi, T.: Simulation methods for system reliability using the survival signature. Reliabil. Eng. Syst. Saf. **167**, 327–337 (2017)
23. Qin, J., Coolen, F.P.A.: Survival signature for reliability evaluation of a multi-state system with multi-state components. Reliab. Eng. Syst. Saf. **218**, 108129 (2022)
24. Reed, S.: An efficient algorithm for exact computation of system and survival signatures using binary decision diagrams. Reliabil. Eng. Syst. Saf. **165**, 257–267 (2017)
25. Reed, S., Löfstrand, M., Andrews, J.: An efficient algorithm for computing exact system and survival signatures of k-terminal network reliability. Reliabil. Eng. Syst. Saf. **185**, 429–439 (2019)
26. Samaniego, F.J.: On closure of the IFR class under formation of coherent systems. IEEE Trans. Reliabil. **34**, 69–72 (1985)
27. Samaniego, F.J.: System Signatures and their Applications in Engineering Reliability. Springer, New York (2007). https://doi.org/10.1007/978-0-387-71797-5
28. Samaniego, F.J., Navarro, J.: On comparing coherent systems with heterogeneous components. Adv. Appl. Prob. **48**, 88–111 (2016)
29. Lopes da Silva, D.B., Sullivan, K.M.: An optimization-based monte-carlo method for estimating the two-terminal survival signature of networks with two component classes. Preprint (2023). https://industrial-engineering.uark.edu/_resources/TwoTerminal-SurvivalSig-MC.pdf(version26.05.2023)
30. Walter, G., Aslett, L.J.M., Coolen, F.P.A.: Bayesian nonparametric system reliability using sets of priors. Int. J. Approx. Reason. **80**, 67–88 (2017)
31. Yi, H., Balakrishnan, N., Li, X.: Joint signatures of two or more semi-coherent systems with shared components. Reliabil. Eng. Syst. Saf. **242**, 109713 (2024)

Using Resizing Layer in U-Net to Improve Memory Efficiency

Lehel Dénes-Fazakas[1,2,3(✉)], Szabolcs Csaholczi[1,3,4], György Eigner[1,2], Levente Kovács[1,2], and László Szilágyi[1,2,4]

[1] University Research and Innovation Center, Physiological Controls Research Center, Óbuda University, Budapest, Hungary
{denes-fazakas.lehel,eigner.gyorgy,kovacs,szilagyi.laszlo}@uni-obuda.hu
[2] John von Neumann Faculty of Informatics, Biomatics and Applied Artificial Intelligence Institute, Óbuda University, Budapest, Hungary
[3] Applied Informatics and Applied Mathematics Doctoral School, Óbuda University, Budapest, Hungary
[4] Sapientia - Hungarian Science University of Transylvania, Cluj-Napoca, Romania

Abstract. The segmentation of medical images is becoming increasingly important in the daily clinical practice, particularly with regard to tumor segmentation. Artificial intelligence, which is becoming more and more widespread, is already able to achieve good results in segmentation, making the task of medical staff easier. The U-net is a widely popular convolutional neural network designed to be involved in image segmentation. However, when training the U-net, we may encounter difficulty not having enough memory due to the large size of the images. The main goal of this paper is to investigate the effects of reducing the resolution of the image data using a resizing layer on the input of the U-net and then convert it back at the output also using a resizing layer. The main question is whether we are able to achieve fine segmentation quality. Also it is investigated, what resolution is needed to achieve acceptable segmentation on the dataset. As a result, we managed to achieve an F1-score of 96.9%. Furthermore, we have demonstrated that it is possible to segment the images well with U-net network at lower resolution. Also, it is shown that neither too large nor too small resolution supports good-quality segmentation.

The project no. 2019-2.1.11-TÉT-2020-00217 has been implemented with the support provided by the National Research, Development and Innovation Fund of Hungary, financed under the 2019-2.1.11-TÉT-2020-00217 funding scheme. On behalf of the AI development for diabetic and brain MRI scans project we are grateful for the possibility to use ELKH Cloud (see Héder et al. 2022; https://science-cloud.hu/) which helped us achieve the results published in this paper. Project no. 2019-1.3.1-KK-2019-00007. has been implemented with the support provided from the National Research, Development and Innovation Fund of Hungary, financed under the 2019-1.3.1-KK funding scheme. This project has been supported by the National Research, Development, and Innovation Fund of Hungary, financed under the TKP2021-NKTA-36 funding scheme. The work of L- Szilágyi was supported by the Consolidator Researcher Program of Óbuda University.

W. Zamojski et al. (Eds.): DepCoS-RELCOMEX 2024, LNNS 1026, pp. 38–48, 2024.
https://doi.org/10.1007/978-3-031-61857-4_4

Keywords: Segmentation · U-net · resizing layer · optimal resolution · memory

1   Introduction

Hundreds of thousand people are diagnosed with brain tumor every year, and approximately 18,000 fatal cases are recorded yearly only in the USA [1]. Chances of survival are higher in case of early diagnosis. However, there are other important factors like the type of the tumor [2]. Automatically detecting the tumor in an early phase does not require a precise segmentation of the tumor, but it is relevant that the procedure is sensitive enough to reliably identify small lesions in the brain volume. Brain tumors can have a great variety of sizes, shapes, positions and appearances, and the detection is obstructed by deformations of normal brain structures and various types of noise as well [3].

During the last decade there has been a great rush in the development of brain tumor segmentation methods, generated and supported by the Brain Tumor Segmentation (BraTS) Challenge [4,5], which was organized yearly since 2012. In the first years of the BraTS era, the whole classical arsenal of machine learning theory and practise was tried out [6–9], while lately the deep learning and various convolutional neural networks (CNN) have totally conquered the whole domain of medical image analysis [10,11].

While the BraTS challenge mostly concentrated on gliomas, there are different types of brain tumor lesions, which demand an accurate classification, if we wish to establish a complex brain tumor diagnosis pipeline. Consequently, the brain tumor segmentation problem, can be considered a spin-off of the BraTS challenges, which had its intensification during the past years. Dozens of very recent CNN-based solutions exist, most of which achieve 97–98% accuracy in a three-class classification problem of brain tumors. Beside being based on deep learning and convolutional networks, they further involve: generative adversarial networks [12], deep hybrid representation learning [13], hybrid deep learning approach and Gabor wavelets [14], adaptive attention mechanisms [15,16], deep residual learning framework [17], recurrent CNN [18], parallel deep CNN [19], or multi-path convolution and multi-head attention network [20].

The main aspect of this publication is the segmentation of tumor tissue. As well as segmentation can often run into memory problems. Therefore, this paper experiments using resizing layer in U-net network to use less memory during training. It investigates how good results can be achieved using metrics at different resolutions. It is deduced that it is worth using resizing layer at the beginning and end of U-net network in order to make memory usage efficient and still achieve good results. In Sect. 2 we describe the dataset and the architecture we use. In addition, the training and testing scenario, and the evaluation metrics. In 3 the results are described. In Sect. 4, the results are explained. In 5, conclusions and opportunities for further improvement are discussed.

2   Materials and Methods

2.1   Data

The brain tumor data set involved in the training and evaluation of the U-net model proposed in this study was published by Cheng et al. [21,22]. Images were collected from 233 patients treated at two different state-owned hospitals in Guangzhou and Tianjin, China, between years 2005 and 2010. It consists of 3064 T1-weighted contrast enhanced brain MRI scans, with a high resolution of 512×512 pixels per image, and pixel size of $0.49\,\mathrm{mm}$ in both directions. The data set contains three distinct types of brain tumors, namely Pituitary, Meningioma, and Glioma, in three different planes, including axial, coronal, and sagittal views. All images are accompanied by the tumor mask provided by human experts.

2.2   Network Architectures

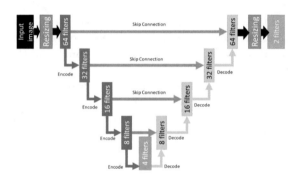

Fig. 1. The proposed network U-net architectures

Our U-net architecture closely aligns with the traditional U-net network, with a key modification to accommodate the original 512×512 resolution of images. Given hardware limitations, we introduced additional resizing layers to manage the challenge posed by the large 512×512 resolution. These resizing layers effectively reduced the resolution to more manageable levels for both training and testing, encompassing dimensions such as 16×16, 32×32, 64×64, 128×128, and 256×256 (Figs. 1 and 2).

The U-net network itself comprises 4 encoder blocks with filter numbers of 64, 32, 16, and 8, respectively. The bridge block represents the smallest form of our network, and after each encoder block, the image size is halved. The filter number in the bridge block is 4. Following the encoder blocks, the decoder branch mirrors the structure of the encoder, with filter numbers in reverse order (8, 16, 32, 64). Resizing layers employ bilinear interpolation.

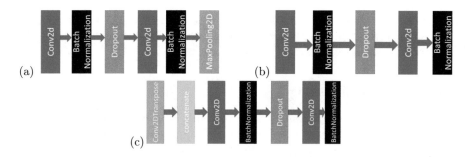

Fig. 2. The proposed network U-net architectures' blocks: (a) encoder, (b) decoder, (c) bridge.

The output layer is a Conv layer consisting of two filters. In our segmentation task, we focused on differentiating backgrounds and tumors within the dataset. Notably, the input images were treated as color images despite having 3 color channels, reflecting the nature of MRI images.

The construction of encoder blocks involves layer-by-layer assembly. Each encoder block's first layer is a Conv2D layer with a filter number matching the block's filters. This layer incorporates ELU activation, employs a 3×3 filter matrix with fill factor, and crucially maintains the image size. Subsequently, a batch normalization layer normalizes the data, followed by a dropout layer (0.2) designed to prevent overfitting.

Continuing, a second Conv2D layer with identical parameters to the first is introduced, followed by another batch normalization layer. The block concludes with a MaxPooling2D layer utilizing a 2×2 pooling matrix, responsible for halving the image dimensions. Importantly, the skip connection mapping originates from the second batch normalization layer.

Our next block, the bridge, comprises fewer layers than the encoder blocks and does not involve dimension reduction. Initiated with a convolution layer featuring 4 filter numbers and a 3×3 filter matrix, it employs ELU activation with a fill factor to maintain image size. A subsequent batch normalization layer normalizes the data, succeeded by a dropout layer (0.2) to prevent overlearning. The block concludes with another convolution layer mirroring the first, followed by a batch normalization layer for data normalization.

Finally, the decoder block represents the most intricate component of u-Net networks, leveraging skip connection inputs from the layer below. These blocks, combining the output of the bridge or decoder block below them, consist of numerous layers. The initial layer is either an upsampling layer or, as in our case, a Conv2Dtranspose deconvolution layer, tasked with doubling the incoming image size due to previous MaxPooling layers halving it.

Subsequently, a concatenate layer combines the enlarged image with the skip connection, followed by a Conv2D convolution layer with a filter number matching the current block. The filter matrix is 3×3, with ELU activation and a fill factor to maintain image size. A batch normalization layer normalizes the data, succeeded by a dropout layer (0.2) to prevent overlearning. Another Conv2D convolution layer follows with parameters similar to the first, concluding with a batch normalization layer for data normalization.

2.3 Training and Testing

In our comprehensive experiment, we delved into the performance nuances of our U-net network, particularly focusing on various image sizes up to my resolution, necessitating intricate transformations by the resizing layer due to hardware constraints. Our exploration encompassed a spectrum of sizes, namely 16×16, 32×32, 64×64, 128×128, and culminating in 256×256, followed by a restoration to the original 512×512 resolution.

To ensure a robust performance assessment, each configuration underwent 5 tests, adopting an 80% training and 20% testing split. The integration of cross-validation further diversified the training and testing datasets. In this process, every 5th item assumed a role in the test dataset, while the rest contributed to the teaching dataset. Additionally, a slight adjustment to the index of the test dataset, from 0 to 4, determined the starting point for extracting test sample data from the entire dataset.

Our evaluation strategy extended to a custom cost function, amalgamating two crucial components. The first component, CategoricalCrossentropy, conventionally applied in classification problems, aptly suited our scenario of pixel-level classification. To contend with the challenge of unbalanced classes, the second component, Dice loss, was introduced. This component calculated the Dice value for both classes, subtracting the result from 1, given that the maximum value of Dice is 1. While effective in handling unbalanced classes, the Dice loss can be volatile. Therefore, the two cost components were combined with a multiplier of 0.5 each and aggregated to formulate our final cost function.

The learning process spanned 1000 epochs with a batch size of 64. An early stopping mechanism was incorporated, excising the model that achieved the smallest cost on the test data during training. Subsequently, this model was reinstated for in-depth performance analysis. The optimizer employed throughout the process was the Adam [23] algorithm, known for its efficiency in training neural networks.

2.4 Evaluation Criteria

The segmentation problem focuses on the separation of tumor pixels from those representing normal tissues, within 2D MR images. The accuracy of segmentation can be characterized by the extent to which the predicted labels of pixels match their ground truth. In this study we counted for each test image the number of true positive (TP), true negative (TN), false positive (FP) and false negative pixels (FN), and them computed the accuracy benchmarks shown in Table 1. We also employed the AUC metric to indicate the segmentation accuracy, which is extracted as the area under the ROC curve.

Table 1. Accuracy indicators employed in this study.

Indicator	Definition	Indicator	Definition
Sensitivity	$\text{TPR} = \frac{\text{TP}}{\text{TP+FN}}$		
Specificity	$\text{TNR} = \frac{\text{TN}}{\text{TN+FP}}$	F1 score	$\text{F1} = \frac{2\cdot\text{TP}}{2\cdot\text{TP+FP+FN}}$
Precision	$\text{PPV} = \frac{\text{TP}}{\text{TP+FP}}$	Accuracy	$\text{ACC} = \frac{\text{TP+TN}}{\text{TP+TN+FP+FN}}$

3   Results

Tables 2 and 3 report the main accuracy benchmarks obtained in case of various resizing numbers. Table 2 exhibits the precision, recall, and F1 scores, while Table 3 shows the AUC values for both the negative and positive classes, and the overall rates of correct decisions denoted by ACC. In case of each indicator, the median, mean, and standard deviations are reported.

First, let us examine the values shown in Table 2. For the five test cases, these values are calculated once. Examining the precision table, we can say that the resizing of 32 and the resizing of 64 are the best. Both the median and the mean precision values of these resizing values are greater than 97%. However, the scale 32 has a smaller standard deviation, so that it is the best scale in terms of precision. Slightly lower values are obtained for the resizing of 128 as well. In that case, both the median and the mean are 97% accurate. The weakest is the highest resolution. When looking at the recall values, we may notice that here the resolution of 256 is the best. Only at this resolution we obtain a 97% response rate for recall when looking at the mean and median. Also, the standard deviation is the second smallest. However, when examining the 128 resolution series, it is observed that it yields the second best recall values for mean and median, as well as the second best standard deviation. In terms of recall, the 32 and 64 resolutions perform much worse, while the worst results are obtained for resolution 16. While precision penalizes only the false positives and recall only the false negatives, the F1 score applies a penalty in case of both types of mistakes. The highest value of the F1 score can most effectively indicate, which is the best performance. The best accuracy of segmentation is obtained by the resolutions of 64 and 128, which have approximately 97% in both recall and precision, and consequently, the F1 scores are situated similarly. One of these two resolutions has a small lead in the median, the other in the mean value of the F1 score. For the standard deviation, we get the lowest value with the resolution of 128. The third best resolution is 32, but with this setting the results are around 96% and the variance is quite high. For the 256 and 16 resolutions, not even 96% is reached by either the median or average values, while the variance is very minimal.

Next, let us examine the AUC metric values reported in Table 3. Class 0 refers to the non-tumor part of the brain while class 1 refers to the focal lesion. AUC metrics were extracted for both of these classes separately, the median, the mean and the standard deviation of the metrics, for the five test cases. The mean

Table 2. Precision, Recall and F1 score mean, median and std values

Resizing number	Precision			Recall			F1 score		
	mean	median	std	mean	median	std	mean	median	std
16	0.967	0.967	0.002	0.932	0.933	0.001	0.950	0.950	0.001
32	0.978	0.978	0.002	0.947	0.943	0.013	0.962	0.961	0.007
64	0.978	0.979	0.006	0.956	0.962	0.016	0.967	0.971	0.007
128	0.970	0.970	0.004	0.968	0.967	0.004	0.969	0.969	0.001
256	0.934	0.934	0.004	0.976	0.975	0.004	0.955	0.955	0.001

Table 3. AUC by classes mean, median and std values

Resizing number	AUC class 0			AUC class 1			ACC		
	mean	median	std	mean	median	std	mean	median	std
16	0.971	0.970	0.003	0.973	0.971	0.005	0.971	0.971	0.001
32	0.986	0.985	0.011	0.988	0.987	0.008	0.978	0.978	0.004
64	0.994	0.985	0.014	0.994	0.982	0.018	0.983	0.981	0.004
128	0.996	0.996	0.001	0.996	0.996	0.001	0.982	0.982	0.001
256	0.995	0.995	0.001	0.995	0.995	0.001	0.973	0.973	0.001

AUC values grow together with the resolution from 16 up to 128, and slightly reduce above that. However, the standard deviations a highest at the resolution of 64, and lower at bothe extremities of the resolution scale. The median AUC values do not always correlate with the mean value, for example the resolution 64 has almost as high mean values as the best resolution of 128, while its median AUC values are considerably lower, the difference being visible also in standard deviation.

Finally, let's examine the last three columns of Table 3 where the correct decision rates are reported for the various resolutions involved in the experiment. Here again, the resolutions of 128 and 64 yield the highest mean and median scores above 98%, indicating than approximately one out of 55 pixels is misclassified in average during the segmentation. The lowest and the highest resolution lead to considerably lower accuracy rates, scoring slightly above 97% in both mean and median value. Also here we observe that the variance is higher in the middle of the resolution scale.

4 Discussion

It can be stated that the lowest resolution performs the worst when all metrics are examined. However, it is also fair to say that its performance is not so bad as to be unworthy of evaluation. Furthermore, since the lowest resolution is too compressed, the images lose important information for the network and cannot perform as well as other resolutions. However, another positive thing about this resolution to consider in terms of variance is that it has almost the minimum standard deviation in all cases. So the results are pretty constant compared to other resolution cases. Next is the case of resolution 32. For this resolution, we have already obtained better results than for resolution 16, proving that resolution 16 already over-compresses the information. However, what we can observe in this case is that it has a greater ability to spread its performance. In almost all cases, it has the highest variance. This proves that the higher resolution has already helped enough to get better results in some test cases. However, there are still cases where even a resolution of 32 is too low compression and important information is lost. Next, the most interesting case is the 64 in the same case where we have seen that in terms of accuracy, it is able to bring the best average. However, a case where the classes are not balanced is not only examined in terms of accuracy. The AUC metric shows that although it can produce the best results in terms of precision, it cannot in terms of AUC metric. Also, the AUC metric is much better at showing a classifier works well for unbalanced classes than accuracy. Finally, we examine the values of the precision, recall and F1 score. For precision it can be the best, but for recall it is almost the weakest. This should show that at lower resolution we compress the image so much that the classifiers predict fewer false positives. However, they predict many more false negatives. This is true for all lower resolutions. Since we are talking about tumor segmentation, one of the key metrics is that the fewer false negatives the classifier predicts, the more tumor pixels it hits. As well as looking further into this table, it can be observed that based on the F1 score, this case of 64 has the best median value. However, it is second best in mean value. It shows that 3 out of the 5 tests achieve a score of 97% but 2 cases are much weaker. This conclusion is supported by the fact that these 64 cases have the highest variance. Looking at the variance, it can be observed that this resolution value has the highest variance for all metrics. This shows that we have even better results for some test cases than for the resolution of 32, but still not a high enough resolution. The correct resolution is the case of 128, which is the best resolution for this problem. It does not perform best in all cases but is very balanced in all cases with almost the smallest variance. It has the best F1 score on average, but more importantly for AUC metrics it has the best score. This implies that we have achieved the optimal resolution, as there is little variation in the metric scores across the tests. This statement is supported by the fact that the precision and recall values are close to each other, with no bias in either direction. This means the classifier is not shifted to find negative classes better than it does at lower resolutions. Nor in the direction that it does for the 256 resolution, where it finds positive classes much better. However, it does bring in a lot of false positives. This resolution is large enough that important information is not lost to help the classification. Next is the 256 resolution which also does not have a large variance. However, the precision and recall values show that it works the other way round, with both the lower resolutions of 16 to 32 and 64.

It tends to introduce more false positives in classification. It follows that this is already too high a resolution here, it is no longer a lack of information because the image is over-compressed, but unnecessary information is introduced because of the higher resolution.

5   Conclusion

To summarise the results of the tests, it can be said that a smaller resolution than the original image size helps to increase the accuracy when using a U-net network. This is illustrated by the fact that the best results were obtained with a resolution of 128. With a resolution of 256, the values start to deteriorate. It can also be said that if there is a hardware limitation, it is safe to work with lower resolutions, since we still get good results at such resolution. It can also be said that at very low resolutions the classification shifts towards false negatives, while at too high a resolution the reverse happens, i.e. the classification shifts towards false positives. Of course, which is worse depends on the problem. In our case, having a high number of false negatives is worse than having a higher proportion of false positives. Another interesting point is that with higher resolution, although there are more false positives, the variance is much smaller. At low resolutions, we observe a large standard deviation, and then we get to the right resolution. At the same time, we have less information in the image. And at low resolutions it is already too compressed.

The overall conclusion is: you can use a resizing layer before and after U-net if your original images are too large to fit in memory. We need to find a reasonable resolution which in our case is 128. This should show that we really shouldn't compress the images since too much information is lost. Also, you can't scale the images too large either because unnecessary information will be introduced into the system. However, if there is a shortage of memory when using U-net for segmentation, it is safe to use resizing layers to reduce images to other resolutions. Very accurate results can be achieved at lower resolutions as well.

As a further development step, the conclusions drawn in this research can be applied to other data sets. Thus the findings of this work can gain a stronger validation.

References

1. Patel, A.P., Fisher, J.L., Nichols, E., et al.: Global, regional, and national burden of brain and other CNS cancer, 1990–2016: a systematic analysis for the Global Burden of Disease Study 2016. Lancet Neurol. **18**, 376–393 (2019)
2. Mohan, G., Subashini, M.M.: MRI based medical image analysis: survey on brain tumor grade classification. Biomed. Sign. Proc. Contr. **39**, 139–161 (2018)
3. Gordillo, N., Montseny, E., Sobrevilla, P.: State of the art survey on MRI brain tumor segmentation. Magn. Reson. Imag. **31**, 1426–1438 (2013)
4. Menze, B.H., Jakab, A., Bauer, S., et al.: The multimodal brain tumor image segmentation benchmark (BRATS). IEEE Trans. Med. Imag. **34**, 1993–2024 (2015)

5. Bakas, S., Reyes, M., Jakab, A., et al.: Identifying the best machine learning algorithms for brain tumor segmentation, progression assessment, and overall survival prediction in the BRATS challenge. arXiv: 1181.02629v2 (2019)
6. Győrfi, Á., Szilágyi, L., Kovács, L.: A fully automatic procedure for brain tumor segmentation from multi-spectral MRI records using ensemble learning and atlas-based data enhancement. Appl. Sci. **11**, 564 (2021)
7. Lefkovits, S., Szilágyi, L., Lefkovits, L.: Brain tumor segmentation and survival prediction using a cascade of random forests. In: Crimi, A., Bakas, S., Kuijf, H., Keyvan, F., Reyes, M., van Walsum, T. (eds.) BrainLes 2018. LNCS, vol. 11384, pp. 334–345. Springer, Cham (2019). https://doi.org/10.1007/978-3-030-11726-9_30
8. Szilágyi, L., Lefkovits, L., Benyó, B.: Automatic brain tumor segmentation in multispectral MRI volumes using a fuzzy c-means cascade algorithm. In: 12th International Conference on Fuzzy Systems and Knowledge Discovery (FSKD), pp. 285–291 (2015)
9. Szilágyi, L., Iclanzan, D., Kapás, Z., Szabó, Z., Győrfi, Á., Lefkovits, L.: Low and high grade glioma segmentation in multispectral brain MRI data. Acta Univ. Sapientiae - Imformatica **10**(1), 110–132 (2018)
10. Macsik, P., Pavlovicova, J., Goga, J., Kajan, S.: Local binary CNN for diabetic retinopathy classification on fundus images. Acta Polytech. Hung. **19**(7), 27–45 (2022)
11. Szepesi, P., Szilágyi, L.: Detection of pneumonia using convolutional neural networks and deep learning. Biocybern. Biomed. Eng. **42**(3), 1012–1022 (2022)
12. Neelima, G., Chigurukota, D.R., Maram, B., Girirajan, B.: Optimal DeepMRSeg based tumor segmentation with GAN for brain tumor classification. Biomed. Sign. Proc. Contr. **74**, 103537 (2022)
13. Kanchanamala, P., Revathi, K.G., Ananth, M.B.J.: Optimization-enabled hybrid deep learning for brain tumor detection and classification from MRI. Biomed. Sign. Proc. Contr. **84**, 104955 (2023)
14. Rajeev, S.K., Rajasekaran, M.P., Vishnuvarthanan, G., Arunprasath, T.: A biologically-inspired hybrid deep learning approach for brain tumor classification from magnetic resonance imaging using improved Gabor wavelet transform and Elmann-BiLSTM network. Biomed. Sign. Proc. Contr. **78**, 103949 (2022)
15. Mishra, L., Verma, S.: Graph attention autoencoder inspired CNN based brain tumor classification using MRI. Neurocomput. **503**, 236–247 (2022)
16. Reddy, K.R., Dhuli, R.: Segmentation and classification of brain tumors from MRI images based on adaptive mechanisms and ELDP feature descriptor. Biomed. Sign. Proc. Contr. **76**, 103704 (2022)
17. Mehnatkesh, H., Jalali, S.M.J., Khosravi, A., Nahavandi, S.: An intelligent driven deep residual learning framework for brain tumor classification using MRI images. Expert Syst. Appl. **213**, 119087 (2023)
18. Vankdothu, R., Hameed, M.A.: Brain tumor MRI images identification and classification based on the recurrent convolutional neural network. Meas. Sensors **24**, 100412 (2022)
19. Rahman, T., Islam, M.S.: MRI brain tumor detection and classification using parallel deep convolutional neural networks. Meas. Sensors **26**, 100694 (2023)
20. Isunuri, B.V., Kakarla, J.: EfficientNet and multi-path convolution with multi-head attention network for brain tumor grade classification. Comput. Electr. Eng. **108**, 108700 (2023)
21. Cheng, J., Huang, W., Cao, S., et al.: Enhanced performance of brain tumor classification via tumor region augmentation and partition. PLoS ONE **10**(10), e0140381 (2015)

22. Cheng, J.: Brain tumor dataset. Figshare. Dataset (2017). https://doi.org/10.6084/m9.figshare.1512427.v5
23. Kingma, D.P., Ba, J.: Adam: a method for stochastic optimization. arxiv:1412.6980 (2014)
24. Ronneberger, O., Fischer, P., Brox, T.: U-Net: convolutional networks for biomedical image segmentation. In: Navab, N., Hornegger, J., Wells, W.M., Frangi, A.F. (eds.) MICCAI 2015. LNCS, vol. 9351, pp. 234–241. Springer, Cham (2015). https://doi.org/10.1007/978-3-319-24574-4_28

Multiprocessor Task Scheduling with Probabilistic Task Duration

Dariusz Dorota[✉]

Cracow University of Technology, street Warszawska 24, 31-155 Cracow, Poland
ddorota@pk.edu.pl

Abstract. Article concentrate on so-called multiprocessor task scheduling problem with uncertain (random) time of tasks duration. Multiprocessor scheduling can be perceived as a tool for improving dependability of the system by hardware and software redundancy. Our overall aim is develop more precise description of various configuration of embedded systems). Our aim is to evaluate the model of task duration distribution on the result of scheduling. We compare two models of the task: normal distribution and Erlang distribution. The latter model is considered as more suitable for multiprocessor scheduling, which reflects more accurately reality. We use MVA (Mean Value Analysis) methodology in the research with the application of modified Muntz-Coffman algorithm. The article is an extension of previous research of the author, considering uncertain task duration in different models. Computational experiments compared results obtained for both distributions in this stochastic model.

Keywords: Multiprocessor tasks · Uncertain task duration · Scheduling

1 Introduction

Scheduling tasks is a problem we encounter in everyday life. Particular attention should be paid to this issue in the area of embedded systems. More branches have emerged from the area of embedded systems, such as IoT (Internet of Things) and IoV (Internet of Vehicles). In all the above-mentioned areas characterized by an increased level of security, optimization issues can be identified related to increasing the reliability of systems through redundancy. In systems with an increased level of reliability, both hardware and software redundancy can be mentioned. Synchronous execution of identical copies of tasks on more than one processor at the same time allows the system to be trusted by making the right decision based on the selection. Systems of this type are used in various industries and areas of life where factors aimed at increased reliability and resistance to possible errors become important. This quite natural concept of making the results of systems credible allows for their extensive application in new areas of life. In particular, going beyond the area of determinism allows for a more accurate reflection of real systems. One of the most natural approaches is to use

© The Author(s), under exclusive license to Springer Nature Switzerland AG 2024
W. Zamojski et al. (Eds.): DepCoS-RELCOMEX 2024, LNNS 1026, pp. 49–58, 2024.
https://doi.org/10.1007/978-3-031-61857-4_5

stochastic to specify task duration, which is done in this work. The overall aim of the paper is to evaluate the chosen distribution of task processing times on the result of scheduling.

2   Problem Formulation

As a result of many years of evolution of the field, the theory and practice of task scheduling have been formulated and significantly developed, covering various issues of practical importance. Currently, the following are being considered: deterministic, fuzzy, stochastic, single- and multi-criteria problems, with additional resources or constraints, with different task flow structures, with different time/cost models, as well as algorithms for solving them with different numerical features. Identification of each considered problem in the context of the taxonomy of scheduling problems and analysis of its specific features allows for the appropriate selection of an existing solution or development of a new algorithm. Uncertainty in scheduling problems may involve different data and assumptions of the problem, [1]. It should be noted that uncertainty can be static or dynamic and can be perceived differently, as follows: the set of tasks is known a priori, all tasks from this set must be performed, but some task parameters are uncertain, e.g. task execution times, desired completion dates, etc.

In order to systematize the approach, certain notations should be introduced: $T = \{T_1, T_2, T_3, ..., T_n\}$ a set of tasks to be performed on machines (processors) defined as $M = \{M_1, M_2, M_3, ..., M_m\}$ in the deterministic case. Basically, it is assumed that the task T_i contains one atomic operation performed simultaneously and synchronously on a predefined number of processors $a_i \geq 1$. and takes time p_i. If task interruptions are allowed, it is assumed that the task $T_i \in T$ consists of a sequence of a certain number x_i operations $O_i = (o_{i,1}, o_{i,2}, o_{i,3}, ..., o_{i,x_i})$, executed in this order, each requiring the synchronous use of a_i processors. Of course, $\sum_{k=1}^{x_i} |o_{i,k}| = p_i$ holds, where $|o_{i,k}|$ is the duration of the operation $o_{i,k}$.

In this work, the division of tasks into operations is not fixed and is subject to choice. Multi-processor tasks are also denoted by the multi-processor ratio MPR (*Multi-Processor Ratio coefficient*). Therefore, $MPR = 1$, means that all tasks are single-processor, i.e. $a_i = 1$, $T_i \in T$. $MPR = k$, means that $a_i = k \leq m$, $T_i \in T$, and $MPR \leq k$, means that $a_i \leq k \leq m$, $T_i \in T$. In the practice of embedded systems, cases where $a_i = 2..3$ are most common, so this work will also focus on such examples. In order to classify the problems considered in the thesis, we will refer to the three-field notation of Graham et al. [16] used in deterministic scheduling. The book [25] proposes some extension of the α symbol for task parallel processing. It is assumed that $\alpha = \alpha_1\alpha_2\alpha_3$. The value α_1 can be a number (then it defines a finite number of machines in the system) or an empty symbol $\circ$, which signals that the number of machines is arbitrary m, which unfortunately does not allow describing the problem considered in the article, so for [12], an extension of the range of meanings of the symbol β was used, namely: $random(p)$, and an additional parameter mpr was proposed for β, identical to a_i. This proposal is a change from the concept presented in the works [2,14,34], because it introduces the multiprocessor model through the constraints of the problem, and not through the structure of the processing system. Using the given classification, the problems considered in the thesis are of classes $P|\beta|C_{\max}$, where $\beta \in \{\circ, mpr = k, mpr \leq k, pmtn, prec, random(p)\}$. The absence of the mpr parameter is equivalent to $mpr = 1$. The $\beta \in \{random(p)\}$ parameter can be divided into several variants of $random(p) \in \{normal, erlang\}$, where they denote the normal distribution and the Erlang distribution, respectively. Problem representation methods determine the choice of analysis tools, solution algorithm or area of solution search. The basic description methods include [30]: (A) data flow diagram; (B) synchronous data flow diagram; (C) task graph; (D) Petri networks; (E) SystemC. The most frequently used specification method (also in this work) are graph methods. We define a graph as a pair $G = (V, E)$, where V is the set of nodes and $E \subseteq 2^V$ is the set of undirected edges. An edge is represented by a two-element set $\{i, j\}$, $i, j \in V$. A directed graph is defined analogously, i.e. $G = (V, A)$, where $A \subseteq V \times V$ is the set of (directed) arcs, [15]. In this article, a task graph was adopted as the system specification. Exemplary task graph was shown in Fig. 1.

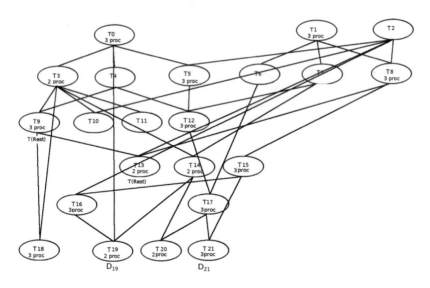

Fig. 1. Exemplary task graph, for $a_i \leq 3$

3 Dependability in Scheduling - Motivation

There are several reasons for considering multiprocessor, dependable models. In classic deterministic scheduling problems, it is typically assumed that the task is performed by a single processor, even if we have a multiprocessor system. In order to increase the reliability of the system, hardware and software redundancy is introduced by synchronously performing identical tasks on several identical processors working in parallel. A typical solution in this regard are systems containing 2 or 3 processors, striving for a certain balance between cost and level of reliability. Note that the use of two processors in the case of conflicting computational results does not allow making a clear decision, resulting in repeated but unpredictable execution of the same task on three processors (the majority votes then). The obvious application is complex embedded system architectures that use multiprocessor tasks. Similar applications include simultaneous access to shared memory, testing VLSI circuits [3–6,9,13,14,21,32,36], or implementing true parallelism [17,26,35]. Embedded system architectures involve some processors, which may seem to limit the problem somewhat, e.g. $mpr \leq 3$. A more interesting interpretation of the $P|mpr|C_{\max}$ problem from $mpr \gg 3$ is the *strip packing/cutting problem, SPP*; an extension of this concept is presented in [12].

4 Application of Probability - Erlang Distribution

One of the most frequently used approaches to modeling and solving is the natural conversion of a deterministic model into a stochastic model, e.g. by analyzing the mean values of appropriate distributions, the MVA method (*Mean Value Analysis*) or through scenario analysis. Stochastic scheduling algorithms can be divided into, (see [23]): (A) two-stage; (B) multi-stage; (C) a constrained programming approach. The article is a continuation of the research from [12], as well as an extension of the [31] research. The work [22] compares scheduling using fuzzy logic and probabilistic. The reports show that the application of fuzzy logic to methods based on PERT analysis (*Program Evaluation and Review Technique*) has a disadvantage related to determining the maximum fuzzy value. The analysis allows us to determine that the fuzzy logic approach does not require, as in the case of stochastic methods, the use of an identical distribution for all considered task times. Additionally, it was established that arithmetic operations in the case of probability distributions are more complicated than when using fuzzy logic. The comparison conducted in the mid-1990s still shows the current trends in the above-mentioned approaches to task scheduling. In the work [29] it was assumed that probabilistic is used to determine the probability of completing a task and, consequently, the impact of this event on subsequent tasks. In this case, SJSS (*Stochastic Job Shop Scheduling Problem*) is considered, in which, unlike usual, it is not one of the parameters in the scheduling process that is uncertain, but the uncertainty includes the success (or lack thereof) of performing the production task. The system model proposed here optimizes the cost through the proposed heuristics. In this case, the 3σ rule is used, closely related to the normal distribution. In [33] the search for concurrency errors in a randomly generated schedule is considered. Testing is performed by repeatedly running the program with data, where the error depth parameter is assumed to be a randomly selected probability distribution. Probabilistic analysis for pre-emptive scheduling priorities for mixed critical systems created in the AMC (*Adaptive Mixed Criticality*) and SMC (*Static Mixed Criticality*) schemes. In the presented analysis, the worst execution time, the worst response time and the worst possible timeout of constraints are described using mixed probability distributions. Typically, it is assumed to use the stochastic concept to specify task durations, as done in [7,10,18]. In this case, the authors assumed a time specification of tasks, represented in the form of a graph, using the Erlang probability distribution. In task scheduling, in this case the research successfully uses an algorithm close to the optimal one, HLF (*Highest Levels First*). Scheduling problems with uncertain data include: (A) unknown task duration; (B) unknown transfer (transmission) time between dependent tasks; (C) unknown task completion time, etc. This work focuses on parameter (A), which can be described using a probability distribution. This is an extension of the research in which the duration was described using the Erlang distribution. According to the current state of knowledge, this is the first work describing the duration of multiprocessor tasks in a multiprocessor architecture organized in a NoC network architecture. To describe the task duration, the geometric distribution of the probability function (discrete time) or the exponential distribution (continuous time), [19], is used.

We recall some definitions. For a sequence of independent random variables $X_1, X_2, \ldots X_n$ such that each variable has the same exponential distribution with parameter λ, the random variable $\Sigma_{i=1}^n X_i$ has the Erlang's distribution with parameters $k = n$, $\theta = \frac{1}{\lambda}$.

5   Description of the Approach

According to the current state of knowledge, using the stochastic approach, it is possible to specify the execution times of [18,31] tasks, the worst execution times of the [20,28] tasks, and can be used in algorithms that guarantee the correct scheduling of periodic tasks [11]. The presented methods are also used for scheduling in systems with increased reliability [24] and in the so-called robust systems [27]. The specification of task duration's using a normal probability distribution is also presented in [12]. Since MC algorithm has been described in details in some my previous papers, [12], because of this paper size we skip formal presentation of MC there, providing only a brief description as follows.

- The system specification is a task graph, where the task duration's are described using a probability distribution. Mean value is crucial for the MVA.
- Priorities are determined according to MVA, then it is deterministic. Tasks with the highest priority values are selected, taking into account the location of each task on the path(s).

- Tasks with the priorities specified above are selected and then these tasks are allocated to the selected processors so that (if possible) there are no unused (idle) processor(s) in a given unit of time.
- If there are no idle processors in a unit of time, or it is not possible to assign any task to unused processors/procesora, the time will be shifted by another time unit, and the algorithm is executed again until they are ranked all tasks.
- If all tasks have been ranked in the assumed probability level, the next values are selected. After this, the algorithm starts prioritization by re-initializing its operation.

The above description of the algorithm also shows the difference between the traditional approach and the use of uncertainty in scheduling. The modification consists in adding elements that consider uncertainty. The specification is loaded, including a task graph description. The difference in this case is the way of representing times in the task graph using probability methods, in this particular case the normal distribution. After loading the system specifications, levels are set for each task. Task levels are determined based on the average value.

6 Computational Experiments

This chapter describe the made computer experiments and provides obtained results. Two benchmark sets have been considered. The former contains single set of dependent task. Table 1 provides specification of tasks shown in Fig. 1, where a_i mean so-called multiprocessor ratio (processors needed to execute task), p_i is a time of task execution, p_i^E value of the Erlang's distribution for p_i. Schedule obtained by MC for Erlang distribution is shown in Fig. 2. The latter benchmark set contains 10 instances with sizes $n \in \{9, 13, 15, 19, 22, 26, 32, 41, 46, 51\}$. Each instance is a mix of independent tasks with various $a_i \in \{1, 2, 3\}$ specified as the "processors" in "task profile". Further details one can find in [12].

Using MC in case of normal distribution we set processing times accordingly to the rule: V1 (40%) - by choosing the mean value from the distribution, V4 (99,7%) - by setting $\pm 3\sigma$ rule. Consequently, V1 and V4 mean the length of the schedule obtained by MC. Tables 2, 3 and 4 presents length of the schedule, where column "Distribution" means makespan $C_{\max}$ (for normal and Erlang's distribution).

Fig. 2. Exemplary scheduling for graph from Fig. 1 and $m = 3$

Table 1. System specification, see also graph in Fig. 1

task	a_i	p_i	p_i^E	$a_i p_i$
T0	3	10	0,8753	30
T1	3	30	1,0000	90
T2	1	10	0,8753	10
T3	2	10	0,8753	20
T4	1	15	0,9797	15
T5	3	15	0,9797	45
T6	1	5	0,2424	5
T7	1	20	0,4562	20
T8	3	25	0,9972	75
T9	3	10	0,9997	30
T10	1	25	0,8753	25
T11	1	10	0,9797	10
T12	2	15	0,9972	30
T13	2	20	0,9797	40
T14	2	15	0,9972	30
T15	3	20	1,0000	60
T16	3	50	0,9797	150
T17	3	15	0,4562	45
T18	3	5	0,9972	15
T19	2	20	0,8753	40
T20	2	10	0,9972	20
T21	3	20	1,0000	60

Table 2. Length of schedule. Stochastic approach. $m = 3$, $\lambda = 2$, $k = 3$

instance		Tasks profile			Distribution		
		processors			Normal		Erlang
nr	n	1	2	3	V1	V4	
1	9	4	2	3	86	69–103	83
2	13	5	4	4	170	148–192	169
3	15	5	5	5	190	172–208	190
4	19	5	7	7	240	222–258	239
5	22	6	6	10	305	276–334	305
6	26	8	7	11	335	307–363	335
7	32	11	8	13	440	404–458	431
8	41	13	10	18	670	631–709	670
9	46	12	10	24	954	906–1002	954
10	51	16	11	24	718	674–762	763

Table 3. Length of schedule. Stochastic approach. $m = 4$, $\lambda = 2, k = 3$

instance		Tasks profile			Distribution		
		processors			Normal		Erlang
nr	n	1	2	3	V1	V4	
1	9	4	2	3	80	63–97	55
2	13	5	4	4	130	107–152	108
3	15	5	5	5	125	107–143	77
4	19	5	7	7	185	167–203	169
5	22	6	6	10	255	225–285	254
6	26	8	7	11	285	257–313	284
7	32	11	8	13	365	330–400	365
8	41	13	10	18	545	506–584	545
9	46	12	10	24	820	722–867	795
10	51	16	11	24	764	720–808	764

Table 4. Length of schedule. Stochastic approach. $m = 5$, $\lambda = 2, k = 3$

instantiation		Tasks profile			Distribution		
		processors			Normal		Erlang
nr	n	1	2	3	V1	V4	
1	9	4	2	3	55	38–72	35
2	13	5	4	4	95	73–117	72
3	15	5	5	5	105	87–123	79
4	19	5	7	7	130	112–148	99
5	22	6	6	10	205	175–235	204
6	26	8	7	11	240	212–268	240
7	32	11	8	13	305	269–341	305
8	41	13	10	18	400	360–439	399
9	46	12	10	24	780	732–828	780
10	51	16	11	24	630	585–674	630

7   Discussion of the Results

The article proposes an approach to using the Erlang distribution to describe task durations for a system specifying multiprocessor tasks. The results were compared with the results for ranking with task durations described by a normal probability distribution. The Erlang distribution was chosen because it describes the tasks more precisely compared to the normal probability distribution [8]. Uncertainty described in this way more closely corresponds to real situations, and therefore the length of the ranking also reflects reality more closely. The adapted

MC algorithm in the version with normal distribution allows you to obtain a schedule in various probability intervals. The consequence of this approach is that the length of the ranking depends on the level of probability, unlike in the case of the Erlang distribution, where we can talk about the same length of the ranking characterized by a certain predetermined (based on the algorithm) probability, in accordance with the formulas presented in the work. Another quite natural approach seems to be the use of more than one primer described using probability, and ultimately the comparison of such results.

References

1. Behnamian, J.: Survey on fuzzy shop scheduling. Fuzzy Optim. Decis. Making **15**(3), 331–366 (2016)
2. Behnamian, J., Fatemi Ghomi, S.: A survey of multi-factory scheduling. J. Intell. Manuf. **27**, 231–249 (2016)
3. Blazewicz, J., Drabowski, M., Weglarz, J.: Scheduling multiprocessor tasks to minimize schedule length. IEEE Trans. Comput. **5**, 389–393 (1986)
4. Błażewicz, J., Drozdowski, M., Ecker, K.: Management of resources in parallel systems. In: Błażewicz, J., Ecker, K., Plateau, B., Trystram, D. (eds.) Handbook on Parallel and Distributed Processing, pp. 263–341. Springer, Heidelberg (2000). https://doi.org/10.1007/978-3-662-04303-5_6
5. Blazewicz, J., et al.: Communication delays and multiprocessor tasks. In: Handbook on Scheduling: From Theory to Practice, pp. 199–241 (2019)
6. Błażewicz, J., Liu, Z.: Scheduling multiprocessor tasks with chain constraints. Eur. J. Oper. Res. **94**(2), 231–241 (1996)
7. Bozejko, W., Hejducki, Z., Wodecki, M.: Flowshop scheduling of construction processes with uncertain parameters. Arch. Civil Mech. Eng. **19**(1), 194–204 (2019)
8. Bozejko, W., Hejducki, Z., Wodecki, M.: Flowshop scheduling of construction processes with uncertain parameters. Arch. Civil Mech. Eng. **19**, 194–204 (2019)
9. Caplan, J., Al-Bayati, Z., Zeng, H., Meyer, B.H.: Mapping and scheduling mixed-criticality systems with on-demand redundancy. IEEE Trans. Comput. **67**(4), 582–588 (2017)
10. Chin, M.K., Kek, S.L., Sim, S.Y., Seow, T.W.: Probabilistic completion time in project scheduling. Int. J. Eng. Res. Sci. **3**(4), 44–48 (2017)
11. Davis, R.I., Cucu-Grosjean, L.: A survey of probabilistic timing analysis techniques for real-time systems. LITES: Leibniz Trans. Embed. Syst. 1–60 (2019)
12. Dorota, D.P.: Szeregowanie zadań wieloprocesorowych w warunkach niepewności (2023)
13. Drozdowski, M.: On the complexity of multiprocessor task scheduling. Bull. Polish Acad. Sci. Techn. Sci. **43**(3), 1–12 (1995)
14. Drozdowski, M.: Scheduling multiprocessor tasks-an overview. Eur. J. Oper. Res. **94**(2), 215–230 (1996)
15. Drozdowski, M.: Scheduling for Parallel Processing. Springer, London (2009). https://doi.org/10.1007/978-1-84882-310-5
16. Graham, R.L., Lawler, E.L., Lenstra, J.K., Kan, A.R.: Optimization and approximation in deterministic sequencing and scheduling: a survey. Ann. Disc. Math. **5**, 287–326 (1979)
17. Ibrahim, H., Salih, M.H.: Design and implementation of embedded true parallelism jammer system using FPGA-SOC for low design complexity. ARPN J. Eng. Appl. Sci. **13**(24), 9410–9420 (2018)

18. Jiang, X., Lee, K., Pinedo, M.L.: Ideal schedules in parallel machine settings. Eur. J. Oper. Res. **290**(2), 422–434 (2021)
19. Mao, H., Chen, Y., Jaeger, M., Nielsen, T.D., Larsen, K.G., Nielsen, B.: Learning deterministic probabilistic automata from a model checking perspective. Mach. Learn. **105**, 255–299 (2016)
20. Maxim, D., Davis, R.I., Cucu-Grosjean, L., Easwaran, A.: Probabilistic analysis for mixed criticality systems using fixed priority preemptive scheduling. In: Proceedings of the 25th International Conference on Real-Time Networks and Systems, pp. 237–246 (2017)
21. Miedema, L., Rouxel, B., Grelck, C.: Task-level redundancy vs instruction-level redundancy against single event upsets in real-time dag scheduling. In: 2021 IEEE 14th International Symposium on Embedded Multicore/Many-core Systems-on-Chip (MCSoC), pp. 373–380. IEEE (2021)
22. Moselhi, O., Lorterapong, P.: Fuzzy vs probabilistic scheduling. In: Proceedings of the 12th Conference "Automation and Robotics in Construction" (ISARC), pp. 441–448 (1995)
23. Ning, C., You, F.: Optimization under uncertainty in the era of big data and deep learning: when machine learning meets mathematical programming. Comput. Chem. Eng. **125**, 434–448 (2019)
24. Pathan, R.M.: Real-time scheduling algorithm for safety-critical systems on faulty multicore environments. Real-Time Syst. **53**, 45–81 (2017)
25. Pinedo, M.: Scheduling. Springer, New York (2015). https://doi.org/10.1007/978-3-319-26580-3
26. Raj, M.D., Gogul, I., Thangaraja, M., Kumar, V.S.: Static gesture recognition based precise positioning of 5-dof robotic arm using FPGA. In: 2017 Trends in Industrial Measurement and Automation (TIMA), pp. 1–6. IEEE (2017)
27. Saint-Guillain, M., Vaquero, T., Chien, S., Agrawal, J., Abrahams, J.: Probabilistic temporal networks with ordinary distributions: theory, robustness and expected utility. J. Artif. Intell. Res. **71**, 1091–1136 (2021)
28. Santinelli, L., Cucu-Grosjean, L.: A probabilistic calculus for probabilistic real-time systems. ACM Trans. Embed. Comput. Syst. (TECS) **14**(3), 1–30 (2015)
29. Shoval, S., Efatmaneshnik, M.: A probabilistic approach to the stochastic job-shop scheduling problem. Procedia Manuf. **21**, 533–540 (2018)
30. Sriram, S., Bhattacharyya, S.S.: Embedded Multiprocessors: Scheduling and Synchronization. CRC Press, Boca Raton (2018)
31. Terekhov, D., Down, D.G., Beck, J.C.: Queueing-theoretic approaches for dynamic scheduling: a survey. Surv. Oper. Res. Manag. Sci. **19**(2), 105–129 (2014)
32. Xin, X., Mou, M., Mu, G.: A polynomially solvable case of scheduling multiprocessor tasks in a multi-machine environment. In: 2017 2nd International Conference on Materials Science, Machinery and Energy Engineering (MSMEE 2017), pp. 1746–1749. Atlantis Press (2017)
33. Xu, M., Kashyap, S., Zhao, H., Kim, T.: Krace: data race fuzzing for kernel file systems. In: 2020 IEEE Symposium on Security and Privacy (SP), pp. 1643–1660. IEEE (2020)
34. Ye, D., Chen, D.Z., Zhang, G.: Online scheduling of moldable parallel tasks. J. Sched. **21**(6), 647–654 (2018)
35. Zahid, Y., Khurshid, H., Memon, Z.A.: On improving efficiency and utilization of last level cache in multicore systems. Inf. Technol. Control **47**(3), 588–607 (2018)
36. Zhao, L., Ren, Y., Sakurai, K.: Reliable workflow scheduling with less resource redundancy. Parallel Comput. **39**(10), 567–585 (2013)

On-Line Scheduling Multiprocessor Tasks in the Non-predictive Environment

Dariusz Dorota[1] and Czeslaw Smutnicki[2]([✉])

[1] Cracow University of Technology, Crakow, Poland
ddorota@pk.edu.pl
[2] Wrocław University of Science and Technology, Wrocław, Poland
czeslaw.smutnicki@pwr.edu.pl

Abstract. The paper considers the problem known in the literature as the multiprocessor task scheduling. It appears in the context of embedded systems and devices working in critical civil and military infrastructure objects, Internet of Things, Internet of Vehicles and so on, with an augmented level of security. Enlarged reliability is achieved through hardware and software redundancy. A number of independent devices (microprocessors) which execute the same programs for identical data, provide independent results to set the final decision by "voting". In the paper we assume that the set of tasks for processing is uncertain while tasks arrive randomly. We provide some on-line scheduling algorithms with the theoretical evaluation of their quality.

Keywords: Multiprocessor tasks · On-line scheduling

1 Introduction

Scheduling algorithms are used frequently in service systems in various applications. We deal with embedded devices with high level of security, working in the Internet of Things (IoT), Internet of Vehicles (IoV), Internet of Robots (IoR). Due to recent reduction of the hardware cost, augmented dependability can be obtained through redundancy of the hardware as well as redundancy of the software. Redundancy has an aim the automatic decision making through the "voting" by replicated identical devices, performing the same functions and running the same programming code for identical data. This leads to the problem called multiprocessor task scheduling. Applications one can find in aviation, space vehicles, military equipment, rockets, means of transporting hazardous materials, nuclear installations, chemical installations, mining, drones. Plainness of the approach allow us to extend it on other areas of human activities.

2 The Problem

Let us denote by $T = \{1, 2, 3, ..., n\}$ the set of computational tasks (programs) dedicated for processing on identical machines (processors, independent embedded devices) defined as $M = \{1, 2, 3, ..., m\}$ used in parallel. Each task $i \in T$

uses $a_i \geq 1$ processors in order to run simultaneously a_i copies of the program. There are given the processing times $p_i > 0$ of the tasks. The schedule is defined by the vector of task starting times $S = (S_1, S_2, \ldots, S_n)$ and allocation of tasks to processors $V = (V_1, V_2, \ldots, V_n)$, $V_i \subseteq M$, $|V_i| = a_i$. We would like to find the schedule for the processors, which satisfies the mentioned constraints and minimizes some scheduling criteria. In this paper we consider chiefly the minimum of the makespan $C_{\max} = \max_{i \in T}(S_i + p_i)$, which also maximizes the level of utilization of processors. The problem is NP-hard.

The above formulation has been provided as a deterministic scheduling problem. Nevertheless, there exist an equivalent formulation in terms of so-called Resource Constrained Project Scheduling (RCPS, [17]). Indeed, processors can be treated as a discrete resource divisible with granularity one unit, required by various tasks in various amount. The scheduling problem has also a link with the Strip Cutting Problem (SCP), see Fig. 4 (right) as an instance and paper [18] for detail. There are also known other specific methods of modeling and solving problems of such type, however differently to the presented here on-line approach, they refer only to the deterministic off-line case, [22].

Starting from the well-known taxonomy $\alpha|\beta|\gamma$ of scheduling problems, [15], there has been proposed in [27] an extension of α to cover parallel processing of tasks for various scheduling problems. Unfortunately, this denotation idea cannot be applied in our case. Therefore, we extend meaning of β. Symbol $a_i = k$ means that all tasks are k-processor, $a_i \leq k$ means tasks own "mixed" but bounded a_i; alone a_i means free a_i. The used denotation is a modification to that used in [3,13,18,20,27], since introduces multiprocessor model via constraints but not by the system structure. Thus, we consider problems $P|\beta|C_{\max}$, where $\beta \in \{\circ, a_i = k, a_i \leq k, a_i, pmtn, prec\}$. Note that the problem differs significantly from the "classical" $P, R, Q||C_{\max}$ because of various a_i, unknown T and p_i.

3   Dependable Scheduling

The paper is a continuation of research from [11] but the approach is completely different with those from [9,10]. Generally, we consider problems with uncertain data of the instance and uncertain task set. It can refer to various parameters and assumptions, see these enumerated in [11]: (A) unknown task notification time; (B) unknown task duration; (C) unknown transfer (transmission) time between dependent tasks; (D) unknown task flow time; (E) unknown processor breakdowns; (F) unknown processor maintenance time; (G) unknown precedence relation between tasks; (H) unknown task resource requirements; (I) unknown number and the set of tasks. Cases (A)...(F) can be modelled and solved by fuzzy and/or stochastic tools, see e.g. [2]. Although such models occur in scheduling area, unfortunately they do not appear in the context of multiprocessor tasks. Therefore the research area is open. Similarly, (G) i (H) do not appear in the literature. Case (I) requires assumptions for the incoming tasks. Infinite stream of incoming tasks rarely appears in closed embedded systems, although has reasonable character. We assume here that tasks are unique taking into account data, arrive randomly, and the stream has no statistical features. Then, in the paper we use assumptions (A), (B), (I) and expand own research from [11] to realize on-line scheduling of multiprocessor tasks.

4 Online Scheduling

Process of task scheduling is performed in the real time, since tasks appear in an unknown time moments and in the unknown sequence of entries. Thus on-line approach is more flexible and more usable than off-line. On-line models are classified as follows: (A) *online list*; (B) *online time*; (C) *machine activity*; see e.g. [1,5,8,12,25]. The fundamental differentiation between scheduling on-line versus off-line is the accessibility of data: (a) offline – all data are available before the process of scheduling; (b) online – data are provided partially step-by-step during the task processing. Additional arguments for (b) can be incompleteness of data, uncertainty or nondeterminism of time events [7,21]. Model *online list* assumes that tasks come accidentally, individually, and each arriving task reveals its characteristics (e.g. duration, involved processors). Immediately after the task arrival, the scheduler plans its processing and this decision cannot be changed in the future; tasks are scheduled in the order of arrivals. Model *online time* assumes that tasks come accidentally, and may come simultaneously. Some parameters of the incoming task may be known (e.g. a_i), others – not known (e.g. p_i). Scheduling decision takes into account all tasks waited in the queue, are dynamical, changeable, usually preemptive since p_i is not known.

Scheduling algorithm can make the decision in the *deterministic* or *random* manner. In the literature, there are known other classifications of on-line algorithms, [19,26]: (A) *clairvoyant*, where task duration is predictable, and (B) *non-clairvoyant*, where task duration is unknown, (C) with various *adversary* for random scheduling. The quality of the on-line algorithm is evaluated by the *competitive ratio*. It compares goal function value $f()$ provided by on-line algorithm A with the optimal goal function value OPT achieved posterior (off-line), for any data instance I. Since formal definitions of competitive ratio are various for various types of algorithms mentioned before, we quote only coefficient r for deterministic on-line algorithm used further in this paper

$$f(A, I) \leq r \cdot f(OPT, I) + \alpha, \tag{1}$$

as minimal value r obtainable for any instance I, where α is some constant. In fact, competitive analysis (worst-case, pessimistic analysis) evaluates algorithms in extreme case from the theoretical point of view, so the measure of algorithms quality is independent on the instance. Alternative commonly used technology of algorithm comparison refers to experimental analysis on the representative sample of instances. Taking into account "no free lunch" theorem of Wolpert and Macready and tendencies of choosing "sample of instances", the experimental research can provide different evaluations of algorithms quality, however subjective. Then the quality of the algorithm should be evaluated from various points of view: worst-case, experimental, probabilistic. In the sequel we provide several algorithms with theoretical evaluation of their properties. Potential experimental research exceed the limited size of this paper, so it has been skipped there.

5   Non-preemptive Independent Tasks

For the online scheduling there have been designed in the submitted paper new algorithms for the case $a_i \leq k < m$. These algorithms can be considered as a development of deterministic off-line algorithms formulated originally for $a_i = 1$, see e.g. [13]. In the paper [16] there has been provided an online-list scheduling algorithm for single-processor scheduling on parallel machines. Tasks are analysed in the order of inflow. There is applied the fundamental rule "if the processor becomes free, then assign next task waiting on the list" (known as the algorithm LIST). Immediate extension of this idea leads to the rule "if a_i processors are empty, assign a_i processors to the task (algorithm m-LIST). Note that m-LIST operates also on the list created in order of tasks arrivals.

Algorytm 1. *Let C_k, $k \in M$, be the time moment so that processor k is continuously idle. We start from $C_k = 0$, $k \in M$. From the list of waiting tasks take the successive task and call it i. Find the earliest time moment S_i so that a_i processors are idle from S_i. Schedule task i starting from S_i and modify suitable C_k. Repeat process for successive task from the list.*

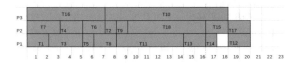

Fig. 1. Exemplary schedule provided by on-line algorithm m-LIST, for $a_i = 1$, $m = 3$

Figure 1 shows exemplary schedule for $a_i = 1$ generated by online-list algorithm m-LIST for the list $(1, 7, 16, 3, 4, 5, 6, 8, 2, 10, 11, 9, 18, 13, 14, 15, 12, 17)$. Task i is denoted in the figure as T_i.

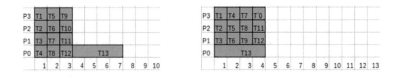

Fig. 2. Case m-LIST $a_i = 1$, $m = 4$, (left). Optimal off-line schedule (right).

Algorithm m-LIST for $a_i = 1$ (namely algorithm LIST) is known in the literature as the best among online-list algorithms for this case of problem [6, 16]. We also refer in Theorem 1 to the fundamental well-known result.

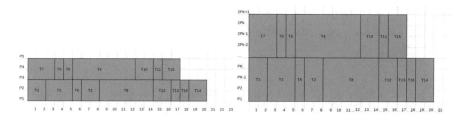

Fig. 3. Case m-LIST: $a_i = 2$, $i \in T$, $m = 5$ (left); $a_i = k$, $i \in T$, any m (right)

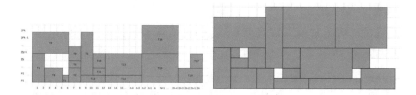

Fig. 4. Case m-LIST. $a_i \leq k$, any m (left); non-guillotined SCP (right)

Theorem 1. *Algorithm LIST for the problem* $P|online - list, a_i = 1|C_{\max}$ *has competitive ratio* $r = 2 - \frac{1}{m}$.

Figure 2 illustrates the theorem by an exemplary on-line schedule for an instance I and the list of incoming jobs $(1, 2, \ldots, 13)$ (left) versus optimal off-line schedule (right). It is clear that $r \to 2$ if $m \to \infty$. In the figure tasks from T are denoted as $T_1, T_2, \ldots$ in order to avoid misunderstanding.

In order to formulate successive properties for m-LIST we introduce the following approximation through data transformation. Let us assume that $a_i = const$ and are the same for all tasks. Then we group processors as follows: (A) for 2-processor tasks, aggregate each 2 processors, (B) for 3-processor tasks, aggregate each 3 processors, (C) for k-processor tasks, aggregate each k processors. Then we apply Algorithm 1 for aggregated processors and artificially defined single-processor tasks. Taking into account previous theorem we obtain

Theorem 2. *Algorithm m-LIST for the problem* $P|online - list, a_i = 2|C_{\max}$ *has competitive ratio*

$$
r = \begin{cases} 2 - \frac{2}{m} & \text{if } m \text{ is even} \\[2mm] 2 - \frac{2}{m-1} & \text{if } m \text{ is odd} \end{cases} \tag{2}
$$

Proof. Since tasks utilize processors in a synchronous manner, there exist at most $\lceil m/2 \rceil$ pairs of 2-processors simultaneously available for processing. Without losing generality, one can assume that these are pairs $(1, 2), (2, 3), \ldots, (m - 1, m)$ if m is even and $(1, 2), (2, 3), \ldots, (m - 2, m - 1)$ if m is odd. Note that for case "odd" there are no possibility of using processor m because it has no pair.

Next we transform problem $P|online - list, a_i = 2|C_{max}$ with data n, m, $(p_j, j = 1, 2, \ldots n)$ to the auxiliary problem $P|online - list, a_i = 1|C_{max}$ with data n', m', $(p'_j, j = 1, 2, \ldots, n)$ in the following manner: $n' = n$, $m' = [m/2]$, $(p'_j = p_j, j = 1, 2, \ldots, n)$. Accordingly to Theorem 1 we have

$$r = 2 - \frac{1}{m'}. \tag{3}$$

Since $m' = [m/2] = m/2$ for even m and $m' = [m/2] = (m - 1)/2$ for odd m, substituting m' into (3) we obtain (2). It ends the proof. ∘

By using the idea of the proof shown the above, we can prove by analogy the following theorem (proof as obvious will be omitted). Let $[x]$ denotes the floor integer from the real x.

Theorem 3. *Algorithm m-LIST for the problem $P|online - list, a_i = k|C_{max}$ has competitive ratio $r = 2 - \frac{1}{[\frac{m}{k}]}$.*

Theorem 3 applied for the case $a_i = 2$ provides result the same as in Theorem 2. Function $r = 2 - \frac{1}{[\frac{m}{k}]}$ is decreasing with increasing integer k and reaches maximum for $k = 1$. For the case $1 \leq a_i \leq k$ (mixed tasks), competitive ratio should work for any instance I, in particular for $a_i = 1, i \in T$. Thus, competitive ratio for m-LIST with various a_i is not less than $r = 2 - \frac{1}{m}$. Figure 3 shows two various instances with processor aggregation for constant a_i obtained by m-LIST. Figure 4 (left) is for the case of various a_i.

Algorithm m-LIST derives from its precursor LIST. Seeking for other inspirations we refer to algorithm MR from [14]. MR uses special technology of balancing loads of processors depending on the type of tasks and partial schedules. It has been proved in [14] that competitive ratio of MR approximates $1 + \sqrt{\frac{1 + \ln 2}{2}} < 1.921$ for $m \to \infty$. Since this is currently the best on-line algorithm, we will provide some its key elements. The incoming tasks are not known a priory and they arrive on the list in the order $(1, 2, \ldots, n)$. Let us assume that tasks $(1, 2, \ldots, t)$ have been already scheduled. We denote by L_i^t the workload of processors after iteration t and assume that $L_1^t \geq L_2^t \geq \ldots \geq L_m^t$, which corresponds to processors $M_1^t \ldots M_m^t$, respectively. Processor M_1^t has the highest load. The workload is the sum of processing times of tasks assigned to the processor, so L_1^t is the makespan. We define the *average load* of processors $M_j^t, \ldots, M_m^t$

$$D_j^t = \frac{1}{m - j + 1} \cdot \sum_{k=1}^{m} L_k^t \tag{4}$$

at step t and denote $D^t = D_1^t$. Next we choose some "magic" numbers according to Authors' of the algorithm

$$c = 1 + \sqrt{\frac{1 + \ln 2}{2}}, \quad i = \left\lceil \frac{5c - 2c^2 - 1}{c} \cdot m \right\rceil - 1, \quad k = 2i - m. \tag{5}$$

We call the schedule *flat* at iteration t if

$$L_k^{t-1} < \frac{2(c - 1)}{2c - 3} \cdot D_{i+1}^{t-1}. \tag{6}$$

Inequality (6) means that, before scheduling task t, the load of k-th highly loaded processor is not too much higher then average load of $m - i$ the least loaded processors. If condition (6) does not hold we called the schedule *steep*. The algorithm MR works as follows.

Algorytm 2. *We have already scheduled tasks $(1, 2, \ldots, t - 1)$ and we want to schedule task t having processing time p^t. If the current schedule is steep or if $p^t + L_i^{t-1} > c \cdot D^t$ then schedule task t on the least loaded processor M_m^{t-1}, otherwise schedule it on the processor M_i^{t-1}.*

Despite the complexity of presentation of MR one can propose advanced version m-MR for multiprocessor task scheduling case. The methodology of the algorithm design and analysis is similar to m-LIST, so we skip in the sequel this subject because of the limited size of the paper.

6   Preemptive Independent Tasks

Independent tasks means $E = \emptyset$ in $G(T, E)$. Our aim is to design an online algorithm for the problem denoted $P|online - list, pmtn, a_i|C_{\max}$ in extended Graham's taxonomy. Nevertheless, at the begin we will provide a few helpful facts. First, there exists the polynomial-time offline algorithm for the problem $P|pmtn, a_i = 1|C_{\max}$, followed from McNaughton, [23]: the preemptive optimal makespan is $C_{\max} = \max(\frac{1}{m} \sum_{j=1}^n p_j, \max_{1 \le j \le n} p_j)$. Secondly, the problem $P|pmtn, a_i = 1|C_{\max}$ is NP-hard.

In the sequel we deal with the problem $P|online - list, pmtn, a_i = 1|C_{\max}$ and its online-list algorithm (called next Algorithm 3, or CVW) for the case of single-processor preemptive task, [6]. The incoming tasks are not known a priory and they arrive on the list in the order $(1, 2, \ldots, n)$. Task j becomes known already after job $j - 1$ has been scheduled. Scheduling of a task means to set assignments to time slots and processors, and the made decision cannot be changed in the future. Let us assume that t tasks $(1, 2, \ldots, t)$ have been already scheduled. We denote by L_i^t the workload of processors after iteration t and assume that $L_1^t \le L_2^t \le \ldots \le L_m^t$, which corresponds to processors $M_1^t \ldots M_m^t$, respectively. The current optimal makespan in iteration t is denoted OPT^t, while the sum of task processing times $S^t = \sum_{j=1}^t p_j$.

The strategy of the algorithm is to maintain lightly loaded machines still lightly loaded, because some longer jobs can arrive in the future. The algorithm provides for any t the schedule satisfying the following properties

$$L_m^t \le t \cdot OPT^t \tag{7}$$

$$\sum_{i=1}^k L_i^t \le \frac{\alpha^k - 1}{\alpha^m - 1} S^t, \quad 1 \le k \le m \tag{8}$$

where $\alpha = \frac{m}{m-1}$. Let us pass to the description of the algorithm, assuming that

$$r \stackrel{\text{def}}{=} \frac{\alpha^m}{\alpha^m - 1} = \frac{m^m}{m^m - (m-1)^m}. \tag{9}$$

Algorytm 3. *We have already scheduled tasks* $(1, 2, \ldots, t)$ *and we want to schedule task* $t+1$. *Calculate* OPT^{t+1} *by using McNaugton rule. If* $L_m^t + p_{t+1} \leq r \cdot OPT^{t+1}$, *then schedule task* $t+1$ *on the processor* M_m^t. *Otherwise find* $\ell = \min\{1 \leq i \leq m : L_i^t + p_{t+1} \geq r \cdot OPT^{t+1}\}$. *Schedule* $r \cdot OPT^{t+1} - L_\ell^t$ *part of task* $t+1$ *on the processor* M_ℓ^t, *and schedule the remain part (if it exists) of task* $t+1$ *on the processor* $M_{\ell-1}^t$.

In the paper [6] and in the review in [5] there has been mentioned that for $m \geq 2$ there are no algorithms with better than CVW worst-case guarantee. Details of the formal proofs one can find in [6], so we skip them. We have.

Theorem 4. *Algorithm CVW for the problem* $P|pmnt, online - list, a_i = 1|C_{\max}$ *has competitive ratio* $r = \frac{1}{1-(1-\frac{1}{m})^m}$.

Note that r which appear in the theorem equals that in (9). For the number of processors $m \to \infty$ competitive ratio $r \to \frac{e}{e-1} \approx 1, 58$.

The extension of CVW on the case $a_i = const > 1$ can be made by the aggregation of processors (like in m-LIST) with the use of "artificial tasks". Such transformation provides new algorithm called further m-CVW. Because of the clear analogy to technology used in m-LIST, we skip formal description of the new version of the algorithm leaving only the following quality evaluation.

Theorem 5. *Algorithm m-CVW for the problem* $P|online - list, a_i = 2|C_{\max}$ *has competitive ratio*

$$r = \begin{cases} \dfrac{1}{1 - \sqrt{(1 - \frac{2}{m})^m}} & \text{if } m \text{ is even} \\ \dfrac{1}{1 - \sqrt{(1 - \frac{2}{m-1})^{m-1}}} & \text{if } m \text{ is odd} \end{cases} \tag{10}$$

Proof. Tasks utilize two processors in the synchronous way, therefore we have at most $[m/2]$ pairs of processors used simultaneously. Without losing generality, one can assume that these are pairs $(1, 2)$, $(2, 3)$, $\ldots,(m-1, m)$ if m is even and $(1, 2)$, $(2, 3)$, $\ldots,(m-2, m-1)$ if m is odd. In case "odd" processor m is useless. The transformation from the problem $P|online - list, a_i = 2|C_{max}$ with data n, m, $(a_j = 2, j = 1, 2, \ldots n)$, $(p_j, j = 1, 2, \ldots n)$ to the auxiliary problem $P|online - list, a_i = 1|C_{max}$ with data n', m', $(a_j' = 1, j = 1, 2, \ldots n)$, $(p_j', j = 1, 2, \ldots, n)$ is as following: $n' = n$, $m' = [m/2]$, $(p_j' = p_j, j = 1, 2, \ldots, n)$. Applying m-CVW algorithm for transformed problem with primes' we have from the Theorem 10

$$r = \frac{1}{1 - (1 - \frac{1}{m'})^{m'}}. \tag{11}$$

Since $m' = [m/2] = m/2$ for odd m and $m' = [m/2] = (m-1)/2$ for even m, substituting m' we get (10). ∘

The next result is clear, so we skip the proof.

Theorem 6. *Algorithm m-CVW for the problem P|online − list, pmtn, $a_i =$ k|$C_{\max}$ has competitive ratio*

$$r = \frac{1}{1 - (1 - \frac{1}{[\frac{m}{k}]})^{[\frac{m}{k}]}}, \tag{12}$$

where [x] is the floor integer from real x.

Function $\frac{1}{1-(1-\frac{1}{[\frac{m}{k}]})^{[\frac{m}{k}]}}$ is decreasing with increasing integer k and reaches maximum for $k = 1$. For the case $1 \leq a_i \leq k$ (mixed tasks), competitive ratio should work for any instance I, in particular for $a_i = 1$, $i \in T$. Thus, competitive ratio for m-CV with various a_i is not less than $r = \frac{1}{1-(1-\frac{1}{m})^m}$.

7   Conclusions

We proposed and evaluated several online algorithms for scheduling multiprocessor tasks in the cases of independent preemptive and non-preemptive tasks with the evaluations of competitive ratios. These pessimistic evaluations have theoretical character and are formulated as theorems. Obtained results do not need to carry out numerical experiments on random sample of instances. The reasons of this attitude has been explained earlier. In fact, pessimistic instances appear statistically rarely in real systems. On the other side, numerical tests usually provide results average for the sample of instances. Therefore this findings will be closer to probabilistic analysis in the average sense than to competitive analysis. Nevertheless, equally competitive analysis as well as experimental analysis provides aggregated numerical characteristics of the algorithm.

References

1. Albers, S.: Online Scheduling. Introduction to Scheduling. CRC Press (2009)
2. Behnamian, J.: Survey on Fuzzy Shop Scheduling, Fuzzy Optimization and Decision Making. Springer, Cham (2016)
3. Błażewicz, J., et al.: Handbook on Scheduling: From Theory to Applications. Springer, Cham (2007)
4. Błażewicz, J., Liu, Z.: Scheduling multiprocessor tasks with chain constraints. Eur. J. Oper. Res. **94**, 231–241 (1996)
5. Bożejko, W., Gawlińska, E.: Algorytmy szeregowania online. W: Bożejko, W., Pempera, J. (red.): Optymalizacja dyskretna w informatyce, automatyce i robotyce, Oficyna Wydawnicza Politechniki Wrocławskiej, Wrocław (2012)
6. Chen, B., Van Vliet, A., Woeginger, G.J.: An optimal algorithm for preemptive on-line scheduling. Oper. Res. Lett. **18** (1995)
7. Chen, X.: Selected problems of online scheduling on parallel machines. Rozprawa doktorska, Politechnika Poznańska, Poznań (2014)

8. Davis, R., Burns, A.: A survey of hard real-time scheduling for multiprocessor systems. ACM Comput. Surv. (CSUR) **43**, 1–44 (2011)
9. Dorota, D.: Scheduling tasks in a system with a higher level of dependability. In: Zamojski, W., Mazurkiewicz, J., Sugier, J., Walkowiak, T., Kacprzyk, J. (eds.) DepCoS-RELCOMEX 2019. AISC, vol. 987, pp. 143–153. Springer, Cham (2020). https://doi.org/10.1007/978-3-030-19501-4_14
10. Dorota, D.: Scheduling tasks with uncertain times of duration. In: Zamojski, W., Mazurkiewicz, J., Sugier, J., Walkowiak, T., Kacprzyk, J. (eds.) DepCoS-RELCOMEX 2020. AISC, vol. 1173, pp. 197–209. Springer, Cham (2020). https://doi.org/10.1007/978-3-030-48256-5_20
11. Dorota, D.: Szeregowanie zadań wieloprocesorowych w warunkach niepewności. Rozprawa doktorska, Politechnika Wrocławska (2023)
12. Drozdowski, M.: Scheduling for Parallel Processing. Springer, London (2009)
13. Drozdowski, M.: Scheduling multiprocessor tasks - an overview. Eur. J. Oper. Res. **94**, 215–230 (1996)
14. Fleischer, R., Wahl, M.: Online scheduling revisited. In: Paterson, M.S. (ed.) ESA 2000. LNCS, vol. 1879, pp. 202–210. Springer, Heidelberg (2000). https://doi.org/10.1007/3-540-45253-2_19
15. Graham, R.L., Lawler, E.L., Lenstra, J.K., Rinnooy Kan, A.H.G.: Optimization and approximation in deterministic sequencing and scheduling: a survey. Ann. Discret. Math. **5** (1979)
16. Graham, R.: Bounds for certain multiprocessing anomalies. Bell Syst. Tech. J. **45**, 1563–1581 (1966)
17. Herroelen, W., De Reyck, B., Demeulemeester, E.: Resource-constrained project scheduling: a survey of recent developments. Comput. Oper. Res. **25**(4), 279–302 (1998)
18. Hurink, J.L., Paulus, J.J.: Online algorithm for parallel job scheduling and strip packing. In: Kaklamanis, C., Skutella, M. (eds.) WAOA 2007. LNCS, vol. 4927, pp. 67–74. Springer, Heidelberg (2008). https://doi.org/10.1007/978-3-540-77918-6_6
19. Im, S.: Online scheduling algorithms for average flow time and its variants. University of Illinois at Urbana-Champaign (2012)
20. Johannes, B.: Scheduling parallel jobs to minimize the makespan. J. Sched. **9**, 433–452 (2006)
21. Karp, R.M.: On-line algorithms versus off-line algorithms: how much. In: Algorithms, Software, Architecture: Information Processing 92: Proceedings of the IFIP 12th World Computer Congress, Madrid, Spain, 7–11 September 1992 (1991)
22. Makuchowski, M.: Problemy gniazdowe z operacjami wielomaszynowymi. Własności i algorytmy. Rozprawa doktorska. Raporty Instytutu Cybernetyki Technicznej PWr, Ser. PRE, no. 37, p. 196 (2004)
23. McNaughton, R.: Scheduling with deadlines and loss functions. Manag. Sci. **6**, 1–12 (1959)
24. Muntz, R.R., Cofmann, E.G., Jr.: Preemptive scheduling of real-time tasks on multiprocessors systems. J. ACM **17**(2), 324–338 (1970)
25. Pinedo, M.: Scheduling. Springer, New York (2015)
26. Pruhs, K., Jiri, S., Torng, E.: Online scheduling (2004)
27. Smutnicki, C.: Algorytmy szeregowania zadań. Oficyna Wydawnicza Politechniki Wrocławskiej (2012)

Solving a Vehicle Routing Problem for a Real-Life Parcel Locker-Based Delivery

Radosław Idzikowski$^{(\boxtimes)}$, Jarosław Rudy , and Michał Jaroszczuk

Department of Control Systems and Mechatronics, Wrocław University of Science and Technology, Wrocław, Poland
{radoslaw.idzikowski,jaroslaw.rudy,michal.jaroszczuk}@pwr.edu.pl

Abstract. In this paper a problem modeled after a real-life system of delivering orders to parcel lockers is considered. The problem is formulated as an extension of the Vehicle Routing Problem with total travel time as the goal function. Several constraints, including vehicle capacity, travel time limits as well as parking and order service times, are formulated. Two heuristics are proposed as solving methods: a Genetic Algorithm population metaheuristic and a greedy method, which is also used as a reference point. A computer experiment is performed using large-size problem instances based on real-life parcel locker locations for an existing delivery company in the city of Wrocław, Poland. The results indicate that the proposed Genetic Algorithm method vastly outperforms the greedy method, obtaining solutions with several times less total travel time and runs under 5 min even for instances with 10000 orders. The influence of the number of orders and parcel locations on the result are discussed as well.

Keywords: vehicle routing · parcel lockers · metaheuristics · ready times · capacity-constrained · time-constrained

1 Introduction

Recent years have seen a gradual change in the way couriers and delivery companies are conducting their services. Previously, parcels were delivered directly to the customer's house and each delivery location corresponded to a single order served by a single vehicle. Such a system could be modeled well enough with classic Vehicle Routing Problems (VRP). Nowadays, the popularity of electronic shopping has grown, resulting in an increased number of deliveries. Moreover, the delivery process changed. Instead of delivering goods directly to customers, a system of parcel lockers is often used. Smaller vehicles (small cars, bicycles etc.) are used for deliveries more commonly, stressing the need to take their limited capacity into account. Furthermore, often the goods have to first be shipped from a different city, before it can be delivered to the parcel locker. As a result, it becomes necessary to consider that a single parcel locker might be visited by

© The Author(s), under exclusive license to Springer Nature Switzerland AG 2024
W. Zamojski et al. (Eds.): DepCoS-RELCOMEX 2024, LNNS 1026, pp. 69–79, 2024.
https://doi.org/10.1007/978-3-031-61857-4_7

multiple vehicles in the same day. Moreover, multiple customers might be served at the same locker, taking more time and violating the previously used one-customer-per-location rule. Finally, time taken to perform the deliveries have to be taken into consideration to minimize customer waiting time, fuel and working costs as well as to observe the working time limits imposed by law.

Due to the above real-life considerations, classic approach to VRP is no longer sufficient. As such, in this paper we propose Time and Capacity Constrained Vehicle Routing Problem with Ready Times (TCVRP-RT) to model and solve the problem of delivering orders to parcel lockers while observing all mentioned limitations. We propose a problem formulation and several heuristic solving methods. The proposed methods are evaluated using problem instances based on real-life data.

The remainder of this paper is structured as follows. In Sect. 2 we perform a brief literature review of similar approaches. In Sect. 3 we present a mathematical model of the problem, including the input data, objective function, all constraints as well as solution representation. Next, in Sect. 4 we describe the proposed solving methods. In Sect. 5 we describe the computer experiment carried out using real-life parcel locker data from the city of Wrocław, Poland and discuss its results. Finally, Sect. 6 contains conclusions as well as planned directions of future research on this topic.

2   Literature Review

VRP is a well-known problem researched for many years with a wide range of practical applications. Due to different applications having different detailed assumptions, a great number of modifications and extensions to VRP has been proposed in the literature. However, the topic of applying VRP to the problem of delivering parcels to parcel lockers is not well-researched with only a handful of closely-related papers, which we will review here.

The most similar problem was considered in the paper by Orenstein et al. [6]. The authors assumed both vehicle capacity and travel time limits as well as non-uniform capacity limit of parcel lockers. Moreover, parcel locker locations were modeled after location of real-life gas stations, making this paper half-practical as gas stations are not as densely placed as parcel lockers. The authors considered both a MILP formulation and a Tabu search metaheuristic. However, it should be noted that MILP was restricted to 1 h time limit and other methods took 30 min for instances of 1500 orders, making this approach of limited use in practice. The authors also did not consider the time parcels arrive in the depot.

A less similar problem was considered by Grabenschweiger et al. [2]. The authors proposed a problem of delivering to parcel lockers with vehicle time limit and non-uniform parcel locker limits, but with no vehicle capacity limits. The authors also consider parcel service and locker unloading times, but did not consider parcel arrival time. Moreover, the algorithm was only tested on extended Mancini and Gansterer benchmarks with no real-life data and used Adaptive Large Neighbourhood Search (ALNS) as a solving method. Next, Pan et al. [7]

considered a multi-trip time-dependent VRP with time windows. Both time- and capacity-constraint were considered as well as order release time with ALNS and Variable Neighborhood Descend as solving methods. However, the paper does not directly consider parcel lockers and uses instances based on Solomon benchmarks instead of a real-life case. A similar approach was done by Li et al. [4] with assumption of time windows, time- and capacity-constraints as well as order release times, but once again with no direct connection to parcel lockers. The solving method was once again ANLS for Solomon benchmarks.

Moving onto the last batch of papers, Yu et al. [10] considered a parcel lockers delivery system with capacity constrains on both lockers and vehicles, but with no time limit on vehicle travel time. Simulated Annealing solving methods was proposed and tested on instances based on Solomon benchmarks and compared to a Gurobi solver formulation. Order arrival times were not considered. Parcel lockers with both pick-up and delivery was considered in the paper by Sitek et al. [9]. The authors considered time travel limit and parcel locker capacity limit, but not vehicle capacity limits or order release times. A hybrid solution method with the use of the solver was employed. Next, Reed et al. [8] considered a VRP problem where multiple orders might be delivered to the same customer. While not directly matching a parcel locker systems it is similar. The authors took both time travel and vehicle capacity limits into account. Gurobi formulation and a heuristic algorithm were used to research the influence of the parking time on the results. Finally, Dondo et al. [1] considered an optimal approach to VRP with multiple depots, time windows and time- and capacity-constraints. Unfortunately, not only there is no direct connection to parcel lockers, but the authors stopped before performing any computer experiments.

To summarize, only the paper by Orenstein et al. [6] is relatively close to the approach we consider in this paper. All other papers either do not consider parcel lockers directly, deviate from our assumptions of TCVRP-RT too much or fail to test their methods on instances based on real-life parcel locker systems.

3   Problem Formulation

The problem is stated as follows. Let $\mathcal{O} = \{1, 2, \ldots, n\}$ and $\mathcal{V} = \{1, 2, \ldots, n\}$ be set of n delivery orders and n vehicles respectively. Next, let $\mathcal{L} = \{0, 2, \ldots, m\}$ be a set of $m + 1$ locations, where 0 is depot location and 1 through m are m possible delivery locations (i.e. parcel lockers).

For each order $i \in \mathcal{O}$ let $r_i \in \mathbb{N}$, $w_i \in \mathbb{N}$ and $d_i \in \mathcal{L} \setminus \{0\}$ be the order ready time, weight (or volume) and destination respectively. Next, for each pair (l, k) of locations such that $l, k \in \mathcal{L}$ let $t_{l,k} \in \mathbb{N}$ be the time needed to travel from l to k. Let $C \in \mathbb{N}$ and $T \in \mathbb{N}$ be the constant vehicle capacity and time limit respectively. Let $S \in \mathbb{N}$ be the constant time of servicing a single order (retrieving the package, placing it in the package locker etc.). Similarly, let $P \in \mathbb{N}$ be the constant time needed for the vehicle to park and take off from a location.

Decision variables are as follows. $x_{v,l,k}$ determines whether vehicle v travels from location l to k while on its tour. Similarly, $y_{v,i}$ determines whether order i

is served by vehicle v. Finally, auxiliary variables q_v determine whether vehicle v is being used.

The task is to minimize the following goal function which is the total time taken by vehicles while delivering orders:

$$\sum_{v \in \mathcal{V}} q_v \left(\max_{i \in \mathcal{O}} r_i y_{v,i} + \sum_{l \in \mathcal{L}} \left(\sum_{k \in \mathcal{L}} (t_{l,k} + P) x_{v,l,k} \right) - P \right) \tag{1}$$

with the following constraints:

$$q_v \sum_{l \in \mathcal{L}} \left(\sum_{k \in \mathcal{L}} (t_{l,k} + P) x_{v,l,k} - P x_{v,l,0} \right) +$$

$$+ \sum_{i \in \mathcal{O}} y_{v,i} S \leq T \qquad \forall v \in \mathcal{V}, \tag{2}$$

$$\sum_{i \in \mathcal{O}} w_i y_{v,i} \leq C \qquad \forall v \in \mathcal{V}, \tag{3}$$

$$\sum_{k \in \mathcal{L}} x_{v,0,k} = q_v \qquad \forall v \in \mathcal{V}, \tag{4}$$

$$\sum_{l \in \mathcal{L}} x_{v,l,0} = q_v \qquad \forall v \in \mathcal{V}, \tag{5}$$

$$\sum_{v \in \mathcal{V}} y_{v,i} = 1 \qquad \forall i \in \mathcal{O}, \tag{6}$$

$$\sum_{l \in M} \sum_{k \in \mathcal{C}(v) \setminus M} x_{v,l,k} \geq 2 \qquad \forall M \subsetneq \mathcal{C}(v) \setminus \{0\}, v \in \mathcal{V}, \tag{7}$$

$$q_v = \begin{cases} 1 & \text{if } \sum_{i \in \mathcal{O}} y_{v,i} > 0 \\ 0 & \text{otherwise} \end{cases} \qquad \forall v \in \mathcal{V}, \tag{8}$$

$$x_{v,l,k} \in \{0,1\} \qquad \forall v \in \mathcal{V}, l, k \in \mathcal{L}, \tag{9}$$

$$y_{v,i} \in \{0,1\} \qquad \forall v \in \mathcal{V}, i \in \mathcal{O}, \tag{10}$$

where $\mathcal{C}(v)$ is the set of all locations that were assigned to vehicle v i.e. $\mathcal{C}(v) = \{d_i : i \in \mathcal{O} \wedge y_{v,i} = 1\}$.

For each vehicle v, function (1) takes into the account the following terms:

1. $q_v \max_{i \in \mathcal{O}} r_i y_{v,i}$ – the time the vehicle leaves the depot (0 if v is not used),
2. $q_v \sum_{l \in \mathcal{L}} \sum_{k \in \mathcal{L}} t_{l,k} x_{v,l,k}$ – travel time between locations (including from and into the depot),
3. $q_v \sum_{l \in \mathcal{L}} \sum_{k \in \mathcal{L}} P x_{v,l,k}$ – vehicle parking time after arriving at location k (including the return to the depot),
4. $-q_v P$ – subtracting the parking time once as it should not be counted when arriving back at the depot.

Such goal function was chosen as it allows to directly or indirectly optimize 1) man-hours, 2) fuel usage, 3) delivery time, 4) number of vehicles used. The goal function does not consider the service time S, as all orders need to be delivered anyway, so this will amount to nS, no matter what values the decision variables take. However, the service time is important for the first constraint below.

Constraint (2) specifies time constraint, namely that the tour of vehicle v (meaning traveling through edges, parking at all visited locations and service time of all its orders) cannot be larger than T. This constraint can be used to obey laws regarding allowed working time of drivers. Constraint (3) enforces that the weight (volume) of all orders served by vehicle v does not exceed its capacity. Constraints (4)–(5) enforce that vehicle's tour starts and ends at the depot. Constraint (6) ensures that each order is served by exactly one vehicle. Constraint (7), based on typical cut constraint for VRP problems, ensures that vehicle v visits all locations assigned to it and that its tour is connected properly. Constraint (8) ensures that vehicle is used only if it has at least one order assigned to it. Finally, constraints (9)–(10) define the allowed domain of decision variables.

As the formulation (1)–(10) is cumbersome in practice, we will use the well-known Giant Tour Representation (GTR) instead. GTR can be thought of as an array storing a permutation of orders with additional zeroes inserted, including at the beginning and the end of the array. Zeroes separate vehicles and represent going out of and into the depot. For n orders and V vehicles, GTR will contain n orders and $V+1$ zeroes. For example, assume we have $\mathcal{O} = \{1, 2, 3, 4, 5, 6, 7, 8, 9\}$ and $|\mathcal{V}| = 5$. Then a GTR for a possible solution could be as follows:

$$(0, 4, 7, 1, 9, 0, 6, 3, 0, 0, 2, 0, 5, 8, 0). \tag{11}$$

Here the orders assigned to vehicle 1 and their sequence is $(4, 7, 1, 9)$. Orders for vehicles 2, 4 and 5 are $(6, 3)$, (2) and $(5, 8)$ respectively. Vehicle 3 has no orders, meaning that it will not leave the depot.

If needed, a transformation from GTR to the original formulation (1)–(10) can be done as follows. For vehicle v we set $y_{v,i}$ to 1 for all orders assigned to v and 0 for the remaining orders. Next, we use the sequence of orders (their d_i values to be exact) to obtain sequence of locations to be visited by v. Then variables $x_{v,l,k}$ are set appropriately, remembering to add going out of and into the depot. For example, if sentence of orders for v is $(4, 7, 1)$ then we set $x_{v,0,d_4} = x_{v,d_4,d_7} = x_{v,d_7,d_1} = x_{v,d_1,0} = 1$ and remaining $x_{v,l,k}$ are set to 0. It should be noted that interpretation of a valid GTR always results in a solution that automatically meets constraints (4)–(10). Solutions obtained through GTR can still, however, violate constraints (2)–(3) and be infeasible.

4   Solving Methods

One further issue with the formulation (1)–(10) is the exponential number of constraints (7), making exact methods infeasible for real-life sized problem instances. As such, we propose two solving methods: a greedy method as well as Genetic Algorithm (GA) metaheuristic.

4.1 Greedy Method

The greedy method first normalizes r_i and w_i of all orders and then sorts them by the normalized $r_i w_i$ product in descending order. The initial GTR is set to $n + 1$ zeroes i.e. n empty vehicles. Next, we proceed iteratively. In each iteration we consider next order i from the sorted list and try to insert it in all currently available positions. For each position we evaluate the goal function and ultimately insert the order into the position that yields the lowest goal function value in this partial solution. After all orders are inserted, the algorithm ends. We also considered using only r_i or only w_i for sorting, but the obtained results were slightly worse.

4.2 Genetic Algorithm

GA is a well-known population metaheuristic. In each iteration the GA search process is guided by genetic operators applied to specimens in the population. The crossover operator is responsible for promoting promising genes, leading to intensification of search in the promising areas of the solution space. On the other hand, the mutation operator introduces small random changes in specimens to introduce new genes and increase gene diversity, leading to diversification of the search process and preventing premature convergence of the method.

In our implementation each specimen (chromosome) is simply a GTR. The starting population contains *popSize* random specimens. In each iteration specimens are first evaluated. Next, *popSize* parent specimens are chosen using a tournament selection. In a tournament of size *tourSize*, we take *tourSize* specimens from the population at random and select the best one as the winner. Tournaments are repeated until necessary number of parents are chosen, repetitions are allowed. Once parents are selected, the crossover operator is applied for each parent or parents pair (depending on crossover operator type) with probability *crossProb*, creating offspring. If the operator is not applied, then the offspring becomes a copy of the parent.

We consider three crossover operators: OX, random and greedy. OX is a well-known crossover operator that considers two parents to create two offspring. For the first offspring, we copy a random section of the first parent. The remaining elements from the first parent are added in the order they are in the second parent. The second offspring is created in the same way, but the order of parents is reversed. Random operator uses a single parent. At first, random section of the parent is copied into offspring. The remaining elements of the parent are put into the offspring in random order. Greedy operator (previously used for Travelling Salesman Problem, which is closely related to VRP) works similarly, but remaining elements are inserted in the greedy order as described in the earlier subsection. In all operators the offspring is constructed as to not create infeasible solutions.

After the crossover is complete, each offspring has mutation operation applied to it with probability *mutProb*. Two types of mutation operators are considered. In the swap operator two random GTR elements are interchanged, while in the reverse operator, the order of elements in a random portion of GTR is reversed. After all mutations are complete, the old population is replaced with the children (thus keeping the population size constant throughout the process). The algorithm stops after *maxTime* time has elapsed. *popSize*, *tourSize*, *crossProb*, *mutProb*, *maxTime* and operator types are algorithm parameters.

5   Computer Experiment

The testing instances were prepared as follows. We have used real-life parcel lockers locations of one delivery company in Wrocław, Poland as a basis for our location set $\mathcal{L}$. After removing duplicate parcel lockers with the same locations (e.g. two lockers across the street from each other were often reported as having the same coordinates) 344 locations were obtained, meaning that we can cover problem instances with $m \leq 344$. Moreover, there are 3 possible depot locations. Next, for each pair of locations we have used Open Street Map [5] to obtain the travel time between those locations. The city of Wrocław was used due to its geographical and infrastructure features that make the problem more challenging (many rivers and waterways, a lot of bridges and one-way streets).

The procedure for generating instances uses 8 input parameters n (number of orders), m (number of locations), maxW (maximal weight), maxR (maximal ready time), C (vehicle capacity constant), T (time capacity constant), S (service time constant), P (parking time constant). For each order i values r_i, w_i and d_i are chosen from an uniform integer distribution. Ranges for each parameter are $r_i \in \{1, 2, \ldots, \text{maxR}\}$, $w_i \in \{1, 2, \ldots, \text{maxW}\}$, $d_i \in \{1, 2, \ldots, m\}$. Finally, depot location is chosen randomly from the 3 possible locations.

For our experiment we consider 10 problem sizes $n \in \{1000, 2000, \ldots, 10000\}$. For each n three numbers of locations were used $m \in \{100, 200, 344\}$. For each $n \times m$ combination we generate 10 instances, meaning 330 instances in total. We have assumed $maxW = 10$ and $C = 1000$ (i.e. a parcel can weight up o 10 kg and a vehicle can hold 1000 kg of cargo). Likewise, we set $maxR = 240$, $T = 480$, $S = 1$, $P = 2$ (i.e. service and parking time is 1 and 2 min respectively, vehicle time limit is 8 h and parcels arrive in the first 6 h of the working day). The instance generator is available on GitHub [3].

All algorithms were coded in Julia 1.10 and their source code is available on GitHub as well [3]. Main tests were run on a DGX machine with AMD EPYC 7742 3.2 GHz with 64 physical cores and 503 GiB of RAM running under Ubuntu Linux 5.4. We have performed small-scale preliminary tests to calibrate the proposed algorithms. As a result for GA we have assumed the following parameters $popSize = 40$, $tourSize = \lfloor \sqrt{n} \rfloor$, $crossProb = 0.8$, $mutProb = 0.2$. As crossover and mutation operators we assume greedy and swap respectively. Finally, $maxTime$ was set to $\lfloor 0.025n \rfloor$ seconds. In other words, the time limit is 25 s for $n = 1000$ and 250 s for $n = 10000$.

Table 1. Aggregated values of $Q(I)$ for different n and m values

n	$m = 100$	$m = 200$	$m = 344$	All m
1000	3.84	3.21	2.73	3.20
2000	4.57	4.33	3.70	4.16
3000	4.99	4.75	4.32	4.67
4000	5.32	5.18	4.77	5.08
5000	5.24	5.25	4.97	5.15
6000	5.33	5.38	5.23	5.31
7000	5.47	5.51	5.41	5.46
8000	5.34	5.63	5.47	5.48
9000	5.37	5.68	5.60	5.54
10000	5.57	5.77	5.64	5.66
All n	5.05	4.92	4.56	4.83

As for quality measure we compare the results of GA to the reference solution. Specifically, for instance I we define $Q(I)$ simply as:

$$Q(I) = \frac{REF(I)}{GA(I)}, \tag{12}$$

where $GA(I)$ is the value of the goal function (i.e. total time) for the solution obtained by GA for instance I, while $REF(I)$ is the reference solution for instance I. As such $Q(I) = 2$ indicates that GA provided a solution with two times smaller (better) total time than the reference solution. Ideally, we would use the optimal value as $REF(I)$, but due to instance sizes we were unable to obtain it. Moreover, to our knowledge no benchmarks for this problem exist in the literature. As such, we have used the solution provided by the greedy algorithm as $REF(I)$. Finally, due to GA being a probabilistic method, we run it 5 times for each instance and report the average. Full results are available on GitHub [3].

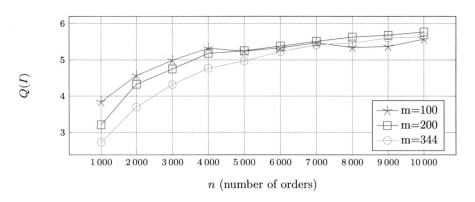

Fig. 1. $Q(I)$ in function of n for different m values

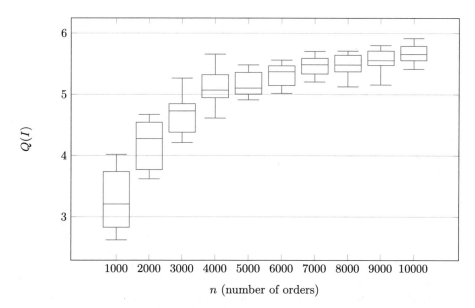

Fig. 2. Boxplot for various numbers of orders n (boxplots are constructed from the 5th, 25th, 50th, 75th and 95th percentiles respectively)

The aggregated results of the computer experiment are shown in Table 1 and Fig. 1, with each value being average results for that problem size and number of locations. We immediately conclude that the proposed GA outperforms the greedy method and does so consistently. In fact, the resulting $Q(I)$ ranged from 2.61 to 6.12 with the average of 4.83 and standard deviation of 0.78. This means that the results provided by GA were usually almost 5 times better and almost always at least 3 times better than the ones obtained by the greedy method.

We can also observe that the advantage of GA increases with the increase of n. In fact, from $n = 1000$ to $n = 10000$, the value of $Q(I)$ increased by 77% on average. This effect is more pronounced the bigger m is. The general effect on increasing m on results is more complex. For smaller n the advantage of GA over the greedy method decreases as m increases (although GA is still vastly superior). The effect slowly diminishes as n approaches 6000. For $n \geq 7000$ the advantage of GA is the largest when $m = 200$ and smallest when $m = 100$, though the differences become small and $Q(I)$ becomes stable with little influence from m.

It should also be noted that the greedy method works in time $O(n^3)$ since for each of n orders, we have to check $O(n)$ possible insert positions and for each we need to check if constraints are met, which takes $O(n)$ time as well. As such, the greedy method no only loses in quality, but does not provide a big speed advantage over GA either.

Finally, as additional analysis a boxplot for various number of orders is shown in Fig. 2. The results prove that higher value of n not only improves on the median value, but also reduces the spread compared to smaller n.

6   Conclusions and Future Works

In this paper we considered a modified Vehicle Routing Problem based on real-life delivery system with parcel lockers with total travel time as an objective function. We presented a mathematical model taking into account travel time limit, vehicle capacity, order arrival times as well as order service and vehicle parking times. Due to large problem size, we have proposed a Genetic Algorithm solving method together with a greedy heuristic as a reference point. We have performed a computer experiment using instances based on real-life parcel locker locations and showed that the Genetic Algorithm managed to obtain several times better solutions than a greedy method with running time under a few minutes.

In future works we plan to extend the research presented in this paper as follows. First, we intend to model the problem in more detail (multiple depots, retrieving parcels from lockers, non-uniform vehicle capacity etc.). Second, we plan to establish theoretical properties of the problem. Third, we plan to propose efficient parallel algorithms with the use of a high-end DGX machine (equipped with multiple General-Purpose Computing on Graphics Processing Units and capable of running 128 threads in parallel).

References

1. Dondo, R., Mendez, C.A., Cerda, J.: An optimal approach to the multiple-depot heterogenous vehicle routing problem with time window and capacity constraints. Lat. Am. Appl. Res. **33**, 129–134 (2003)
2. Grabenschweiger, J., Doerner, K.F., Hartl, R.F., Savelsbergh, M.W.P.: The vehicle routing problem with heterogeneous locker boxes. Central Eur. J. Oper. Res. **29**, 113–142 (2021)
3. Idzikowski, R.: Github repository (2024). Accessed 05 Feb 2024
4. Li, W., Li, K., Kumar, P.N.R., Tian, Q.: Simultaneous product and service delivery vehicle routing problem with time windows and order release dates. Appl. Math. Model. **89**, 669–687 (2021)
5. Luxen, D., Vetter, C.: Real-time routing with OpenStreetMap data. In: Proceedings of the 19th ACM SIGSPATIAL International Conference on Advances in Geographic Information Systems, GIS 2011, pp. 513–516. ACM, New York (2011)
6. Orenstein, I., Raviv, T., Sadan, E.: Flexible parcel delivery to automated parcel lockers: models, solution methods and analysis. EURO J. Transp. Logist. **8**, 683–711 (2018)
7. Pan, B., Zhang, Z., Limb, A.: Multi-trip time-dependent vehicle routing problem with time windows. Eur. J. Oper. Res. **291**, 218–231 (2020)
8. Reed, S., Campbell, A.M., Thomas, B.W.: Does parking matter? The impact of parking time on last-mile delivery optimization. Elsevier (2024)

9. Sitek, P., Wikarek, J., Rutczyńska-Wdowiak, K.: Capacitated vehicle routing problem with pick-up, alternative delivery and time windows (CVRPPADTW): a hybrid approach. In: Herrera-Viedma, E., Vale, Z., Nielsen, P., Martin Del Rey, A., Casado Vara, R. (eds.) DCAI 2019. AISC, vol. 1004, pp. 33–40. Springer, Cham (2020). https://doi.org/10.1007/978-3-030-23946-6_4
10. Yu, V.F., Susanto, H., Jodiawan, P., Ho, T.-W., Lin, S.-W., Huang, Y.-T.: A simulated annealing algorithm for the vehicle routing problem with parcel lockers. IEEE (2022)

Hammering Test for Tile Wall Using AI

Atsushi Ito[1]($\boxtimes$) , Yuma Ito[1], Masafumi Koike[2], and Katsuhiko Hibino[3]

[1] Chuo University, 742-1 Higashi Nakano, Hachioji, Tokyo 192-0351, Japan
{atc.00s,a20.r3yg}@g.chuo-u.ac.jp
[2] Utsunomiya University, 7-1-2 Yoto, Utsunomiya, Tochigi 321-8585, Japan
koike@is.utsunomiya-u.ac.jp
[3] PORT DENSHI Corporation, 5-8-2 Miyanishi, Fuchu, Tokyo 183-0022, Japan
4hibino@port-d.co.jp

Abstract. Social infrastructure, such as bridges and tunnels, is extremely important for economic activities. Among inspection services, it is possible to inspect abnormal areas that cannot be visually observed, such as the interior of concrete walls, by using sound changes caused by hammering. In this inspection, called hammering test, the detection of abnormal sounds depends on the sensory perception of the person performing the hammering test, and the results may differ from one inspector to another. In addition, since the hammering is performed manually, there is the problem that it takes time to perform a wide range of inspections and that the supply of inspections themselves cannot keep up due to the retirement of skilled inspectors. Therefore, this paper describes a hammering test system that uses artificial intelligence to analyze and judge hammering sounds, enabling even unskilled workers to judge abnormal areas. Furthermore, the effectiveness of AI-based sound inspection for adhesive-applied tile walls, which is currently the mainstream, will be discussed.

Keywords: Hammering test · Non-destructive testing · Adhesive-applied tile wall · Deep Learning

1 Introduction

I Various social infrastructures have been aging in recent years, and the demand for infrastructure inspection services is expected to increase [1]. Infrastructure inspection involves various maintenance and management tasks and requires focused diagnosis and inspection related to structural strength and seismic resistance. Inspection robots and drones are being used to target areas that require scaffolding and are difficult for people to access, such as large bridges and high-rise walls [2].

Although there are a wide variety of methods for inspecting anomalies that cannot be seen, such as the interior of concrete walls, including infrared inspection [3] and electromagnetic wave inspection [4], hammering test is still often used to estimate the interior condition by the sound generated when the wall surface is struck with a hammer.

© The Author(s), under exclusive license to Springer Nature Switzerland AG 2024
W. Zamojski et al. (Eds.): DepCoS-RELCOMEX 2024, LNNS 1026, pp. 80–89, 2024.
https://doi.org/10.1007/978-3-031-61857-4_8

Hammering test is a low-cost and highly accurate inspection method, but it requires experienced and skilled inspectors to distinguish abnormal sounds. However, the number of skilled inspectors is decreasing due to the aging of the workforce.

The objects of hammering test can be broadly divided into concrete structures, such as bridges and high-rise buildings, and decorative tiles, such as condominiums and low-rise buildings. Tile decoration has been used around the world since the invention of pottery 5,000 years ago. Many apartment buildings use tiles for their exteriors from the standpoint of creating aesthetics. In Japan alone, there are approximately 6.65 million apartments [5].

There are two main methods of attaching tiles: mortar and adhesives. Tile adhesion with mortar has been used since the time of the Egyptian dynasty, but it is prone to peeling, and when peeling occurs, a large area is peeled off. For this reason, it is recommended to use adhesives with muscular adhesive strength and localize the peeling area. For this reason, adhesives are already used in more than half of all new tiled buildings in urban areas, but they still need to be used in local cities due to cost issues.

We have been researching and developing a technique that enables even inexperienced workers to inspect hammering sounds efficiently by using deep learning to distinguish hammering sounds [6]. However, the target was a tile wall using mortar.

Even if peeling occurs in adhesive-backed tile walls, the extent of peeling is limited to a single tile, and because the adhesive absorbs and weakens the sound of impact and vibration caused by the peeling, it is difficult to distinguish the sound, and even skilled inspectors have difficulty in making a judgment. Therefore, ICT has not been applied to the sound impact inspection of adhesive-bonded tile walls, and other non-destructive testing methods have been used, such as peeling off a tile sample for inspection or requiring large-scale equipment.

Therefore, we used our previous experience in developing a technique for utilizing neural networks for concrete wall sound inspection. We aimed to develop an AI for sound judgment of adhesive-applied tile walls, which is difficult even for skilled inspectors to judge.

The structure of this paper is as follows. Section 2 discusses previous research; Sect. 3 describes the research environment for a method of hammering test of adhesive-applied tiles; Sect. 4 evaluates the generated AI model for hammering test; and Sect. 5 provides the conclusion.

2 Related Works

2.1 Testing of Adhesive-Applied Tile Wall

Various studies have been conducted on the bonding of tiles to exterior walls. For example, it was found from reference [7] that the stress at the bonding area exceeds the bond strength, causing the tiles to delaminate, and reference [8] found that tile walls with adhesive bonding have a larger relative maximum amplitude at the sound area than those with mortar bonding. Reference [9] found that two frequency peaks exist when there is a cavity under the tile, resulting in a muddy sound, and reference [10] found that it is difficult to determine the presence of a float when the area of the float is small

(1)Rolling hammer   (2)Hitting rod

Fig. 1. Equipment for hammering test

Fig. 2. Hammering test on a tile wall

in the sound inspection of a tile wall. However, there has been no research using neural networks yet, so a study was conducted to examine the practical feasibility of such a system.

2.2 Hammering Test

Various studies have been conducted on ICT-based hammering test. Studies in the literature [11] and [12] propose an efficient method of inspecting concrete without using hammers, but it requires expensive equipment. In addition, no study has been made for tile walls. In a study in [13], the same as ours, the authors study sound hammering inspection, but there is no study for tiled walls.

We have also previously conducted a study of hammering sound inspection for mortar-applied tile walls using neural networks [14], and we used the findings of this study in our research.

3 Non-destructive Test for Adhesive-Applied Tile Wall

3.1 Hammering Test for Adhesive-Applied Tile Wall

We have been using a rolling hammer (Fig. 1, left) for hammering test of tile walls [5]. However, a rolling hammer is a tool for concrete walls, so it may break a tile wall that uses thin ceramic tile. Therefore, a hitting rod shown in Fig. 1 (right) generates a friction sound by rubbing with an inspection stick with a sphere attached to its tip (Fig. 2).

3.2 Test Specimen

Figure 3 shows the specimen used in this study. The specimen consists of a concrete block with 18 tiles in three horizontal rows and six vertical columns. The dimensions of each tile are 4.5 cm long and 9.5 cm wide, and the total length of a row is 29.5 cm, including joints (Fig. 3).

Fig. 3. Test specimen

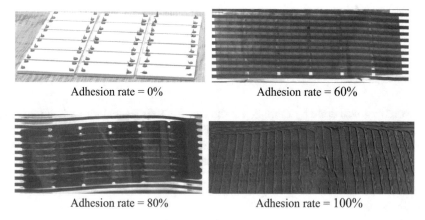

Adhesion rate = 0%   Adhesion rate = 60%

Adhesion rate = 80%   Adhesion rate = 100%

Fig. 4. Adhesion rate

The Ministry of Land, Infrastructure, Transport, and Tourism's standard [7] and tile industry guidelines [8] stipulate that at least 60% of the bonded area must be bonded for an abnormal area in a tile wall. Therefore, we used four specimens whose adhesion rates were 100%, 80%, 60%, and 0%. Figure 4 shows samples to display the difference in the adhesion ratio of tiles.

3.3 Experiment Flow

The flow of the hammering test in this study is shown in Fig. 5. The sound of a strike is acquired from a microphone, and the sound of each tile is extracted by separating the sound at the joints. The frequency components are extracted by FFT and put into a model created by DNN to determine the degree of peeling. The procedure for creating the data for this experiment is as follows.

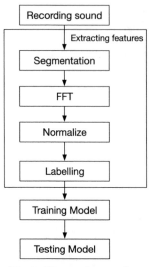

Fig. 5. Hammering test flow

Table 1. The number of segments

Adhesion rate	The number of segments
100%	2313
80%	2375
60%	2435
0%	2389

Recording was done by rubbing the tile surface with a stick with a 2 cm diameter sphere attached, as shown in Fig. 1 (right). This process was recorded using a smartphone from a fixed distance; 12 recordings of approximately 30 s were made for each specimen. The recorded audio data was extracted from the video files and converted to mono format, and silent portions were manually removed.

Recording Sound
Recording sounds were made by rubbing the tile surface with a hitting rod with a 2 cm diameter sphere, as shown in Fig. 1 (right). The behavior was recorded using a smartphone from a fixed distance, with 12 recordings lasting approximately 30 s for each specimen. The recorded audio data was extracted from the video files and converted to mono format, and silent portions were manually removed. Since it is essential to accurately assess the bonding condition of individual tiles in the inspection of adhesive-laminated tile walls, it was necessary to accurately isolate the sounds corresponding to individual tiles from a series of friction sounds. Therefore, segmentation was an essential element in this study. The segmentation was performed using Python in the steps below.

1. Load audio data with the librosa library
2. Define the segment length L and overlap rate O by analyzing the audio waveform
3. Perform segmentation. Segmentation is generated according to the following formula where Si is the i-th segment, audio is the audio data, and n is the starting sample index.

$$Si = audio[n : n + L], n = i \times (L - L \times O)$$

4. Each segmented segment is saved as an independent file for later processing.

Table 1 shows the number of segments by adhesion rate segmented by the above procedure.

Extracting Features of Sound
The Fast Fourier Transform (FFT) was applied to extract features from speech data. The FFT effectively analyzes the frequency components of speech signals by transforming signals in the time domain into the frequency domain. The following equation represents the FFT.

$$X(K) = \int_{n=0}^{N-1} x(n)e^{-i2\pi kn/N}$$

where $X(n)$ is the signal in the time domain, $x(k)$ is the signal in the frequency domain, and N is the number of samples, this process used in this study was performed using the np.fft.fft [16] function of the NumPy library in Python. The study also computed a 2048-point FFT, which is widely used in general speech processing applications and is also suitable for the 44.1 kHz sample rate used for the recordings in this study. The calculation of this FFT was based on the following formula.

$$X(K) = \int_{n=0}^{2047} x(n)e^{-i2\pi kn/2048}$$

Because of the symmetry of the FFT results, only the first 1024 points of the obtained frequency components were used to represent the frequency components. This process extracted the frequency domain features obtained from each voice segment. These features were used for subsequent data analysis and model training. The FFT was applied to all data in this study because previous studies have shown that frequency domain data after applying the FFT is more accurate than time domain data before the FFT was applied.

Normalize
The features extracted by FFT were subjected to a normalization process. This process was performed using the fit transform method [15] of the Standard Scaler library for each adhesion rate data separately. Normalization was performed based on the following equation where x is the original feature, u is the mean of the feature, and s is the standard deviation of the feature.

$$Z = (x - u)/s$$

This transformation normalized the data so that each feature had a mean of 0 and a standard deviation of 1 for each adhesion rate. This constructed a dataset for training and evaluation of the deep learning model. The split ratio of the datasets in this study is 60% for the training set, 20% for the validation set, and 20% for the test set. The data set created by the above procedure was used to create a 4-value classification model that aims to discriminate data with each adhesion rate from each other.

Table 2. Hyper-parameters of Model_1

Activation (hidden)	Relu
Activation (Output)	Softmax
Optimizer	Adam
Epoch	75

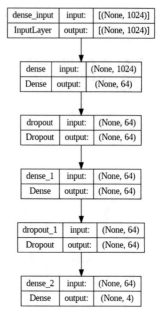

Fig. 6. Model_1

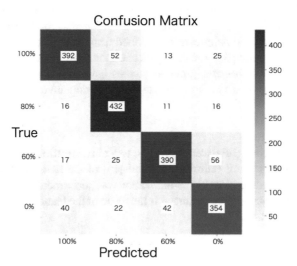

Fig. 7. Confusion Matrix

4 Learning Model and Evaluation

4.1 Our Learning Model

For each adhesion rate data, this study used TensorFlow2 and Keras to construct a deep learning model (hereafter referred to as Model_1) that classifies adhesion rates into four classes: 100%, 80%, 60%, and 0%.

The specific structure of Model_1 is shown in Fig. 6. The model consists of an input layer followed by a 64-unit all-coupled layer, a 50% dropout layer, another 64-unit all-coupled layer, another 50% dropout layer, and finally a 4-unit output layer. Dropout layers placed between each all-coupled layer are used to prevent overlearning.

4.2 Hyper Parameters

The hyperparameters are listed in Table 2. The model was compiled using the Adam optimizer and a sparse categorical cross-entropy loss function. The training was performed with a batch size of 32 and 75 epochs. The training setup was chosen based on an analysis of the learning curve from prior experiments.

Table 3. Precision, Recall and F1

Adhesion rate	Precision	Recall	F1
100%	84.30%	81.33%	82.79%
80%	81.36%	90.95%	85.88%
60%	85.53%	79.91%	82.62%
0%	78.49%	77.29%	77.89%

4.3 Experiment

Model_1 achieved an accuracy of approximately 82.40% for the test set. Table 3 shows each adhesion rate's precision, recall, and F1 scores, and Fig. 7 shows the confusion matrix. Table 3 shows each adhesion rate's precision, recall, and F1 score. Although some mistakes were observed in identifying tiles with 0% adhesion, the precision, recall, and F1 scores for all adhesion rates were over 77%. This result is accurate enough for practical use.

The confusion matrix shown in Fig. 5 also reveals a trend of misclassification.

(1) the data misclassified with 100% adhesion, 54.79% were misclassified as 0%
(2) the data misclassified with 80% adhesion, 52.53% were misclassified as 100%
(3) the data misclassified with 60% adhesion, 63.64% were misclassified as 0%
(4) the data misclassified with 0% adhesion, 57.73% were misclassified as 60%
(5) the data misclassified with 100% adhesion, 63.64% were misclassified as 0%

(6) the data misclassified with 0% adhesion, 63.64% were misclassified as 0% adhesion rate data, and 57.73% were misclassified as 60%

(7) the data misclassified with all adhesion rates, 50% were misclassified at a specific adhesion rate.

However, as more than half of the misclassified data with a 100% adhesion rate were determined to be 0%, the data were not necessarily misjudged to the closest adhesion rate. However, since there was only a tiny amount of misjudged data, we would like to examine the tendency of misclassified data in detail in the future.

These results showed that it is possible to use a neural network to inspect hammering sound by utilizing the friction sound of adhesive-applied tile walls. In addition, it was confirmed that the friction sound of each adhesion rate has unique characteristics that the neural network can discriminate.

5   Conclusion

This study describes the research results of using neural networks to perform hammering tests for adhesive-applied tile walls, which are difficult even for skilled inspectors to find problems. In previous studies, we have confirmed the effectiveness of neural networks for the hammering test for mortar tile walls. We confirmed that neural networks are also effective for judging from the friction sound generated by rubbing tile walls with an inspection stick with a sphere on the tip, which is commonly used in workplaces. In addition, the neural network was able to classify the friction sounds recorded from four different samples (Adhesion rate: 100%, 80%, 60%, and 0%). We also confirmed the existence of unique features that can be discriminated by the neural network for each adhesion rate. These results show the possibility of creating a hammering inspection system that can handle a wide range of adhesion rates by training the model with a wide range of adhesion rate data. In the future, we would like to examine the characteristics of each adhesion rate in detail. For example, we would like to test a specimen that has an adhesion rate of 85%.

The confusion matrix analysis also revealed that the misjudged data also tended to be misjudged to one adhesion rate. However, it was not necessarily the closest adhesion rate that would be misjudged. This result suggests that factors other than the adhesion rate may be deeply involved in the mis-judgment. In the future, we would like to study this misjudged data in detail.

References

1. Situation Concerning Maintenance and Management of Social Infrastructure. Ministry of Land, Infrastructure, Transport and Tourism. http://www.mlit.go.jp/hakusyo/mlit/h25/hakusho/h26/html/n1131000.html. Accessed 31 Jan 2024
2. TERRA DRONE. https://terra-drone.net/service/inspection. Accessed 31 Jan 2024
3. Guidelines for Exterior Wall Surveys by Infrared Surveys (Including Infrared Surveys by Unmanned Aerial Vehicles) under the Periodic Reporting System. Ministry of Land, Infrastructure, Transport and Tourism. https://www.mlit.go.jp/jutakukentiku/build/content/001474154.pdf. Accessed 31 Jan 2024

4. Non-destructive testing by electromagnetic waves. Just corporation. https://www.just-ltd.co.jp/service/existing/concreteexploration/electrowave/. Accessed 31 Jan 2024

5. Fukumura, T., Aratame, H., Ito, A., Koike, M., Hibino, K.: Improvement of sound classification method on smartphone for hammering test using 5G network. Int. J. Netw. Comput. **12**(2), 359–371 (2022). https://doi.org/10.15803/ijnc.12.2_359. https://www.jstage.jst.go.jp/article/ijnc/12/2/12_359/_article/-char/en

6. Seminar on construction problems seen in tile peeling and problem solving from a legal point of view (2016). http://fukukenkyo.org/slide-koga.pdf. Accessed 31 Jan 2024

7. Soeda, T., Mikami, T.: Basic study into a diagnostic system for exterior tile debondig. https://doi.org/10.3130/aijs.81.1779

8. Yokoi, M.: A study on diagnosis method using hammering test. J. Osaka Sangyo Univ. Nat. Sci. (131), 59–69 (2021)

9. Makatamura, M., Tanano, H., Nemoto, K.: Research on the diagonostic accuracy of the tile finishing external wall by various measuring methods. AIJ J. Technol. Des. **21**(49), 919–924 (2015)

10. Louhi Kasahara, J.Y., Fujii, H., Yamashita, A., et al.: Unsupervised learning approach to automation of hammering test using topological information. Robomech J. **4**, 13 (2017). https://doi.org/10.1186/s40648-017-0081-7

11. Kurita, K., Oyado, M., Tanaka, H., Tottori, S.: Active infrared thermographic inspection technique for elevated concrete structures using remote heating system. Infrared Phys. Technol. **52**, 208–213 (2009)

12. Akamatsu, R., Sugimoto, T., Utagawa, N., Katakura, K.: Proposal of non contact inspection method for concrete structures using high-power directional sound source and scanning laser Doppler vibrometer. Jpn. J. Appl. Phys. (2013). https://doi.org/10.7567/JJAP.52.07HC12

13. Survey Methods for Exterior Finish Materials of Exterior Walls, etc. in Periodic Building Inspection Reports. Ministry of Land, Infrastructure, Transport and Tourism (2018). https://www.tilenet.com/documents/20180523s.pdf. Accessed 31 Jan 2024

14. Study on rationalization of periodic survey methods for wet exterior walls. Ministry of Land, Infrastructure, Transport and Tourism (2015). https://www.mlit.go.jp/common/001129988.pdf. Accessed 31 Jan 2024

15. Yang, J., Ito, Y., Koike, M., Hibino, K., Ito, A.: Live demonstration: hammering test on a wall using AI. IEEE Sensors **2022**, 1 (2022). https://doi.org/10.1109/SENSORS52175.2022.9967264

16. NumPy Developers. https://numpy.org/doc/stable/reference/generated/numpy.fft.fft.html#numpy.fft.fft. Accessed 31 Jan 2024

Models of Resilient Systems with Online Verification Considering Changing Requirements and Latent Failures

Vyacheslav Kharchenko[1] [ID], Yuriy Ponochovnyi[2(✉)] [ID], Sergiy Dotsenko[3] [ID], Oleg Illiashenko[1,4] [ID], and Oleksandr Ivasiuk[1] [ID]

[1] National Aerospace University KhAI, Kharkiv, Ukraine
{v.kharchenko,o.ivasiuk}@csn.khai.edu, o.illiashenko@khai.edu
[2] Poltava State Agrarian University, Poltava, Ukraine
yuriy.ponch@gmail.com
[3] Ukrainian State University of Railway Transport, Kharkiv, Ukraine
docenko@kart.edu.ua
[4] The Institute of Informatics and Telematics of the National Research Council (IIT-CNR), Pisa, Italy
oleg.illiashenko@iit.cnr.it

Abstract. The paper presents the outcomes of developing and implementing a cybernetic approach to creating and deploying resilient systems. It discusses the concepts and principles of real-time evolution for resilient systems, considering changing requirements, environments, and latent failures. This increased resilience stems from the necessity of integrating additional channels into the control system to address shifts in requirements, environments, or unspecified faults and failures. The general structure of resilient systems is founded on the principle of partitioning and managing channels, as well as system reconfiguration based on functional and non-functional characteristics. A case study of a space resilient system with online verification is presented. Three scenarios of system behavior to ensure resilience are proposed, considering various tolerance and management mechanisms. The paper analyzes the results of developing and researching Markov models for these scenarios, offering insights for enhancing the availability of resilient systems.

Keywords: Information and Control System · Security · Safety · Resilience · Online Verification

1 Introduction

Safety and security-critical systems, such as railway control systems, aerospace onboard systems, and instrumentation and control systems of nuclear power plants, operate in demanding physical and informational environments [1]. The requirements for their functional and non-functional characteristics are stringent, considering their prolonged operational lifespans, potential evolution, changing environmental conditions, and various cyber and physical threats [2]. Consequently, these systems must exhibit self-adaptation and resilience throughout use [3].

© The Author(s), under exclusive license to Springer Nature Switzerland AG 2024
W. Zamojski et al. (Eds.): DepCoS-RELCOMEX 2024, LNNS 1026, pp. 90–99, 2024.
https://doi.org/10.1007/978-3-031-61857-4_9

Numerous proceedings and normative documents address the concepts of resilience and resilient systems. Standards and reports from organizations such as NIST [3–5], ASIS [6], CNSS, CSRC, ECSS, etc. The definition of resilience, as highlighted in sources [7], underscores the capacity to minimize the impact and duration of disruptive events on critical systems or infrastructure. The efficacy of a resilient system hinges on its ability to anticipate, absorb, adapt to, and rapidly recover from potentially disruptive events.

Standard [5] defines resilience as the capability to swiftly adapt and recover from any known or unknown environmental changes through a comprehensive implementation of risk management, contingency, and continuity planning. Formal definitions and models of resilient systems are elaborated in [8, 9]. In [10, 11], examples of applying metrics for assessing the resilience of transport information systems are explored.

In conclusion, ensuring resilience necessitates the development of a management system capable of promptly adapting and recovering amidst changing operational conditions [12]. This entails predicting, mitigating, and swiftly recovering from the impacts of known or unknown environmental changes [13]. To achieve this, it is imperative to analyze the structure and features of resilient systems, enhance a methodology for resilience management [14, 15], and develop and research the availability models of different kinds of resilient systems.

The rest of the paper is structured as follows: Sect. 2 delves into explaining the system architecture and reliability block diagram, including the set of states considering failure combinations. Section 3 presents and examines the systems availability of Markov models for three online verification scenarios, alongside justification for the values of simulation input parameters. Finally, Sect. 4 summarizes our findings, offers guidance on utilizing the developed models, and discusses future work.

2 The Structure and Markov Model of the Single-Channel Space I&C Systems

2.1 Space Control Reconfigurable System with Online Verification

The Space Computer Control Reconfigurable System (SCCRS) comprises onboard (BCS) and ground (GCS) information and control (I&C) subsystems [8, 16]. This configuration enables the execution of additional functions, such as online verification, via a telecommunications channel during automatic flight mode (see Fig. 1). Conducting such verifications on Earth is often challenging, costly, and, at times, impossible due to the inability to replicate all outer space conditions, as well as the necessity to modify and re-engineer software during operations. Presently, most composite manned space complexes are equipped with the capability for hardware repair.

Utilizing software that can be modified provides greater flexibility in distributing verification steps. Consequently, several non-critical software functions can be promptly verified post-launch and during spacecraft operation. Following online verification (OV) procedures, actions can be taken to address identified shortcomings and defects through corrective online verification (COV) [16].

The architecture of the SCCRS with serviced hardware includes a single hardware channel, within which operates one version of the software (see Fig. 2). The symbolic

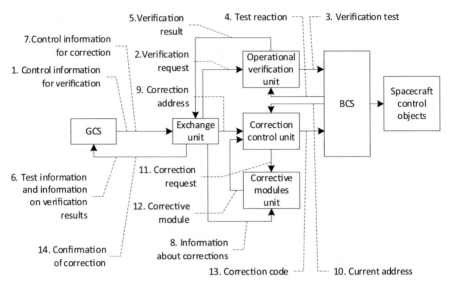

Fig. 1. Space control reconfigurable system with online verification and correction of software [8]

representation of such a system structure is denoted as S_{11}. The model has the following input parameters:

a) hardware channel failure rate λ_{HW} (1/h);
b) system recovery rate after hardware failure μ_{HW} (1/h);
c) software failure rate λ_{SW} (1/h);
d) system recovery rate after software defect occurrence μ_{SW} (1/h).

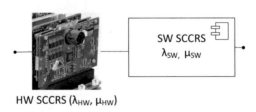

HW SCCRS (λ_{HW}, μ_{HW})

Fig. 2. Reliability block diagram of a single-channel, single-version SCCRS

Contemporary facilities require a mandatory communication link with the terrestrial SCCRS, through which decisions are made to amend the program code. Commands to initiate verification procedures originate from the terrestrial complex and are executed by a dedicated input data processing and decision-making unit (cancellation block).

2.2 The Markov Single-Fragment Model of an Onboard Computer System with Serviced Units

The Markov single-fragment model of the SCCRS with serviced hardware, which allows for the restart of the software after a software failure, is shown in Fig. 3.

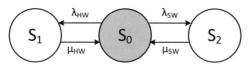

Fig. 3. Graph of states and transitions of a single-channel single-version onboard computer system of the spacecraft

To construct the single-fragment model for evaluating the availability of the SCCRS without considering online verification procedures, the following assumptions were made:

a) the event stream that transitions the system from one functional state to another has properties of stationarity, ordinariness, and absence of aftereffects; accordingly, the input parameters of the model λ_{HW}, λ_{SW}, μ_{SW} are assumed to be constant;
b) at any given moment, each element of the I&C system may be either operational or non-operational;
c) the assumption of the ideal functioning of diagnostic and control tools is accepted. This assumption leads to somewhat inflated availability indicators. However, it allows reducing the dimensionality of the solved problem in terms of reducing the number of considered functional states.

The Kolmogorov-Chapman system of differential equations corresponding to the graph in Fig. 3 and the initial conditions are as follows:

$$\begin{cases} \frac{dP_0}{dt} = -(\lambda_{HW} + \lambda_{SW})P_0 + \mu_{HW}P_1 + \mu_{SW}P_2; \\ \frac{dP_1}{dt} = -\mu_{HW}P_1 + \lambda_{HW}P_0; \\ \frac{dP_2}{dt} = -\mu_{SW}P_2 + \lambda_{SW}P_0; \\ P_0 + P_1 + P_2 = 1; \end{cases} \qquad (1)$$
$$P_0(0) = 1, \ P_1(0) = 0, \ P_2(0) = 0.$$

The availability function corresponds to the probability of the system being in the operational state S0. To simulate the operation of the I&C system under conditions of hardware and software failures and the implementation of online verification procedures, three separate scenarios were developed, based on which separate multi-fragment models were constructed.

3 Markov's Availability Model of Resilient Systems with Online Verification

3.1 The Models of Three Online Verification Scenarios of Space Computer Control Reconfigurable System

The diagram in Fig. 4 illustrates the Markov graph of the SCCRS model depicting the initial online verification scenario.

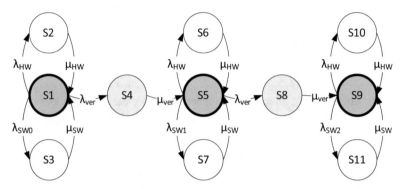

Fig. 4. The graph represents the functionality of SCCRS in ensuring the operational trust of noncritical functions.

The operational process of SCCRS unfolds as follows: Initially, the system executes all planned functions and resides in state S1. During operation, hardware defects arise, causing the system to transition to state S2, followed by restoration to state S1. Subsequently, a software defect occurs, leading the system into state S3. Following the manifestation of this software defect, the system is restored to state S1. After a designated time interval determined by parameter λver, a verification of non-critical functions not completed on Earth due to project time constraints occurs. This transition places the system in state S4 (inoperable). Following corrective online verification procedures, the system progresses to a new model fragment (state S5), characterized by a change in intensity.

The Markov graph representing the resilience SCCRS model for the second corrective online verification scenario is depicted in Fig. 5.

The operational process of SCCRS unfolds similarly: Initially, the system executes all planned functions and resides in state S1. During operation, hardware defects arise, causing the system to transition to state S2, followed by restoration to state S1. Subsequently, a software defect occurs, leading the system into state S3. Following the manifestation of this software defect and information communication to the ground computer control system, corrective measures are developed, and commands to alter the program code are issued. The time interval for these activities is characterized by parameter λver. Upon successfully eliminating the defect, the system advances to a new model fragment (state S5), whereas unsuccessful correction results in a return to state S1. After addressing all potential unspecified defects, the system operates normally, with only hardware failures considered.

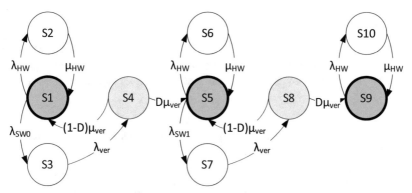

Fig. 5. The graph illustrates the functioning of SCCRS during online verification subsequent to the occurrence of unspecified software defects.

The Markov graph representing the SCCRS model for the third corrective online verification scenario is depicted in Fig. 6.

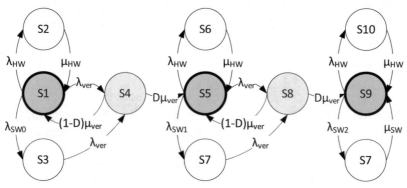

Fig. 6. The graph illustrates the functioning of SCCRS during online verification aimed at specifying environmental changes.

The operational process of SCCRS unfolds as follows: Initially, the system executes all planned functions and resides in state S1. During operation, hardware defects arise, causing the system to transition to state S2, followed by restoration to state S1. Subsequently, a software defect occurs, leading the system into state S3. In this scenario, the defect results from insufficient information about external environmental parameters. Thus, the system remains in a state of software failure until environmental information is clarified during corrective online verification procedures. Consequently, the transition to corrective online verification state S4 occurs from state S1 and state S3 with intensity λver. Successful defect elimination leads to progression to a new model fragment (state S5), whereas unsuccessful correction results in a return to state S1. After addressing all potential defects associated with inaccurate assessment of environmental parameters, the system operates in the event of hardware and software failures. Additionally, after software failures (with known GCS causes), the system recovers with intensity μ_{SW}.

3.2 Research of the Space Resilient Systems Models

Different scenarios require different sets of SCCRS functions. Scenarios 2 and 3 require larger memory cells to load the correction codes. Therefore, the values of input parameters of the Space Resilient Systems models will differ, as shown in Table 1 [8, 16].

Table 1. Resilience indicator assessment models parameter values for various scenarios of carrying out COV.

#	Parameter	Model 1	Model 2	Model 3
1	λ_{HW}	3,00E−05 (1/h)	3,20E−05 (1/h)	3,20E−05 (1/h)
2	μ_{HW}	0,00298 (1/h)	0,00298 (1/h)	0,00298 (1/h)
3	$\lambda_{SW}\,0$	4,00E−04 (1/h)	7,30E−05 (1/h)	2,00E−04 (1/h)
4	μ_{SW}	0,2 (1/h)		0,2
5	$\Delta\lambda_{SW}$	2,00E−06 (1/h)	5,00E−06 (1/h)	5,00E−06 (1/h)
6	λver	0,0104 (1/h)	0,0833 (1/h)	1/3…1/30 (1/h)
7	μver	0,25 (1/h)	0,25 (1/h)	0,25 (1/h)
8	Dver		0,4…0,9	0,8
9	Nver or Nfr	20…100	15	20

The metric for resilience was determined by measuring the deviation of the availability function from its stationary value. The concept of assessing this indicator is demonstrated in the paper [8]. A single-fragment hardware/software model with specific input parameter values highlighted in Table 1 was chosen for the system model without COV. In the second scenario model, which does not allow for simple software recovery but only recovery through verification, the parameter value μ_{SW} for a single-fragment model is set to $\mu_{SW} = \mu\text{ver} = 0.25$ (1/h).

The research outcomes regarding the impact of input parameter values on the resilience index ΔA are depicted in Fig. 7, 8 and Fig. 9.

Figure 7 illustrates the findings from studying the COV model in scenario #1. When transferring verification procedures to the post-launch phase of spacecraft operation, the system's availability decreases initially. As the volume of verification increases, there is a nonlinear increase in the ΔA indicator.

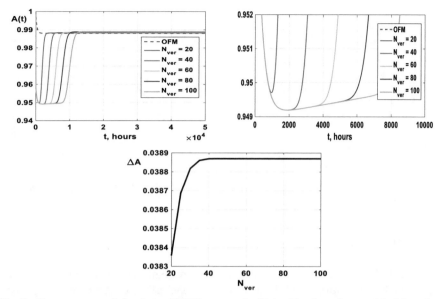

Fig. 7. The outcomes of simulating COV by scenario #1 (verification of non-critical functions after startup).

Figure 8 depicts the findings of the COV model study in scenario #2. As the probability of resolving the identified defect during COV procedures rises, the ΔA indicator value increases linearly.

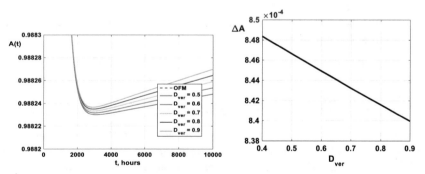

Fig. 8. The outcomes of simulating COV in scenario #2 (performing COV after detecting unspecified defects).

Figure 9 shows the results of the study of the COV model in scenario #3. As the intensity of COV procedures increases, the value of the indicator ΔA increases nonlinearly. This is explained as it is necessary to consider the COV time (parameter μ_{VER}), which is added to the total system downtime.

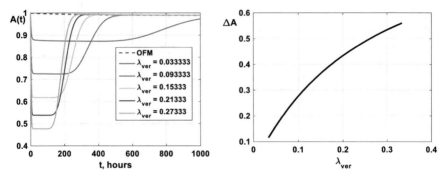

Fig. 9. The results of modeling the COV in scenario #3 (conducting COV to clarify the parameters of the environment).

4 Conclusion

The proposed approach to building and developing resilient control systems is based on a cybernetic scheme with divided channels for functional and non-functional management, considering changes in system requirements, environmental parameters, and unspecified failures. This approach facilitates the real-time adaptation of such systems.

Some findings from research on the developed Markov models of resilient systems include:

- when the number of functions requiring trust increases by fivefold, the resilience index A increases by 1.3%;
- doubling the probability of eliminating unspecified defects during COV leads to a decrease in the resilience index A by 0.95%;
- tripling the intensity of COV procedures to refine environmental parameters results in an 85.7% increase in the resilience index A.

To enhance resilience through COV procedures, efforts should be made to reduce or execute their implementation duration without completely losing the system's primary functions' operability.

Future steps may involve applying a cybernetic approach and models to:

1. Ensure industrial resilience by integrating cybersecurity and safety management systems for enterprises in Industry 4.0/5.0. This integration would combine IT, OT, and ET levels of information security and functional and ecological safety.
2. Enhance resilience by considering strategies for combined and separate maintenance, incorporating a range of dependability (reliability) and cybersecurity assurance policies during the operation of complex robotic and robotic-biological systems [17].

References

1. Mayar, K., Carmichael, D.G., Shen, X.: Resilience and systems - a review. Sustainability **14**, 8327 (2022). https://doi.org/10.3390/su14148327

2. Vogel, E., Dyka, Z., Klann, D., Langendörfer, P.: Resilience in the cyberworld: definitions, features and models. Future Internet **13**, 293 (2021). https://doi.org/10.3390/fi13110293

3. NIST Special Report 800-160, Vol. 2 Developing Cyber-Resilient Systems: A Systems Security Engineering Approach (2021). https://doi.org/10.6028/NIST.SP.800-160v2r1

4. NIST Interagency Report 8074 Volume 2. Interagency Report on Strategic U.S. Government Engagement in International Standardization to Achieve U.S. Objectives for Cybersecurity (2015). https://doi.org/10.6028/nist.sp.800-30r1

5. NIST. Risk Management Framework for Information Systems and Organizations: A System Life Cycle Approach for Security and Privacy. Special Publication 800-37 (2018). https://doi.org/10.6028/NIST.IR.8074v2

6. American National Standards Institute (ANSI) DS 3001:2009. Organizational Resilience: Security, Preparedness, and Continuity Management Systems - Requirements With Guidance For Use (2009). https://webstore.ansi.org/standards/ds/ds30012009

7. Computer Security Resource Center. Glossary. Resilience. https://csrc.nist.gov/glossary/term/resilience/

8. Kharchenko, V., Dotsenko, S., Ponochovnyi, Y., Illiashenko, O.: Cybernetic approach to developing resilient systems: concept, models and application. Inf. Secur. Int. J. **47**, 77–90 (2020). https://doi.org/10.11610/isij.4705

9. Uslu, S., Kaur, D., Durresi, M., Durresi, A.: Trustability for resilient Internet of Things services on 5g multiple access edge cloud computing. Sensors **22**, 9905 (2022). https://doi.org/10.3390/s22249905

10. Feoktistov, A., Edelev, A., Tchernykh, A., Gorsky, S., Basharina, O., Fereferov, E.: An approach to implementing high-performance computing for problem solving in workflow-based energy infrastructure resilience studies. Computation **11**, 243 (2023). https://doi.org/10.3390/computation11120243

11. De Marchi, M., Friedrich, F., Riedl, M., Zadek, H., Rauch, E.: Development of a resilience assessment model for manufacturing enterprises. Sustainability **15**, 16947 (2023). https://doi.org/10.3390/su152416947

12. Johnson, G.: Cyber Robust Systems: The Vulnerability of the Current Approach to Cyber Security (2020). https://www.igi-global.com/chapter/cyber-robust-systems/258678

13. Shelekhov, I., Barchenko, N., Kalchenko, V., Obodyak, V.: A hierarchical fuzzy quality assessment of complex security information systems. Radioelectronic Comput. Syst. **4**(96), 106–115 (2020). https://doi.org/10.32620/reks.2020.4.10

14. Moskalenko, V., Kharchenko, V., Moskalenko, A., Kuzikov, B.: Resilience and resilient systems of artificial intelligence: taxonomy, models methods. Algorithms **16**(3), 165 (2023). https://doi.org/10.3390/a16030165

15. Madni, A.M., Erwin, D., Sievers, M.: Constructing models for systems resilience: challenges, concepts, and formal methods. Systems **8**, 3 (2020). https://doi.org/10.3390/systems8010003

16. Kharchenko, V., Ponochovnyi, Y., Boyarchuk, A., Brezhnev, E.: Resilience assurance for software-based space systems with online patching: two cases. In: Zamojski, W., Mazurkiewicz, J., Sugier, J., Walkowiak, T., Kacprzyk, J. (eds.) DepCoS-RELCOMEX 2016. AISC, vol. 470, pp. 267–278. Springer, Cham (2016). https://doi.org/10.1007/978-3-319-39639-2_23

17. Fedorenko, G., Fesenko, H., Kharchenko, V., Kliushnikov, I., Tolkunov, I.: Robotic-biological systems for detection and identification of explosive ordnance: concept, general structure, and models. Radioelectronic Comput. Syst. **2**(106), 106–115 (2023). https://doi.org/10.32620/reks.2023.2.12

Performance Optimizations of Real World Map Transformations for 3D Realtime Mobile Games

Maciej Kopczynski[(✉)] [ID]

Faculty of Computer Science, Bialystok University of Technology, Wiejska 45A,
15-351 Białystok, Poland
m.kopczynski@pb.edu.pl

Abstract. This paper presents results and describes a solution and techniques allowing for transformation of highly detailed real world maps into simpler maps for using in 3D mobile games applications running on medium and low computing power devices. This approach is especially interesting for 3D apps developers focusing on real world maps that can be implemented in games like tycoons, strategies or geolocation type. The obtained results show possibility of achieving good compromise between level of details extracted from original maps and high performance of realtime 3D graphics generated on mobile device.

Keywords: Map · Mobile game · 3D game · Transformation · Optimization · Mobile device

1 Introduction

Location-Based Mobile Games (**LBMG**) are a subtype of broad-based games, where real world maps data and geographic location is fundamental to the game design process. In these games, mobile devices such as smartphones and tablets are used both to play and, in many cases, to determine the player's position. Example of such game is [14]. The creation and development of LBMG is more complicated than traditional games. The need for a smooth and constant flow of information from the real world maps to the virtual world maps, as well as high FPS rate of game, is a fundamental function of LBMG that affects gameplay. This is difficult, because we are dealing with a very large amount of data to analyse, select and convert. This is also a problem for users who do not have mobile devices with high computing performance.

As part of this work, an optimization solution was proposed that addresses this problem through functionalities in the preprocessing, analysis and visualization of large input real world GIS data sets focused on mobile devices. This leads to a significant reduction in the production time of new real maps based games for mobile devices and contributes to the creation of games that are stable, smooth and have low failure rates.

© The Author(s), under exclusive license to Springer Nature Switzerland AG 2024
W. Zamojski et al. (Eds.): DepCoS-RELCOMEX 2024, LNNS 1026, pp. 100–110, 2024.
https://doi.org/10.1007/978-3-031-61857-4_10

At the moment there are not many comprehensive and ready-to-use solutions for optimization of this data for mobile devices game engines. In the literature one can find mainly descriptions of concepts or partial process optimizations for transformation such type of data into optimized and ready-to-use form. Approach to open geospatial data integration in 3D game engine for urban digital twin applications was presented in [1] and, for industry, in [2]. Description of GIS-based educational game using low-cost virtual tour experience can be found in [3]. Example of location-based framework for mobile games is described in [4,7] and [5]. Planetary-scale geospatial analysis platform based on Google Earth engine was proposed in [6]. Main assumptions concerning gamification and space in video games in general are described in [8] and [9]. Existing approach for maps optimization and performance analysis in games can be found in [10] and [11]. Design process of maps devoted for games is presented in [12] and [13].

The paper is organized as follows. In Sect. 2 some information about the basic definitions are presented. The Sect. 3 focuses on description of implemented solutions and optimizations made in test applications, while Sect. 4 is devoted to presentation of the experimental results.

2 Basic Definitions and Requirements

For the purpose of measuring obtained results related to maps transformations and mobile device application performance, some definitions have to be introduced. Each geographical region will have the following values measured:

- R_{points} - output number of total points defining shape of the road,
- E_{dist} - total combined road distance,
- T_{gen} - time required to generate the area,
- T_{load} - time required to load and process the region in the client environment,
- T_{frame} - time required to render the resulting geometry in the client environment,
- R_{verts} - output number of total vertices of road network mesh,
- R_{tris} - output number of triangles forming a final road network mesh.

To ensure each playable region in the game is engaging, the following requirements have been established for the city and road graph generation algorithm:

1. **Connected Weighted Graph (CWG)** - each region is represented as a connected weighted graph, comprising cities (vertices V) and roads (edges E).
2. **Minimum Number of Cities (MNoC)** - the graph CWG must contain at least V_{count} vertices representing cities or towns.
3. **City Separation** - each city must be located no closer than V_{sep} (geodetic distance) kilometres from any other city.
4. **Road Network (RN)** - minimum total weight of E_{dist} kilometres of roads is required to deem the generation as successful (total summed weight of edges E).

3   Solution Description

Test application was created in .NET technology using C# language and Microsoft Visual Studio 2022 on the PC side. Mobile device running Android operating system part was created in Unity 2021 LTS.

Figure 1 presents general architecture of solution.

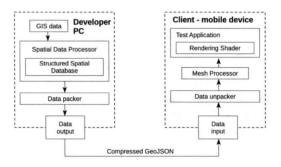

Fig. 1. Solution architecture in general.

System consists of two parts - test application running on mobile device (client side) and real-world map processing application running on PC (developer side). Transformed data is packed alongside game files and embedded into the client application using compressed GeoJSON format [15]. Main functional parts of developer PC's application are:

– *GIS Data Source* - real-world map data source based on OpenStreetMap,
– *Structured Spatial Database* – stores geospatial data using PostgreSQL database with PostGIS extension,
– *Spatial Data Processor* - core geographical data processing unit,
– *Data packer* - responsible for preparing collected and processed data into game data packages.

Main functional parts of mobile device are *Data unpacker* responsible for unpacking the embedded data and preparing it for processing by *Mesh Processor* and further visualisation by *Test application* using *Rendering Shader*.

GIS Data Source is a planetary dataset in PBF format acquired from OpenStreetMap, imported with *imposm3 OSM* import workflow. Some data irrelevant from the point of game needs (e.g. street names) was stripped to keep the database size more manageable. The dataset is structured inside the Postgres database with PostGIS extension to handle spatial indexing and other GIS acceleration structures.

Spatial Data Processor is the key module on developer side that handles extracting only the data, that is essential to client environment. It contains algorithms that decide, how the game world is constructed. It is designed as a

.NET application that exchanges information with a *Structured Spatial Database* to construct the final shape of the game world that contains only necessary data.

The routes and cities set from algorithm implemented in *Spatial Data Processor* are then considered a result representing E and V respectively. Routes are then measured to calculate R_{points}. The entire process stops at this point and T_{gen} is written as the total generation algorithm run time.

Once the resulting set of roads is collected, another geometry aggregation pass is run, where roads close to each other are patched together into one. In addition, as a placeholder for 3D models, each urban city center is turned into a roundabout, and the roads in the center of the roundabout circle are brought to the edge of the circle.

This process creates a playable road network that contains all the essential roads between the most important cities while discarding most of the redundant communication links and excessive geometry caused by the variable quality of the data in the OSM dataset. In doing so, the amount of data is reduced and standardized to the expected set of complexity, which can be presented to players as relatively balanced gameplay maps that, at the same time, can run at the expected level of performance that can be achieved on mobile devices.

Data Packer and *Data unpacker* transform the spatial data to and from the intermediate format that can be embedded into client applications. GeoJSON format is used along with LZMA compression to reduce the payload size of resulting spatial features.

On the client side, the unpackager decompresses data packets and reassembles the GIS data that is then loaded as a navigation graph in *Mesh Processor*, which can be later used for finding paths between points in the represented region. Navigation graph is then used to generate a road network mesh. Generation time is measured and described as T_{load}. Final road mesh is defined by the set of triangles R_{tris}, as well as the set of vertices R_{verts}. Algorithm on mobile device is prepared in a way, that allows the reuse of as many vertices as possible for subsequent triangles to reduce GPU memory usage. Because of page limit of this paper, algorithm is not presented.

Generated mesh can be loaded into the rendering engine to measure T_{frame} for test scenarios defined further.

Generation process is visualized by following graphics presented on Fig. 2. Part a) presents exemplary section of initial road graph, part b) shows first stage of mesh generation corresponding to intersections geometry, while part c) presents final appearance with textures applied.

3.1   Spatial Data Processor Algorithm

Main goal of algorithm on developer PC side is to create a road network that connects a set of cities in a region while meeting some criteria such as the total distance of the roads and the density of the cities. Pseudocode of processing algorithm is presented below.

INPUT: Region-clipped map data including *way* (linear geometry, eg. roads) and *point* (points geometry data)

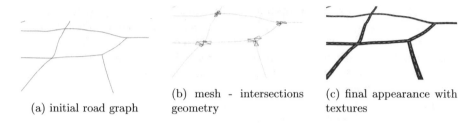

(a) initial road graph

(b) mesh - intersections geometry

(c) final appearance with textures

Fig. 2. Visualisation of mesh network road generation

OUTPUT: List of *routes* and *cities*
1: $allCityClasses \leftarrow [city, town, village]$
2: $roadClassesConsidered \leftarrow [motorway, trunk]$
3: $additionalRoadClasses \leftarrow [primary, secondary, tertiary]$
4: **repeat**
5:     $roadClassesConsidered.push(allRoadClasses.pop())$
6:     $roads \leftarrow CollectRoads(way, roadClassesConsidered)$
7:     $mergedRoads \leftarrow Union(roads)$
8:     $simplifiedRoads \leftarrow Simplify(roads)$
9:     $allCities \leftarrow CollectCities(point, allCityClasses)$
10:    $citiesConnected \qquad\qquad\qquad\qquad\qquad\qquad\qquad \leftarrow$
       $FilterDisconnectedCities(allCities, simplifiedRoads)$
11:    $citiesSorted \leftarrow SortCitiesByPopulation(cities)$
12:    $citiesReduced \leftarrow ReduceDensity(citiesSorted, V_{sep})$
13:    $citiesTrimmed \leftarrow TakeFirst(citiesReduced, V_{count})$
14:    $cities \leftarrow SnapCitiesToRoads(citiesTrimmed, simplifiedRoads)$
15:    **for** each city $cityStart$ in $cities$ **do**
16:       **for** each city $cityEnd$ in $cities$ **do**
17:          $routes.push(CalculateRoute(cityStart, cityEnd))$
18:       **end for**
19:       $totalDistance \leftarrow SumDistances(routes)$
20:    **end for**
21: **until** $totalDist < E_{dist}$ **or** $additionalRoadClasses = \emptyset$

Input to the algorithm is a region clipped world map data in PostgreSQL database containing ways described as *way* representing roads and other linear geometry and points described as *point* containing point geometry. Output is defined as two lists, which are *routes* representing connections between selected *cities*. On the beginning, algorithm defines the classes of cities and roads according to the standardized OpenStreetMap attribute guidelines. Then it starts with the highest road classes (motorway and trunk) and adds lower road classes (primary, secondary, and tertiary) until the conditions are met. Next, the algorithm extracts the roads from a structured spatial database as individual instances for each road segment and merges them with a small tolerance to connect those, that were added in different OpenStreetMap changesets. After that, the algorithm simplifies the road lines by removing excessive points along the lines using the Visvalingam-Whyatt's algorithm. The algorithm then collects all the cities from the region as points, and filters out the cities that are not close to the calculated road network (disconnected from the main and only graph). The algorithm then sorts the inner cities by population to pick the largest cities of the region and removes all lower-ranked cities in a given V_{sep} area to prevent clutter. The algorithm then takes only the first V_{count} cities and snaps them to the points on the road network, creating new nodes on the network at the city location, if necessary. Finally, the algorithm uses Dijkstra's algorithm to find the shortest path between each pair of cities and adds it to the routes list. The algorithm calculates the total distance of the roads in the routes list and compares it to defined stop-condition threshold E_{dist}. The algorithm repeats the steps until the total distance is greater than or equal to E_{dist} or until runs out of roads in the dataset.

4   Experimental Results

Main goal of the presented results is to show differences between raw and processed map data on selected mobile devices and map regions.

Presented results were obtained using a PC equipped with an 64 GB RAM DDR4 running at 3200 MHz and 6-core AMD Ryzen 5600X processor. Database engine was PostgreSQL 14 with its tablespace physically mapped to SSD drive connected via NVMe at PCIe Gen3 speed. Mobile devices used in tests were Samsung Galaxy S20 and Samsung Galaxy S6 updated to newest versions of Android operating system provided by phone manufacturer. Samsung Galaxy S20 is considered as fast mobile device, while Samsung Galaxy S6 has low computing power.

Two different areas were used during obtaining experimental results. Selected regions represent various geographical parts of the world with different properties - vast wilderness of Canada with low road density to the dense urban agglomerations packed on the east coast of the United States. For purposes of this work, the following exemplary areas are considered:

- *sample 1: Northeastern US.* Combined states of New York, Connecticut, Rhode Island, Massachusetts, New Hampshire, Vermont and Maine.
- *sample 2: Saskatchewan.* Sparsely populated Canadian province.

Highways (roads in OpenStreetMap terms) classified with "motorway", "trunk", "primary", "secondary", and "tertiary" were considered roads. Places tagged with "city", "town" and "village" from OSM data were classified as settlements.

Parameters for the algorithms running on the PC computer (developer side) were set in a way, where each final map region after processing targeted $10 \leq V_{count} \leq 20$ cities as resulting city complexity targets and road complexity of $E_{dist} > 8000$.

It should be noted, that all time measurements are average from 100 runs of each process to avoid impact of background operations performed by operating system, both on PC computer, as well as mobile devices. Additionally, in case of test application on mobile device, length of projection and scheme of camera movement was the same for all types of regions.

Table 1 presents results obtained on PC computer related to selected regions processing in terms of preparing maps for mobile device test application. *Input* columns group describe complexity of raw data related to selected region, where *Area* presents region size, *Road network* shows total length of forementioned types of roads and *Settlements* describes total number of settlements of forementioned types. T_{gen} column presents total time from start of reading raw data to the moment, where final data package for mobile device is ready.

Table 1. Results acquired for real map processing

Region	Input			Output		T_{gen}
	Area	Road network	Settlements	E_{dist}	R_{points}	
	$[km^2]$	$[km]$	$[-]$	$[-]$	$[-]$	$[s]$
Northeastern US	348 076	89 020	1 325	19 003	5 143	207
Saskatchewan	504 367	32 319	179	20 017	4 365	14

Figure 3 shows two regions after input map processing.

(a) Northeastern US

(b) Saskatchewan

Fig. 3. Visualisation of two regions after map processing

As can be seen, the generation time varies drastically depending on the complexity of the region in terms of road network and number of cities, but is not significantly dependent on the size of the region itself. This is explained by the fact that the algorithms operate on geometry and graphs, and the Saskatchewan area, despite of its very vast area, do not have much of the roads.

Another fact that can be observed in the Fig. 3 and data in Table 1 is that, despite the input set being very different in characteristics, each output set is relatively similar in appearance and characteristics. Parametrized variations of this algorithm can be applied to any region in the world that has properly classified roads and cities.

Map data generated by PC computer (developer side) was then sent to the mobile device (client side) and test application with rendering part was used to acquire results in terms of time, what has direct impact on mobile application performance. Measurements related to complexity of generated geometry was also collected. R_{verts} describes total number of vertices of road network mesh, while R_{tris} represents number of triangles forming a final road network mesh. It should be noted, that those values are independent on type of mobile device, because geometrical statistics are connected only with map region. Figure 4 shows exemplary screenshot of Northeastern US region from test application.

Fig. 4. Exemplary screenshot of Northeastern US region from test application.

It should be noted, that full raw map with all road tags, especially for Northeastern US region, could not be loaded and rendered at once on current implementation of test application on lower grade device. Most rendering engines by default use 16 bit integer numbers as indices for mesh index buffer data, so the limit of indices in one mesh is 65 536, but high end mobile devices are able to use 32 bit indices. For making transformation algorithm performance comparison between two sets of map data, filtered map data set was prepared in a way, where roads with tags "motorway" and "trunk" were used, while tags for cities remained as previously defined.

Table 2 presents results obtained on different mobile devices depending on selected region using both sets of map data: filtered raw and processed ones. Column T_{load} describes map data loading time measured from the accessing the data, while T_{frame} presents frame rendering time. Column group *Filtered raw map* corresponds to unprocessed raw map data, which was only filtered by previously mentioned roads and cities tags, while column group *Processed map* contains measurements taken on transformed by PC original and full detailed map data.

Table 2. Results on mobile devices for selected regions

Mobile device	Filtered raw map		Processed map	
	T_{load}	T_{frame}	T_{load}	T_{frame}
	[ms]	[ms]	[ms]	[ms]
Northeastern US				
Samsung Galaxy S20	5 478	1.68	106.32	0.03
Samsung Galaxy S6	15 548	7.13	288.58	0.11
Saskatchewan				
Samsung Galaxy S20	244	0.06	68.67	0.02
Samsung Galaxy S6	582	0.26	174.14	0.10

Geometrical results for filtered raw and processed map data is equal to:

- filtered raw map:
  - R_{verts} = 873 224 for Northeastern US; 29 732 for Saskatchewan,
  - R_{tris} = 828 336 for Northeastern US; 25 322 for Saskatchewan.
- processed map:
  - R_{verts} = 12 142 for Northeastern US; 9 866 for Saskatchewan,
  - R_{tris} = 10 592 for Northeastern US; 8 926 for Saskatchewan.

Timing and geometrical results show, that each area was transformed into a similarly performing renderable package that allows the game to achieve similar performance targets across multiple devices with different computing power. Even if there are regions like Saskatchewan, where non processed maps can work properly on mobile devices, this is essential to provide players stable performance across all regions in the world, even on slower devices.

5   Conclusions

Performed research shows, that creating efficient and versatile methods for processing real world data focusing on game development industry, especially for mobile devices is possible. Proposed solution helps in speeding up creating new geolocation games, as well aids creating smooth gameplay. It is recommended to keep vertices count as low as possible. For older devices the reasonable limit is about 100k vertices per scene, going up to 1 million on modern devices. Using unprocessed data would exhaust the entire scene limit. Rendering of such complex shapes would take most of the time intended for the entire scene, leaving no time left for drawing any other 3D objects while keeping reasonable framerate.

Further research will focus on further development and optimization of proposed methods, especially in the field of selecting settlements, creating more general road mesh for bigger cities, adding mechanisms for creating parallel road considering requirements for different road categories, as well as testing more mobile devices with bigger regions.

Acknowledgments. The work was supported by the grant WZ/WI-IIT/3/2023 from Bialystok University of Technology. Research results are based on the project "Geo Game Service - a scalable service that provides useful data about geolocation information in asynchronous multiplayer games" financed by National Center of Research and Development.

References

1. Rantanen, T., Julin, A., Virtanen, J.-P., Hyyppä, H., Vaaja, M.T.: Open geospatial data integration in game engine for urban digital twin applications. ISPRS Int. J. Geo Inf. **12**(8), 310 (2023)
2. Schleich, B., Anwer, N., Mathieu, L., Wartzack, S.: Shaping the digital twin for design and production engineering. CIRP Ann. **66**, 141–144 (2017)
3. Varinlioglu, G., Sepehr, V.A., Eshaghi, S., Balaban, O., Nagakura, T.: GIS-based educational game through low-cost virtual tour experience – khan game. In: Proceedings of the 27th CAADRIA Conference, Sydney, 9–15 April 2022, pp. 69–78 (2022)
4. Ionescu, G., Valmaseda, J.M.D., Deriaz, M.: GeoGuild: location-based framework for mobile games. In: International Conference on Cloud and Green Computing, Karlsruhe, Germany, pp. 261–265 (2013)
5. Predescu, A., Mocanu, M., Chiru, C.: A case study of mobile games design with a real-world component based on Google Maps and Unity. In: 13th International Conference on Electronics, Computers and Artificial Intelligence (ECAI), Pitesti, Romania, pp. 1–6 (2021)
6. Gorelick, N., Hancher, M., Dixon, M., Ilyushchenko, S., Thau, D., Moore, R.: Google Earth engine: planetary-scale geospatial analysis for everyone. Remote Sens. Environ. **202**, 18–27 (2017)
7. Lee, A., Chang, Y.S., Jang, I.: Planetary-scale geospatial open platform based on the Unity3D environment. Sensors **20**(20), 5967 (2020)
8. Deterding, S., Dixon, D., Khaled, R., Nacke, L.: From game design elements to gamefulness: defining gamification. In: Proceedings of the 15th International Academic MindTrek Conference: Envisioning Future Media Environments, pp. 9–15 (2011)
9. Wolf, M.: 3 space in the video game. In: Wolf, M. (ed.) The Medium of the Video Game. University of Texas Press, New York (2021)
10. Abubakar, A., Zeki, A.M., Chiroma, H.: Optimizing three-dimensional (3D) map view on mobile devices as navigation aids using artificial neural network. In: 2013 International Conference on Advanced Computer Science Applications and Technologies, Kuching, Malaysia, pp. 232–237 (2013)
11. Koh, E., Park, G., Lee, B., Kim, D., Sung, S.: Performance validation and comparison of range/INS integrated system in urban navigation environment using Unity3D and PILS. In: Proceedings of the 2020 IEEE/ION Position, Location and Navigation Symposium (PLANS), Portland, OR, USA, 20–23 April 2020, pp. 788–792 (2020)
12. Zagata, K., Medynska-Gulij, B.: Mini-map design features as a navigation aid in the virtual geographical space based on video games. ISPRS Int. J. Geo-Inf. **12**(2), 58 (2023)

13. Toups, Z.O., Lalone, N., Alharthi, S.A., Sharma, H.N., Webb, A.M.: Making maps available for play: analyzing the design of game cartography interfaces. ACM Trans. Comput. Hum. Interact. **26**, 1–43 (2019)
14. Transportico game. https://play.google.com/store/apps/details?id=com.riftcat. transportico. Accessed 20 Jan 2024
15. RFC 7946 standard. https://datatracker.ietf.org/doc/html/rfc7946. Accessed 20 Jan 2024

Preliminary Study on the Detection of Subtle Variations in Image Sequences for Identifying False Starts in Speedway Racing

Jacek Krakowian[1] and Łukasz Jeleń[2(✉)]

[1] Foundation for the Support of Innovation, Research and Development, Na Grobli 12 lok. 0.211, 50-421 Wrocław, Poland
jacek.krakowian@wibir.org
[2] Department of Computer Engineering, Wrocław University of Science and Technology, Wybrzeże Wyspiańskiego 27, 50-370 Wrocław, Poland
lukasz.jelen@pwr.edu.pl

Abstract. Computer Vision algorithms have gained significant popularity and prove to be highly beneficial in detecting motion. These methods have proven to be profitable in motor sports especially. This paper presents a preliminary study on image processing, computer vision, and Long Short-Term Memory (LSTM) networks to detect subtle variations in image sequences for identifying false starts in speedway racing. Traditional methods often rely on manual observation, which lacks the precision to capture delicate movements at the start line. The presented methodology utilizes mentioned methods to enhance image quality, computer vision to detect and track racers' positions, and LSTM networks to analyze temporal sequences for movements indicating a false start. Results indicate that the adopted approach outperforms manual false start detection in accuracy and reliability. This advancement not only ensures fairness in speedway racing by providing a more objective detection of false starts but also demonstrates the potential of these technologies for applications requiring the identification of subtle changes in sequences. The efficiency and precision of our method offer promising implications for sports regulations and beyond, advocating for the broader application of technology in similar contexts.

Keywords: motion detection · speedway · false start · Gaussian mixture model · frame difference · Long Short-Term Memory · LSTM

1 Introduction

In the highly competitive world of speedway racing, the precision of start detection is crucial for guaranteeing equitable competition and upholding the sport's integrity. Identifying false starts accurately is crucial in race officiating as they

W. Zamojski et al. (Eds.): DepCoS-RELCOMEX 2024, LNNS 1026, pp. 111–120, 2024.
https://doi.org/10.1007/978-3-031-61857-4_11

can have a significant impact on race results. Conventional techniques for identifying false starts, which rely mainly on manual observation and basic timing
devices, frequently struggle to detect the very slight movements that are characteristic of a false start. Here, a referee makes challenging and subjective decisions by relying on their observations both a live situation and video recordings.
These techniques lack the necessary sensitivity and precision to reach conclusive
decisions, which can lead to disagreements and discrepancies in race outcomes.
Recent developments in image processing and computer vision has created new
opportunities for movement accuracy improvement detection that can be applied
to start detection in speedway racing [15]. Moreover, evolution of machine learning techniques, particularly Long Short-Term Memory (LSTM) networks, offers
potentially more beneficial capabilities in analyzing temporal sequences of data
for pattern recognition and anomaly detection [13]. This study introduces a
methodology that applies image processing, computer vision, and LSTM networks to detect subtle variations in image sequences indicative of false starts.
The research shows that proposed methodology aims to overcome the limitations of traditional methods, offering a more reliable, objective, and automated
solution for false starts' identification in speedway racing.

The implications of this study go beyond speedway tracks; unfortunately,
numerous sports lack the necessary tools or technologies to assist referees in
the decision-making process. Here, speedway motorsport serves as a prime illus-
tration of this issue. In speedway racing, the starting moment is critical and
significantly influences the entire race. The response time of the rider is a crucial
factor in determining the success of the competition, and the ability to earn
valuable points can depend on that starting moment. This article presents the
adopted methodology for investigating movement detection algorithms applied
to real-world racing scenarios, highlighting its effectiveness in improving the fairness and competitiveness of speedway racing.

2   Materials and Methods

In this section, the materials and methods will be described. In Sect. 2.1 a
brief definition of a false start in speedway sport is described. In the following
sections, we will provide the short literature findings followed by a description
of the applied algorithms.

2.1   False Start Definition in Speedway Racing

The speedway race consists of four riders racing for four rounds from a starting
point. Road cyclists wear helmets of different colors. Usually in team events, red
and blue are designated for the home team while white and yellow are designated
for visitors. At the start of the race, riders must take a position on their respective
starting lines, which is also indicated by the helmet color. All riders must be
present at the start line within a specified period of time, otherwise they will be
disqualified. They should get their bike to the starting area themselves without

external help. The starting area contains a white line and a starting gate with two or more tapes that run through the starting line. It is very important that all riders take their starting position between the white lines defining the grid and within 10 cm of the starting point. Figure 1 shows the correct position of the speedway driver at the start. When the specified time has elapsed, the referee turns on the green light, now riders cannot touch the tape until it tape is released nor move their bikes. If a driver does not comply, it is considered a false start. In such cases, a race is repeated, and the rider who causes a false start may be disqualified. In this case, the referee must analyze the movement of all four drivers simultaneously. Referees can use video recordings to help make decisions. This means that video is the only way to check if the start is correct or not. Unfortunately, throughout the race history, there have been situations where the referee has made wrong decisions without seeing the start of the rider movement.

Fig. 1. Speedway riders on starting positions. Author: Grzegorz Misiak Photography.

2.2 Related Work Review

For decades, image processing and computer vision methods have been an active research fields and raised a lot of interests in object detection [7,8,18]. In this section, we briefly show and review some of the existing computer vision applications for object tracking and moving object detection. In the literature, various applications related to ball detection in soccer images can be found. These algorithms are applied to help in the decision-making process during a game-play. In 2002, D'Orazio *et al.* introduced a modification of a circular Hough transform for this task. Recently, with the introduction of deep learning methods, ball detection applications based on this methodology have been introduced [3,14].

In addition to soccer, computer vision algorithms can also be applied to other sports. One of the examples is a work of Christian José Soler that created a framework for table tennis ball detection and calculation of its trajectory [12]. Table tennis is a very fast sport, making the detection problem difficult. It may even be problematic for coaches and instructors to supervise these rapid actions.

Soler's method solves this problem with common OpenCV image processing and computer vision algorithms. The described system processes the frames in real-time, where each frame is calibrated and preprocessed to remove noise. In the next step, a ball is detected and its trajectory is determined. According to the author, the biggest issue of the framework is a very fast-moving ball. To resolve this issue, a system would require a high-resolution camera with relatively high frame-rate.

Another example of a computer vision application in sports is the work of Burden *et al.* [2]. The authors describe a system that is used to track riders during the changeover in the Olympic cycling discipline. Here, two teams of four cyclists race on a 4000 m track. Each team has a leader that changes periodically with one of the three remaining riders. In this competition, the biggest problem is aerodynamic drag. A team leader fights the greatest drag and other riders fight with the lowest drag. The key to success is to make optimal leader changes during which a cyclist moves from the front to the back. Burden *et al.* proposed a video tracking system that was able to show the leader's moving path during the changeover. Coaches want this system to improve changeover maneuvers. The proposed system was able to detect and track all riders during the race with OpenCV built-in libraries.

In 2013, Song Ailing and Chen Kai described an improved Surendra motion detection algorithm of an athlete during a long jump [1]. The proposed framework was able to detect and track jumper on a so-called technical video. Video analysis is very helpful during training and can reveal important information for training personnel that can draw crucial practice conclusions. The proposed algorithm was able to detect the athlete in long jumping videos and the computer vision methods were able to extract only the athlete's motion and remove unwanted information from the videos.

In addition to the applications mentioned above, in the literature, we can notice numerous descriptions of motion detection algorithms. According to Hussain *et al.*, a frame difference (FD) approach is one of the most popular motion detection methods [5]. Its potential was proved by Singla, who described a frame difference algorithm for moving car detection achieving very promising results [11]. An additional review of the literature revealed various methods to enhance the frame difference technique [6, 17]. Some of them concentrate on the calculation of differences between three consecutive frames. Other improvements introduce additional methods to enhance the difference calculations. For example, Xiaoyang *et al.* introduced the Gaussian mixture model to the moving object algorithm [16] and Zhang *et al.* described a frame difference algorithm based on mathematical morphology. In all these cases, experimental results show that improved algorithms are superior to two frame difference as well as to common methods of background subtraction.

Recently, with the evolution of artificial intelligence and deep learning algorithms, several neural network methods have been introduced. These methods include recurrent neural networks (RNNs) and long-short-term memory (LSTM) networks that can be used to analyze temporal patterns in motion data. From the literature, we learn that they can capture complex dynamics and dependencies over time, improving the accuracy of motion detection and prediction [4, 9, 10].

2.3   Research Methodology

Based on the literature review and the nature of speedway motor-sport, a motion detection framework was created. Here, we describe the systematic approach and techniques employed to detect subtle variations in image sequences. The purpose of the proposed methodology is to identify false starts in speedway racing. This section outlines the specific computer vision algorithms and LSTM network model we implemented to analyze frame-by-frame changes in racer positions, ensuring the methodology's robustness and precision in capturing the critical moments leading to a false start.

The proposed framework takes a sequence of images as input of a video stream and processes them with one of the described methods. The first set of methods of the described scheme is based on background subtraction, where motion is detected by comparing a static background image and moving pixels in the sequence of input images. For this purpose, we compared two- or three-frame difference methods with a Gaussian Mixture Model. Furthermore, we have also evaluated the results of LSTM networks, as they have been proven to be powerful for subtle change detection in sequential data analysis.

1. **Two-frame difference (TFD)** - this algorithm considers two successive video frames F_k and F_{k+1}. Movement is recognized when $(F_{k+1} - F_k) \leq T$, where T is a predefined movement threshold. Subsequently, we propose our own modification of the method. In contrast to the proposed method, the TFD technique uses denoised grayscale images for the calculations. The difference result is then thresholded and a binary image $B(x, y)$ is obtained.

2. **Custom two-frame difference (CTFD)** is the author's modification of a traditional two-frame difference algorithm that also takes two video frames F_k and F_{k+n} as input. Here, k is a frame number, $K + n$ is a number of the next frame, and n is a frame step. The proposed algorithm not only does not restrict the images to two consecutive frames, but also calculates a difference between the foreground and background values, not as in a predefined OpenCV *absdiff* method. Our algorithm determines these values with thresholding of the input frame converted to gray-scale and blurred. Motion is detected if the result of the subtraction of binary foreground values from the background is lower than 0.

3. **Three-frame difference (ThFD)** - here the difference between the sequence of frames F_{k-1}, F_k and F_{k+1} is calculated. Similarly as for the TFD, a threshold for moving object detection is defined. Values greater than or equal to T determine a moving object. In this algorithm, the denoised gray-scale frames F_k and F_{k-1} are subtracted to provide a binary image $B_1(x, y)$. Furthermore, the frames F_{k+1} and F_k are also subtracted, producing another binary image $B_2(x, y)$. Motion is determined as a logical "AND" operation of images B_1 and B_2.

4. **Gaussian mixture model (GMM)** is an improved version of the background subtraction method based on a Gaussian Mixture Model. It uses a probability density function and background representation as a weighted sum of the densities of the Gaussian components. The background model is calculated from a preprocessed image. This involves gray-scale conversion, blurring, and thresholding. In consecutive steps, the model is continuously updated with information from a new video frame. This methodology ensures that the background model is resistant to changes and is not related to moving objects. The final motion is determined by comparing the pixel with the background model.

5. **Long-Short-Term Memory (LSTM)** along with Recurrent Neural Networks (RNNs) are neural network types created to process sequential data. LSTM architecture is built with additional components called memory cells or memory blocks that help with the problem of vanishing gradients. This problem is overcome by allowing the network to retain information over long sequences. This property allows them to be a good choice in situations where capturing long-term dependencies is crucial, like speech recognition, machine translation, sentiment analysis, and motion detection. On the contrary, RNNs are frequently applied in tasks that deal with sequential data, like language modeling, forecasting time series, and generating sequences.

2.4   Database of Speedway Videos

In this preliminary study, we utilized a specially curated database to detect subtle variations in image sequences, aimed at identifying false starts in speedway racing. The database comprises high-resolution image sequences captured during the Official International Competitions that were made publicly available by the Official *Speedway GP* Channel. Each video was recorded with a frame-rate of 30 FPS. In order to train and test the proposed methodology, the videos were carefully investigated to determine rider movement during the starting procedure. Out of 20 real-life streams, we managed to segment 75 starts out of which 11 (14.5%) were treated as misclassified and 64 (85.5%) classified as false start. The misclassified cases are a situation where the rider movement was visible, but the referee did not make a decision on the false start. Each starting procedure was recorded from the YouTube stream using Open Broadcaster Software and saved with resolution of 960×540 pixels.

In addition, a database of 40 test videos recorded during practice at the Opole Speedway track and in the laboratory was collected. These videos were also recorded with 30FPS but with a resolution of 1920×1080 and were used to verify the correctness of the proposed method.

3   Results

Identifying false starts in speedway racing is a challenging tasks mostly due to the close distance between competitors at the starting line. Traditional methods are highly dependent on human judgment, which, although very valuable, can be prone to errors when dealing with the high-pressure circumstances of the beginning of a race. In response to this challenge, our study sought to develop and evaluate an automated framework capable of identifying false starts by analyzing subtle variations in image sequences captured at the start of the race. In this section, we present the results of the performance of the proposed framework. The performance of the algorithms was compared against traditional human-based assessments to highlight its relative strengths and areas for improvement. To achieve this, we have compared five object detection methods described in Sect. 2.3 applied to videos from the collected database. For the first experiment, we applied the methods to the practice sequences from Opole track and the laboratory setup. In this case, the proposed methods have proven to have excellent results with accuracy reaching 99% for all methods. Such high precision is the effect of concentrating only on a single object with fairly uniform background. The second experiment was shown to be less effective but still demonstrates the possibility of false start detection during real competition. For this experiment, we used the 70 image sequences from our database. As mentioned previously, 14.5% of the false starts were noticed by a referee and repeated, while the remaining starts were not.

In Fig. 2 a sample of motion detection results are presented, with the motion marked in green. In Fig. 2a results obtained with the TFD method. Here, one can notice that the entire rider in the yellow helmet is detected, while the reaming riders are not. Figure 2b shows the results of the customary CFTD method, where it can be noticed that the described algorithm identified the front and back wheels of a speedway rider wearing a red helmet, as well as an advertisement rotation.

The results of the three-frame difference are depicted in Fig. 2c, where it can easily be seen that the rider in red has moved. In addition, a motion of the clutch, spokes and part of the engine was also marked. In the last figure, the results of the GMM algorithm are shown (Fig. 2d). Here, we can see that the yellow rider with part of a front wheel is marked as moving.

The results for the entire database showed that GMM methods was able to detect riders motion in most of the cases, reaching 95% of all false starts. The proposed CTFD method did not perform as well as GMM but outperformed two other background subtraction methods. The best noticeable false start detection was recorded at 80%. To complite the flase start detection image, we introduced the term "Half detected" which represents a situation where not all riders committing a false start were detected by the algorithm. In this situation, the GMM algorithm performed best and the TFD worst. The results are gathered in Table 1.

In addition to the methods mentioned above, we tested the performance of LSTM networks to verify their applicability to movement detection and false start classification. For this purpose, the network was trained to classify image sequences as false start or normal race start. During the test, the LSTMs proved their high accuracy in detecting motion and correctly classifying all false starts. During the experiments, the confidence level of the network was observed to be

(a) Two Frame Difference.

(b) Custom Two Frame Difference.

(c) Three Frame Difference.

(d) Gaussian Mixture Model.

Fig. 2. Motion Detection Results.

as high as 53% for false starts and 48% for normal starts. This shows that the difference between the false start and normal procedure is very subtle and could cause confusion for a referee to make a correct judgement during the start of the race.

Table 1. False start detection results.

Method	CTFD	TFD	ThFD	GMM
Detected (%)	65%	70%	50%	75%
Half detected (%)	15%	5%	15%	20%
Not detected (%)	20%	25%	35%	5%

4 Conclusions and Future Work

In this study, we explored the effectiveness of advanced image analysis techniques for detecting subtle variations in sequences of images to identify false starts in speedway racing. The aim was to develop a reliable, objective and automated framework that could assist a referee in making critical decisions, which often have significant implications for the outcome of races and the fairness of competition. Our preliminary study contributes to this evolving field, offering a solution to a long-standing problem in speedway racing. In this paper, we describe the results of moving object detection methods applied to speedway video sequences. These results were compared against traditional human-based assessments. The results described in Sect. 3 clearly show that the proposed methodology is capable of detecting subtle changes in the videos and classifying them as false starts. The experimental results clearly show that the LSTMs were able to detect false starts correctly, but their confidence level could be improved. Other methods also proved to be a good choice for movement detection and the proposed CTFD algorithm was able to detect false starts with better accuracy than TFD and ThFD. The GMM method outperformed all background subtraction methods reaching 95% of accuracy.

Furthermore, future research should focus on database enlargement and the collection of better image sequences, as well as on refining the methodology and exploring its application in various conditions and settings. A larger data base should help train LSTMs to provide classifications with a better confidence levels. Further studies could also investigate the integration of such systems with other technological advances in sports analytics and monitoring, creating a more comprehensive and fair framework for competition evaluation.

In conclusion, this study demonstrates the potential of using advanced image processing and machine learning techniques to detect false starts in speedway racing.

References

1. Ailing, S., Kai, C.: OpenCV detection of athletes in long jumping videos. In: International Conference on Information, Business and Education Technology (ICIBIT), pp. 10–13. Atlantis Press (2013)
2. Burden, J., et al.: Tracking a single cyclist during a team changeover on a velodrome track with Python and OpenCV. In: 8th Conference of the International Sports Engineering Association (ISEA), pp. 2931–2935. Elsevier (2010)
3. Buric, M., Pobar, M., Ivasic-Kos, M.: Ball detection using YOLO and mask R-CNN. In: 2018 International Conference on Computational Science and Computational Intelligence (CSCI), pp. 319–323 (2018). https://doi.org/10.1109/CSCI46756.2018.00068
4. Carrara, F., Elias, P., Sedmidubský, J., Zezula, P.: LSTM-based real-time action detection and prediction in human motion streams. Multimed. Tools Appl. 1–23 (2019). https://api.semanticscholar.org/CorpusID:190637711
5. Hussain, M.Z., Naaz, M.A., Uddin, M.N.: Moving object detection based on background subtraction & frame differencing technique. IJARCCE 5, 817–819 (2016)
6. Li, S., Liu, P., Han, G.: Moving object detection based on codebook algorithm and three-frame difference. Int. J. Signal Process. Image Process. Pattern Recogn. 10(3), 23–32 (2017)
7. Padilla, R., Netto, S.L., Da Silva, E.A.: A survey on performance metrics for object-detection algorithms. In: 2020 International Conference on Systems, Signals and Image Processing (IWSSIP), pp. 237–242. IEEE (2020)
8. Prasad, D.K.: Survey of the problem of object detection in real images. Int. J. Image Process. (IJIP) 6(6), 441 (2012)

9. Rego Drumond, R., Dorta Marques, B.A., Nader Vasconcelos, C., Clua, E.: Peek - an LSTM recurrent network for motion classification from sparse data. In: Proceedings of the 13th International Joint Conference on Computer Vision, Imaging and Computer Graphics Theory and Applications - GRAPP, pp. 215–222. INSTICC, SciTePress (2018). https://doi.org/10.5220/0006585202150222

10. da Silva, R.E., Ondřej, J., Smolic, A.: Using LSTM for automatic classification of human motion capture data. In: VISIGRAPP (2019). https://api.semanticscholar.org/CorpusID:88489943

11. Singla, N.: Motion detection based on frame difference method. Int. J. Inf. Comput. Technol. **4**(15), 1559–1565 (2014)

12. Soler, C.J.: Table tennis ball tracking and bounce calculation using OpenCV. Technical report, Universitat de Barcelona (2017)

13. Sun, S., Mu, L., Wang, L., Liu, P.: L-Unet: an LSTM network for remote sensing image change detection. IEEE Geosci. Remote Sens. Lett. **19**, 1–5 (2020)

14. Teimouri, M., Delavaran, M.H., Rezaei, M.: A real-time ball detection approach using convolutional neural networks. In: Chalup, S., Niemueller, T., Suthakorn, J., Williams, M.-A. (eds.) RoboCup 2019. LNCS (LNAI), vol. 11531, pp. 323–336. Springer, Cham (2019). https://doi.org/10.1007/978-3-030-35699-6_25

15. Wiley, V., Lucas, T.: Computer vision and image processing: a paper review. Int. J. Artif. Intell. Res. (2018)

16. Xiaoyang, Y., Yang, Y., Shuchun, Y., Yang, S., Huimin, Y., Xifeng, L.: A novel motion object detection method based on improved frame difference and improved Gaussian mixture model. In: 2nd International Conference on Measurement, Information and Control, pp. 309–313. IEEE (2013)

17. Xu, Z., Zhang, D., Du, L.: Moving object detection based on improved three frame difference and background subtraction. In: International Conference on Industrial Informatics - Computing Technology, Intelligent Technology, Industrial Information Integration, pp. 79–82. IEEE (2017)

18. Zou, Z., Shi, Z., Guo, Y., Ye, J.: Object detection in 20 years: a survey. arXiv preprint arXiv:1905.05055 (2019)

Artificial Intelligence in Renewable Energy: Bibliometric Review of Current Trends and Collaborations

Paweł Kut[1]([✉]) [iD], Katarzyna Pietrucha-Urbanik[1] [iD], Martina Zelenakova[2] [iD], and Hany F. Abd-Elhamid[2,3] [iD]

[1] Faculty of Civil, Environmental Engineering and Architecture, Rzeszow University of Technology, Al. Powstańców Warszawy 6, 35-959 Rzeszów, Poland
`pkut@prz.edu.pl`
[2] Faculty of Civil Engineering, Technical University of Kosice, Vysokoškolská 4, 04200 Kosice, Slovakia
[3] Faculty of Engineering, Zagazig University, Zagazig 44519, Egypt

Abstract. The article presents a bibliometric analysis of publications on the use of artificial intelligence (AI) in the area of renewable energy sources (RES). The study is based on data downloaded from the Web of Science database, using the keywords "artificial intelligence" and "renewable energy". The aim of the analysis is to identify the main trends, research areas, as well as leading institutions and authors in this interdisciplinary field. The CiteSpace tool was used to process and visualize the data. The analysis covers publications from the last two decades, enabling understanding of the evolution and development directions of the combination of AI and renewable energy. Particular attention was paid to interactions between various branches of science, which allows for the identification of potential new research areas and interdisciplinary cooperation. The results of the analysis show how AI contributes to the development of renewable energy efficiency, management and integration, also highlighting challenges and opportunities for future research in this field.

Keywords: CiteSpace · Bibliometric · Artificial Intelligence · Renewable Energy

1 Introduction

In the face of global challenges related to climate change and the growing demand for energy, renewable energy sources (RES) are gaining importance as a key element of sustainable development. At the same time, the dynamic development of artificial intelligence (AI) opens new possibilities for optimizing and integrating energy systems based on renewable energy sources [1–3]. The interpenetration of these two fields is the subject of intensive research, which is reflected in a growing number of scientific publications. This article aims to conduct a bibliometric analysis of publications on the use of AI in the context of renewable energy, using the Web of Science database. An introduction to

© The Author(s), under exclusive license to Springer Nature Switzerland AG 2024
W. Zamojski et al. (Eds.): DepCoS-RELCOMEX 2024, LNNS 1026, pp. 121–131, 2024.
https://doi.org/10.1007/978-3-031-61857-4_12

the topic requires an understanding of the importance of both fields and their interrelationships. Artificial intelligence, understood as the ability of machines to perform tasks requiring human intelligence, has contributed to significant progress in many areas of science and technology in recent years [4, 5]. The use of AI in energy systems enables, among other things, more effective management of energy networks, forecasting energy production and demand, as well as optimizing the use of renewable energy sources. On the other hand, renewable energy, which includes solar, wind, hydropower, geothermal and other forms of energy obtained from renewable sources, is a key element in the pursuit of reducing greenhouse gas emissions and achieving climate neutrality. Integration of AI with renewable energy systems can contribute to a significant increase in the efficiency of these systems, enabling more precise adjustment of energy production to current demand and better management of energy resources. Bibliometric analysis of publications on this topic allows for the identification of key trends, main research topics, as well as leading research centers dealing with these issues. The use of data from Web of Science, one of the most renowned scientific databases, guarantees the comprehensiveness and reliability of the analysis [6].

The use of the CiteSpace bibliometric analysis tool allows for visualization and detailed analysis of the collected data. CiteSpace is an advanced network analysis tool that enables you to identify trends, critical nodes and the evolution of topics in scientific research. Thanks to this, it is possible not only to understand the current state of research, but also to predict future development directions and identify potential new research areas. Figure 1 shows the stages of bibliometric analysis using the CiteSpace program [7–9].

2   Methodology

In the article, the focus is on examining the impact of artificial intelligence (AI) on the development and utilization in the renewable energy sector. The methodology of the study is based on a detailed bibliometric analysis using the Web of Science database as the main data source. The key search terms used as the foundation were "artificial intelligence" and "renewable energy". Then the highly cited papers filter was selected to use the most frequently cited publications for analysis. Publications from the last 10 years were selected for analysis. A total of 3896 articles were identified. Figure 1 presents the distribution of publications over the years.

After preparing the data for analysis, a detailed bibliometric analysis was conducted using the CiteSpace tool. This allowed for the visualization of collaboration and citation networks, identification of key authors and publications, and understanding of the main research trends and thematic evolution in the field of artificial intelligence and renewable energy. This analysis formed the basis for a deeper understanding of how AI technologies are currently used in the renewable energy sector and what new opportunities and challenges might arise in the future. Through this, the main areas where AI has the potential to impact the development of renewable energy sources were identified. Particular attention was paid to case studies and empirical research, demonstrating practical applications of AI, such as the optimization of energy networks, forecasting energy production from renewable sources, and managing energy demand and supply.

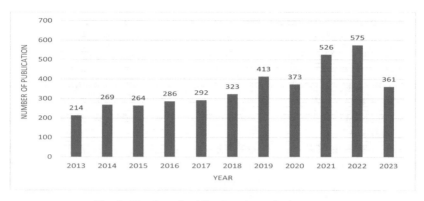

Fig. 1. Number of publications in particular years.

3 Bibliometric Analysis of Literature

The first aspect of bibliographic analysis is the analysis of the geographical origin of the publication. This cross-section allows you to identify global trends, research leaders and regions that are at the forefront of innovation and development in this field. Analysis of the countries where publications come from provides valuable information about the international distribution of research and international cooperation. The performer analysis is presented in Fig. 2.

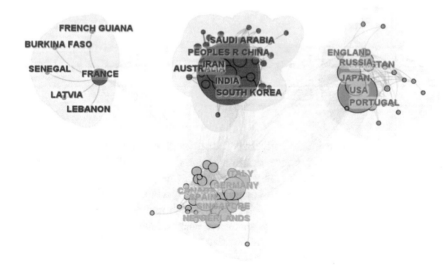

Fig. 2. Cooperation between countries.

During the analysis, special attention was paid to the centrality index, which is an important tool for identifying a country that is a center of scientific cooperation in a given field. It is a metric used to determine the importance of nodes in a network. In practice, nodes with a higher centrality coefficient are perceived as key in the network, acting as connectors between different groups of nodes. In the context of bibliometrics, such nodes may represent authors, institutions or countries that play a central role in a given research area, connecting different research streams or collaborations. A high centrality index suggests that a given country plays a key role in scientific exchange in a given field. The analysis of this parameter allows to identify scientific leaders and focus attention on countries that play an important role in developing research related to AI and renewable energy.

Table 1 shows the 10 countries with the highest centrality index that stand out in the context of scientific cooperation related to the use of AI in renewable energy. Additionally, the table contains information on the number of publications, which allows us to assess both the quality and quantity of a given country's contribution to the development of this interdisciplinary field.

Table 1. Top 10 countries with highest centrality index.

Country	Number of publications	Centrality index
France	116	0.17
China	2023	0.12
USA	769	0.10
England	235	0.10
Germany	230	0.09
Australia	289	0.09
Pakistan	77	0.07
South Korea	181	0.06
India	252	0.05
Spain	118	0.05

The results show considerable diversity in the role that individual countries play in the global research network. France, despite the small number of publications, plays a central role, which may indicate its high involvement in international cooperation and coordination of research. China is also characterized by a high centrality coefficient and the largest number of highly cited publications, which indicates its high involvement in research on artificial intelligence in renewable energy sources. The results show that it is not always the quantity of publications that is most important, but also their high quality, which has an impact and importance in the development of the discipline.

The next stage of the analysis focused on examining cooperation between universities and institutes in the area of using artificial intelligence in the context of renewable energy sources. This segment of research is extremely important because academic and research institutions play a key role in generating new knowledge and technological innovations. The aim of the analysis is to identify leading universities and institutes involved in given research through the analysis of the centrality index. The results of the analysis are presented in Fig. 3.

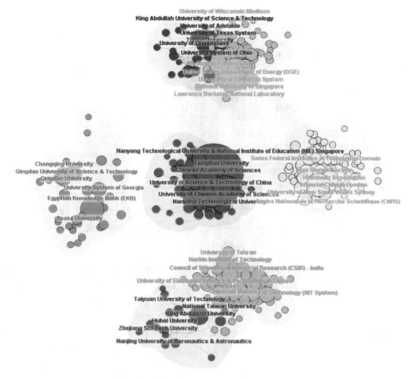

Fig. 3. Cooperation between institutes.

Table 2 shows the top 10 institutes with the highest centrality index and the number of highly scored publications with their affiliation.

The Chinese Academy of Sciences stands out in terms of both the number of publications and the centrality coefficient. The high centrality coefficient (0.36) indicates its key role in the cooperation network between universities and institutes, which proves its importance as a center of innovation and knowledge in this field. The University of California, CNRS and the United States Department of Energy also play a significant role in research, as indicated by their high centrality index. The high centrality coefficients of the above institutions emphasize their role in creating bridges between different research areas. This indicates the importance of inter-institutional cooperation in generating breakthrough innovations and knowledge in the field of using artificial intelligence in renewable energy sources.

The next important stage of the analysis is the identification of key research topics in the field of the use of AI in the context of renewable energy sources. This stage focuses on recognizing and understanding the main research areas that currently dominate the field, and on identifying new, emerging trends that may shape future technological and scientific developments. The use of bibliometric analysis methods allows for the identification

Table 2. Top 10 institutes with highest centrality index.

Country	Number of publications	Centrality index
Chinese Academy of Sciences	482	0.36
University of California	96	0.15
Centre national de la recherche scientifique (CNRS)	71	0.12
United States Department of Energy	150	0.10
National University of Singapore	72	0.08
Tsinghua University	90	0.05
Swiss Federal Institutes Of Technology	67	0.05
Max Planck Society	56	0.05
University of Science and Technology of China	107	0.04
Helmholtz Association	39	0.04

and categorization of the most frequently researched topics. In this way, it is possible to identify the areas that are most intensively researched and those that are gaining importance. Understanding these trends is not only crucial for scientists and researchers to guide their future research, but also for policymakers and funding institutions to better direct resources and support. Analysis of key research topics also reveals relationships between different fields of knowledge, which can inspire interdisciplinary research and collaboration. This allows you to understand how different concepts and technologies in the field of artificial intelligence are applied to solve problems related to renewable energy sources, which can lead to innovative solutions and accelerate development in this field. The results of the analysis are presented in the Fig. 4.

The largest node in the analyzed case is the node called "photo-catalytic hydrogen production". It consists of 443 nodes. The second largest node is "hydrogen evolution reaction" with 388 nodes, and the third is "liquid cellulosic biomass". Of the 13 nodes, 5 are hydrogen research. Over the last 10 years, hydrogen research has accelerated significantly, focusing mainly on the development of electrolysis technology and the use of renewable energy to produce clean hydrogen. Interest in this fuel has increased with the falling costs of solar and wind energy, as well as with growing ecological awareness. Hydrogen, which is currently widely used in industry, has become an important element in the efforts to decarbonize many sectors, including transport, energy production and heavy industry. Going forward, hydrogen has the potential to play a key role in a clean, secure and affordable energy future. Further research and technology development, combined with government policy and international cooperation, aim to reduce costs and accelerate the implementation of hydrogen technologies. Hydrogen can help solve many critical energy challenges, offering ways to decarbonize various sectors, including long-distance transport, for which electric propulsion may be ineffective due to the high mass of batteries, limited range and relatively long charging times.

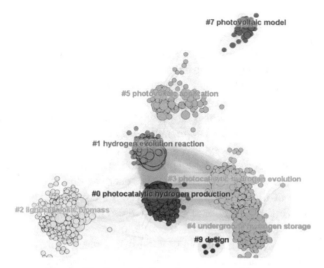

Fig. 4. Citation analysis and major area of interest

Much of the research also concerns applications in photovoltaics. AI is used to better predict solar energy production, which is especially important due to changing weather conditions affecting the efficiency of solar panels. Currently, neural network models are used to forecast hourly energy production by photovoltaic modules. These models are able to predict weather and sky characteristics, allowing for more precise forecasts and increasing the efficiency of solar power plants.

In the area of maintenance and maintenance of photovoltaic systems, AI also plays a key role. The use of machine learning algorithms to analyze data collected using drones or other tools allows for quick and accurate identification and classification of solar panel defects. This makes it possible to manage the maintenance and upkeep of solar farms more effectively, which translates into reduced costs and increased system efficiency.

AI is also used in mitigating pollutants, such as dust or pollen deposits on photovoltaic panels, which may affect the efficiency of the panels. AI-powered platforms can analyze data from solar panels to determine when and how the panels should be cleaned, increasing efficiency and reducing maintenance costs.

In photovoltaic production, AI is used to analyze data from production plants in real time, which allows for continuous optimization of production processes. For example, in Germany, research is being carried out on the use of self-learning photovoltaic factories that use process data to continuously optimize production. This approach can significantly contribute to the development and optimization of the production of high-technology solar cells and modules.

AI also plays a significant role in predicting energy demand and supply, which is essential in next-generation energy systems, especially for renewable technologies where energy production can be unstable due to weather conditions. Machine learning enables variable supply to better match rising and falling demand, maximizing the financial value of renewable energy and enabling it to be more easily integrated into the grid.

One example of the use of AI is the forecasting of wind energy production, which can be enhanced using weather models and information about the location of turbines. The models can accurately predict future energy production up to 36 h in advance. AI is also crucial in preventing network failures, which increases reliability and security. Using AI to continuously monitor and analyze the performance of energy assets enables potential faults to be identified in advance.

In addition to better forecasting energy supply and demand and predictive maintenance of physical infrastructure, AI applications may include managing and controlling networks, facilitating demand response, and delivering improved or expanded services to consumers. The development and implementation of AI in renewable energy is crucial for the transition to more sustainable and reliable energy systems.

The final stage of the analysis is to determine the top 5 articles with the highest citation burst rate (Fig. 5). CiteSpace citation burst refers to a sudden increase in the number of citations to a specific article or topic over a specific period of time. This is an indicator that may indicate growing interest in a given topic or discovery in the scientific community. This feature is often used to identify research trends, key papers, or breakthroughs in a given scientific field.

References	Year	Strength	Begin	End	2017 - 2023
Qing X, 2018, HOURLY DAY-AHEAD SOLAR IRRADIANCE PREDICTION USING WEATHER FORECASTS BY LSTM @ ENERGY, V0, P0	2018	2.83	2021	2023	
Zhang D, 2018, REVIEW ON THE RESEARCH AND PRACTICE OF DEEP LEARNING AND REINFORCEMENT LEARNING IN SMART GRIDS CSEE J. POWER ENERGY SYST. @ 4, V3, P362-370	2018	2.43	2021	2021	
Ghiasi M, 2019, DETAILED STUDY MULTI-OBJECTIVE OPTIMIZATION AND DESIGN OF AN AC-DC SMART MICROGRID WITH HYBRID RENEWABLE ENERGY RESOURCES @ ENERGY, V169, P496-507	2019	2.38	2020	2021	
Kow KW, 2016, A REVIEW ON PERFORMANCE OF ARTIFICIAL INTELLIGENCE AND CONVENTIONAL METHOD IN MITIGATING PV GRID-TIED RELATED POWER QUALITY EVENTS @ RENEW. SUSTAIN. ENERGY REV., V56, P334-346	2016	2.38	2020	2021	
Borhanazad H, 2014, OPTIMIZATION OF MICRO-GRID SYSTEM USING MOPSO @ RENEW. ENERGY, V71, P295-306	2014	2.38	2020	2021	

Fig. 5. Top 5 references with strongest citation burst.

The first article presents a novel method for predicting solar radiation intensity based on weather forecast data. The authors focused on the use of artificial intelligence, in particular the LSTM (Long Short-Term Memory) network, to more accurately predict hourly solar radiation intensity a day in advance. Their method proved to be much more accurate compared to other algorithms such as back propagation neural networks (BPNN), and also showed better generalization ability and lower risk of model overfitting [10].

The second article focuses on the application of deep learning and reinforcement learning in smart grids. It provides an overview of current research and practice in this area, highlighting the potential of these AI techniques to predict energy demand, manage demand, detect anomalies and optimize the operation of energy networks. [11].

The third paper examines the use of hybrid renewable energy systems, such as wind turbines and photovoltaic panels, in smart microgrids using the MOPSO algorithm. Particular emphasis was placed on the use of artificial intelligence to optimize energy design and management to increase network availability and reduce costs. The article also raises issues related to the security and stability of the energy system and the impact of these technologies on the environment and economy [12].

The fourth article focuses on analyzing the impact of grid-connected photovoltaics on power quality and comparing artificial intelligence and traditional methods in mitigating power quality problems. The article discusses in detail various AI approaches such as machine learning in the context of energy system monitoring, controlling inverters and other compensation devices, highlighting the improved performance and responsiveness of AI compared to conventional methods [13].

The fifth article focuses on the optimization of micro-grid system using MOPSO. The study concerns the design of a hybrid system using wind, solar energy and diesel generators, taking into account the costs and reliability of the system. Particular emphasis was placed on the use of the MOPSO algorithm as an AI tool for effective sizing of system components and energy management to optimize performance and costs [14].

4 Conclusion

The article presents a bibliometric analysis of scientific publications on the use of artificial intelligence (AI) in the context of renewable energy sources (RES). The study, based on data from the Web of Science database and conducted using the CiteSpace tool, reveals key trends, main research areas and leading research centers dealing with this interdisciplinary topic.

The analysis showed that the integration of AI with renewable energy systems can significantly increase their efficiency, enabling more precise adjustment of energy production to current demand and better management of energy resources. France and China, as countries with a high centrality index, play key roles in international cooperation and coordination of research in this area.

Moreover, the analysis reveals a growing interest in AI in the context of renewable energy over recent years, with an emphasis on practical applications such as energy network optimization, energy production forecasting and energy demand management. It is worth paying special attention to the growing importance of hydrogen as a key element in the future energy industry. Hydrogen, which is seen as the clean fuel of the future, has the potential to play an important role in the decarbonization of various economic sectors, including transport, energy and heavy industry. In the context of research on the use of AI in the renewable energy sector, hydrogen is an interesting field for the use of modern technologies, including machine learning algorithms and data analysis. The use of AI in photovoltaics is also important, where modeling and forecasting of solar energy production is becoming more and more advanced thanks to machine learning algorithms.

The development of AI and its integration with renewable energy reflects the global desire to increase the efficiency and sustainability of energy systems. In the context of climate change, the integration of AI with renewable energy can contribute to a faster energy transformation, enabling more efficient use of renewable energy and reducing greenhouse gas emissions.

Additionally, the development of AI technology in renewable energy opens up new opportunities for smart grids, which can better manage energy distribution and storage, contributing to greater stability and reliability of energy systems. The use of AI to analyze and process data from various sources, including sensors and production monitoring, allows for more effective planning and response to changing conditions, which is key to optimizing the use of renewable energy sources.

The transition to sustainable energy sources, driven by global initiatives, has accelerated the use of AI in managing complex, multi-source energy operations. AI facilitates real-time optimization to cope with the variability of renewable energy sources such as wind and solar by adjusting energy production to maintain grid stability. This dynamic management is key to integrating diverse energy assets, ensuring operational efficiency and compliance with regulatory standards.

Moreover, AI-powered predictive maintenance is revolutionizing the renewable energy sector. By using AI and machine learning algorithms, energy companies can predict equipment failures before they occur, reducing downtime and maintenance costs. This predictive approach extends the life of key infrastructure such as solar panels and wind turbines, increasing their efficiency and reliability.

AI is also key to optimizing energy storage and use. By anticipating fluctuations in demand and supply, AI enables more efficient storage and use of renewable energy. This not only helps with grid sustainability, but also reduces dependence on non-renewable energy sources. The application of AI in energy trading and market operations further increases the profitability and sustainability of renewable energy projects.

Taken together, our observations highlight the transformative role of AI in renewable energy systems. AI integration not only strengthens the efficiency and reliability of renewable energy sources, but also drives the global transition to a more sustainable and resilient energy future. Therefore, further research and development at this intersection is essential to overcome the contemporary challenges posed by climate change and unleash the full potential of renewable energy resources.

References

1. Chen, C., Hu, Y., Karuppiah, M., Kumar, P.M.: Artificial intelligence on economic evaluation of energy efficiency and renewable energy technologies. Sustain. Energy Technol. Assess. **47**, 101358 (2021). https://doi.org/10.1016/j.seta.2021.101358
2. Smadi, T.A., Handam, A., Gaeid, K.S., Al-Smadi, A., Al-Husban, Y., Khalid, A.S.: Artificial intelligent control of energy management PV system. Results Control Optim. **14** (2024). https://doi.org/10.1016/j.rico.2023.100343
3. Zhang, L., Ling, J., Lin, M.: Artificial intelligence in renewable energy: a comprehensive bibliometric analysis. Energy Rep. **8**, 14072–14088 (2022). https://doi.org/10.1016/j.egyr.2022.10.347
4. Lateef, A.A.A., Ali Al-Janabi, S.I., Abdulteef, O.A.: Artificial intelligence techniques applied on renewable energy systems: a review. In: Bashir, A.K., Fortino, G., Khanna, A., Gupta, D. (eds.) Proceedings of International Conference on Computing and Communication Networks. Lecture Notes in Networks and Systems, vol. 394, pp. 297–308. Springer Nature, Singapore (2022). https://doi.org/10.1007/978-981-19-0604-6_25
5. Hannan, M.A., et al.: Impact of renewable energy utilization and artificial intelligence in achieving sustainable development goals. Energy Rep. **7**, 5359–5373 (2021). https://doi.org/10.1016/j.egyr.2021.08.172
6. Kut, P., Pietrucha-Urbanik, K.: Most searched topics in the scientific literature on failures in photovoltaic installations. Energies **15**, 8108 (2022). https://doi.org/10.3390/en15218108
7. Chen, C.: Visualizing and exploring scientific literature with CiteSpace: an introduction. In: Proceedings of the 2018 Conference on Human Information Interaction&Retrieval - CHIIR 2018, pp. 369–370. ACM Press, New Brunswick (2018). https://doi.org/10.1145/3176349.3176897

8. Ding, X., Yang, Z.: Knowledge mapping of platform research: a visual analysis using VOSviewer and CiteSpace. Electron. Commer. Res. **22**, 787–809 (2022). https://doi.org/10.1007/s10660-020-09410-7

9. Zhao, X., Nan, D., Chen, C., Zhang, S., Che, S., Kim, J.H.: Bibliometric study on environmental, social, and governance research using CiteSpace. Front. Environ. Sci. **10**, 1087493 (2023). https://doi.org/10.3389/fenvs.2022.1087493

10. Qing, X., Niu, Y.: Hourly day-ahead solar irradiance prediction using weather forecasts by LSTM. Energy **148**, 461–468 (2018). https://doi.org/10.1016/j.energy.2018.01.177

11. Zhang, D., Han, X., Deng, C.: Review on the research and practice of deep learning and reinforcement learning in smart grids. CSEE J. Power Energy Syst. **4**, 362–370 (2018). https://doi.org/10.17775/CSEEJPES.2018.00520

12. Ghiasi, M.: Detailed study, multi-objective optimization, and design of an AC-DC smart microgrid with hybrid renewable energy resources. Energy **169**, 496–507 (2019). https://doi.org/10.1016/j.energy.2018.12.083

13. Kow, K.W., Wong, Y.W., Rajkumar, R.K., Rajkumar, R.K.: A review on performance of artificial intelligence and conventional method in mitigating PV grid-tied related power quality events. Renew. Sustain. Energy Rev. **56**, 334–346 (2016). https://doi.org/10.1016/j.rser.2015.11.064

14. Borhanazad, H., Mekhilef, S., Gounder Ganapathy, V., Modiri-Delshad, M., Mirtaheri, A.: Optimization of micro-grid system using MOPSO. Renew. Energy **71**, 295–306 (2014). https://doi.org/10.1016/j.renene.2014.05.006

Impact of Learning Data Statistics on the Performance of a Recommendation System Based on MovieLens Data

Urszula Kużelewska[(⊠)] and Michał Falkowski

Bialystok University of Technology, Wiejska 45a, 15-351 Bialystok, Poland
u.kuzelewska@pb.edu.pl, 74726@student.pb.edu.pl

Abstract. Recommendation systems are an efficient way to improve user satisfaction on the internet and personalise web services. They use various algorithms to analyse large amounts of data and offer customers the most useful products. The accuracy of the offered propositions is the most important feature of all recommendation systems, which is closely related to the principle of the methods. However, the input data also influences their performance. The objective of this study is to determine whether there is a correlation between data statistics, such as density, shape and skewness, and the performance of the algorithm as measured by RMSE. Additionally, we examined the coverage of both items and users with respect to the input data. The verification was performed using classical item and user-based methods with cosine similarity.

Keywords: Recommender systems · Evaluation of recommender systems · Data statistics

1 Introduction

In the realm of digital content, the volume of information available can be overwhelming. Recommender systems act as digital advisors, offering personalised suggestions to users based on their preferences and behaviours. Collaborative filtering is a prominent technique used by these systems, relying on the collective wisdom of users to improve recommendations [11,15,16].

Recommendation systems operate on customer and product data within the service. The system offers each customer recommendations based on customer ratings, product and customer similarity, or product popularity. It may interest them and potentially persuade them to make a purchase. However, a recommendation system that performs well in one context may not be equally effective in a system with a different data structure [14].

The aim of this study was to determine whether the data characteristics affect the quality of recommendations generated by a collaborative recommendation system. The selected data features, as outlined in [3,8], include data density, data shape in relation to the number of users or items, and the popularity factors

© The Author(s), under exclusive license to Springer Nature Switzerland AG 2024
W. Zamojski et al. (Eds.): DepCoS-RELCOMEX 2024, LNNS 1026, pp. 132–142, 2024.
https://doi.org/10.1007/978-3-031-61857-4_13

of items. The system's accuracy was used to measure the impact. The most meaningful index, RMSE, was used to evaluate the results. However, we also examined the coverage of both items and users with respect to the input data. The tests were performed on classical item and user-based methods using cosine similarity. Although other similarity measures, such as Pearson correlation or Euclidean distance, were tested, the results were similar.

This article was inspired by the work presented in [1,4]. However, the difference consists in the datasets used in the experiments and the performance measures used for the recommender system's evaluation. In [1], the authors only analysed error-based metrics, while in [4] the authors also considered fairness, but the MovieLens datasets used in the experiments were smaller: they contained 100 000 and 1 million ratings. The experiments presented in this paper are based on the MovieLens dataset with 25 million ratings. In addition, the other quality metrics related to item and user coverage were investigated.

The main contributions of this work are the following:

- The accuracy of recommender systems can be improved by investigating data characteristics and appropriately preparing learning data.
- The quality of generated recommendation lists can be improved by eliminating popular items, which reduces the Skewness and Gini values.

The article is divided into several sections. The first section provides an analysis of related work on the topic. The following section describes the measures used to evaluate the performance of the recommender system in the experiments. Section 4 presents the procedures for calculating the data characteristics, while Sect. 5 presents the results of the experiments. Finally, the paper concludes by summarising the key findings and discussing the study's implications.

2   Related Work

Recently, researchers have investigated the impact of data characteristics on the performance of classical recommender systems [4,6]. Their findings have yielded fruitful results regarding the relationship between data characteristics and the accuracy of recommendations.

Hsu [9] identified skewness as a characteristic that reduces the accuracy of collaborative filtering methods. Experiments conducted on a naturally skewed real dataset, which contained clickstream data from an online advertising agency, confirmed this conclusion.

In [17] a novel approach to data sparsity was applied. The authors propose a process of data augmentation and refinement to improve data characteristics, thereby increasing the accuracy of recommendations. The process of data augmentation uses some of the ratings predicted with high confidence in every iteration to augment the training set. The refinement phase removes the ratings predicted with low confidence from the training set.

Data management has had a positive impact also within methods. In [19] study, the accuracy of a recommender system was improved compared to the traditional Matrix Factorisation approach by adding the item's variance minimisation regularisation term.

3   Evaluation of Recommender Systems

The evaluation of generated recommendations can be conducted online, when users interact with them immediately after generation, or offline, when the systems are assessed using specially prepared input data [5,11,13]. Despite its imperfections, the offline evaluation is faster and easier to carry out. The process involves dividing the input data into training and test sets. The system is trained on a specific data set and then evaluated on a separate test set to simulate real user interactions. A cross-validation method is used to reduce the impact of random selection on the results. The dataset is divided into k equidistant subsets, then k iterations of the evaluation are performed, each hiding a different subset of the data. The final evaluation results are calculated as the average of the values from the iterations. Cross-validation can further reduce the random factor by creating different subsets of the data in each approach [12].

In the experiments described in this paper, the evaluation criteria were related to the following standard main metrics [2,3,8]:

- Root Mean Squared Error ($RMSE$) - RMSE is a baseline way to measure the error in model evaluation studies. It is a square root of an arithmetic mean of the squares of the predictions between the model and the observations. The lower value of RMSE refers to a better prediction ability.
- $ICoverage$ - Data coverage measures the relationship between the content of the recommendation lists and the learning set. It estimates the ability of the recommendation system to solve the long tail problem. The ICoverage value is typically used in the products dimension, defined as the percentage of recommended products compared to all system products. A low value indicates that many products are never offered to the user, which is known as the long tail problem.
- $UCoverage$ - The user dimension also defines algorithm coverage, which is the percentage of users who receive relevant recommendations. The recommendation system can generate recommendations for every user, but the algorithm's confidence in the presented products is taken into account. A decrease in UCoverage when new users are added to the system indicates a cold start problem. The analysis of coverage also enables the definition of the minimum requirements for a user that can be served by the algorithm.

4   Characteristics of Data

This section presents the characteristics determining how the recommender system evaluation results are impacted. The data metrics of the recommender system are based on a rating matrix known as the User Rating Matrix (URM). The URM is a matrix with columns and rows corresponding to the system's number of items and users. Each cell in the matrix contains the rating given by user u to the product v. If no rating is given, the cell remains empty. The shape of the rating matrix is a key feature, indicating the ratio of users to products in the system, as presented in Eq. 1.

$$Shape(URM) = \frac{|U|}{|V|} \tag{1}$$

where U is a set of users and V a set of items.

The literature reports cases where the similarity measure depends on the Shape value [4]: similarity between users is more useful for Shape values greater than 1 (the number of users exceeds the number of products), while otherwise, it is preferable to use similarity between products [3].

Another crucial aspect of a rating matrix is its data density, which indicates the proportion of grades given concerning the total size of the matrix (see Eq. 2).

$$Density(URM) = \frac{n_r}{|U| \cdot |V|} \tag{2}$$

where n_r is a number of all ratings in the matrix URM.

Recommender systems often suffer from low data density, which is caused by the large amount of processed data and the limited capability of users to rate all the products in the system. Furthermore, the constant addition of new products and users to the system has a negative impact on the density. Recommendation algorithms may have lower accuracy when working with low-density data [1,4].

The impact of popular products on the efficiency of the recommendation algorithm is a significant characteristic [18]. It is expected that products with a high number of ratings will be recommended to users more often, resulting in reduced system efficiency, lower product space coverage, and less diversity in recommendation lists [1,4]. Popularity is determined by sorting products in descending order of their ratings. When representing this arrangement as a histogram, the diagram shows a small number of large values at the beginning of the set, followed by a long tail of numerous products that are rarely rated. Several measures are available to quantify the frequency of popular products in the grade distribution. One such measure is the skewness coefficient, which determines the asymmetry of the data distribution (see Eq. 3).

$$Skewness(URM) = \frac{|V| \sum_{k=1}^{|V|} (x_k - \mu)^3}{(|V| - 1) \cdot (|V| - 2) \cdot s^3}, s = \sqrt{\frac{\sum_{k=1}^{|V|} (x_k - \mu)^2}{|V| - 1}} \tag{3}$$

where x_k is the popularity of k-the items, μ is an average overall popularity of all items, and s is a standard deviation.

Another way to detect an uneven ratings distribution is to calculate the Gini coefficient. It measures the concentration of ratings to capture the rating frequency distribution. A Gini index equal to 0 indicates equal popularity of all products, while 1 value indicates the presence of a few highly popular products (see Eq. 4).

$$Gini(URM) = 1 - 2 \cdot \sum_{k=1}^{|V|} \frac{|V| + 1 - k}{|V| + 1} \cdot \frac{|R_k|}{n_r} \tag{4}$$

where R_k is a number of ratings associated with an item k.

5   Experiments

The experiments were conducted using MovieLens, a well-known benchmark dataset that includes movie ratings. Subsets were generated from the original set of 25 million ratings, with assumed values for selected statistics such as Shape, Density, Skewness, and Gini. The purpose was to investigate the relationship between individual statistics and recommendation quality.

The first part of the experimental data focuses on the shape of the rating matrix. The generated subsets are presented in Table 1. The objective of the selection process was to obtain data with varying ratios of users to products. This was achieved by randomly selecting 2000 movie identifiers, with repetitions, and adding their ratings to the initial dataset. The Shape value of the resulting rating matrix was then checked, and additional identifiers were generated if necessary. The selection process was restarted if the number of unique movie or user identifiers exceeded 13,000. The limit was determined based on the computational capabilities of the equipment used. As a result, collections with fewer films and users were generated. User identifiers were randomly selected, and their ratings were added to the collection. Finally, 12 datasets were obtained, ranging in size from 6 to 1/6.

Table 1. Characteristics of datasets with different Shape values

Dataset	Number of users	Number of items	Number of ratings	Shape value
items1	12460	2231	326 760	5.58
items2	12496	4182	548 306	2.99
items3	12473	6670	897 697	1.87
items4	12474	8250	1 181 486	1.51
items5	12532	10317	1 444 267	1.21
items6	12525	11916	1 621 194	1.05
users1	1989	12439	821 995	0.16
users2	3948	12391	720 372	0.32
users3	5917	12223	608 924	0.48
users4	7813	12155	469 955	0.64
users5	9704	12095	594 568	0.33
users6	11588	12981	288 884	0.80

The following data set is related to the density of the evaluation matrix. The dataset *user*6, which had a high-density value, was used as the basis for generating the samples. Data filtering was performed to remove as many ratings as possible without altering the size of the rating matrix. Rows and columns of the evaluation matrix containing more than one evaluation were analysed. The values were removed in the following way that each film and user had at least one rating assigned. The initial number of ratings, 821,994, was reduced to 20,610 through a series of nine intermediate sets. Each set contained 80,000 fewer ratings than the previous one, resulting in a total reduction of approximately 800,000 ratings. Assessments not appearing in the final set were selected from the initial set to create the intermediate sets. This data generation method enables the creation of sets characterised by a regular change in density. The resulting subsets are presented in Table 2.

Table 2. Characteristics of datasets with different Dense values

Dataset	Number of users	Number of items	Number of ratings	Dense value
dense1	11588	12439	20 610	0.00014
dense2	11588	12439	101 994	0.00071
dense3	11588	12439	181 994	0.00126
dense4	11588	12439	261 994	0.00182
dense5	11588	12439	341 994	0.00237
dense6	11588	12439	421 994	0.00293
dense7	11588	12439	501 994	0.00348
dense8	11588	12439	581 994	0.00404
dense9	11588	12439	661 994	0.00459
dense10	11588	12439	741 994	0.00515
dense11	11588	12439	821 994	0.00570

The data sets were then generated taking into account the popularity of the products. Products were first sorted by their popularity. The removal of product ratings was then performed iteratively to remove as few as possible without changing the number of users. The resulting sets were then analysed in terms of the skewness coefficient, and sets whose changes in the coefficient were significant were selected. An analysis of the Gini coefficient of the data obtained was also carried out, which showed that the original set was less balanced than some of the intermediate sets. The characteristics of the data from this set and the skewness and Gini coefficients are given in Table 3.

Table 3. Characteristics of datasets with different popularity values: Skewness and Gini

Dataset	Number of users	Number of items	Number of ratings	Skewness	Gini
pop1	11588	12439	26 328	108.43	0.240
pop2	11588	12439	36 444	64.63	0.174
pop3	11588	12439	56 840	40.58	0.112
pop4	11588	12439	95 682	26.52	0.067
pop5	11588	12439	163 380	18.22	0.040
pop6	11588	12439	274 079	13.20	0.026
pop7	11588	12439	434 490	10.34	0.023
pop8	11588	12439	617 812	9.12	0.029
pop9	11588	12439	821 994	8.98	0.060

The prepared datasets were used to generate and evaluate recommendation lists. Cross-validation with a split of 25 sets was used. Recommendations were obtained using collaborative filtering algorithms (implementations from the Surprise library [10]). It included item-base and user-based method with similarity based on cosine value.

Then, the recommendation lists were evaluated according to the following indices: RMSE, ICoverage, UCoverage.

Table 4 shows the results of the evaluation of the recommendations for the Shape factor. In the table, a horizontal line separates the sets according to the number of products and the number of users. It is evident that there are no significant changes in all metrics based on the data matrix's shape, whether for item-based or user-based systems. However, an increase in accuracy is observed as the number of items and users increases. Furthermore, it has been noticed that UCoverage shows a positive correlation a number of items in both the items-based and user-based methods. Although there is also a correlation between the number of users and both coverage indices, it is not particularly strong. In both types of recommenders, ICoverage does not show any relationship with the number of items.

The evaluation results for the Density index and generated recommendations are presented in Table 5. As expected, there is a distinct relationship between data density and accuracy of recommendation lists. RMSE decreases as data density grows. However, the item-based approach produced more accurate outcomes, with an RMSE of 0.943 compared to 0.984. A similar relationship can be observed in the case of coverage indices, with their values being greater for dense data. The ICoverage was higher with the item-based approach, while the user-based method resulted in a higher UCoverage.

Table 6 presents the evaluation results for the recommendation lists and the popularity based indices: Skewness and Gini. Lower recommendation accuracy is associated with greater Skewness and Gini values. The propositions generated from the skewed data were evaluated with an RMSE of 1.134, whereas the data

Table 4. Evaluation results for the recommendation lists and the Shape index. The best values are in bold.

Shape value	Item-based recommender			User-based recommender		
	RMSE	ICoverage	UCoverage	RMSE	ICoverage	UCoverage
5.58	0.972	**0.400**	0.317	0.994	0.288	0.373
2.99	0.947	0.367	0.426	0.980	0.271	0.489
1.87	**0.930**	**0.386**	0.545	0.975	**0.290**	0.608
1.51	**0.933**	0.378	0.622	**0.967**	**0.291**	0.698
1.21	**0.926**	0.371	0.653	**0.967**	**0.295**	0.735
1.05	**0.931**	0.361	**0.689**	0.972	0.280	**0.767**
0.93	**0.942**	0.223	0.538	0.986	0.174	0.612
0.78	**0.947**	0.207	0.568	0.992	0.157	0.637
0.64	0.953	0.218	0.546	0.978	0.169	0.621
0.49	0.952	0.196	0.540	0.975	0.150	0.626
0.33	0.960	0.291	**0.689**	0.986	0.231	**0.799**
0.15	0.963	0.186	0.662	0.970	0.151	**0.781**

Table 5. Evaluation results for the recommendation lists and the Density index. The best values are in bold.

Density value	Item-based recommender			User-based recommender		
	RMSE	ICoverage	UCoverage	RMSE	ICoverage	UCoverage
0.00014	1.159	0.000	0.000	1.160	0.000	0.000
0.00071	1.003	0.074	0.054	1.019	0.060	0.061
0.00126	0.982	0.108	0.104	0.991	0.088	0.121
0.00182	0.969	0.132	0.157	0.991	0.105	0.181
0.00237	0.959	0.151	0.211	0.985	0.120	0.245
0.00293	0.958	0.167	0.269	0.988	0.131	0.307
0.00348	0.950	0.179	0.323	0.985	0.139	0.370
0.00404	0.950	0.191	0.376	0.987	0.149	0.428
0.00459	0.947	0.203	0.430	**0.984**	0.157	0.491
0.00515	**0.943**	**0.214**	**0.481**	**0.984**	**0.166**	**0.547**
0.00570	**0.943**	**0.222**	**0.537**	0.986	**0.174**	**0.610**

with the lowest Skewness values resulted in an RMSE of 0.914. This suggests that popular items in the input data have a negative impact on the performance of recommendation systems, resulting in less accurate propositions. The coverage indices have a negative correlation with the skewness-based index values. It has been observed that UCoverage significantly increases as the Gini values decrease. The outermost values are 0 and 0.610 for the item-based recommender and 0 and 0.584 for the user-based method. ICoverage also increases, but its values remain lower compared to the former index. The extreme values for the item-based recommender are 0 and 0.176, and for the user-based method are 0 and 0.162.

Table 6. Evaluation results for the recommendation lists and the Skewness and Gini indices. The best values are in bold.

Gini value	Skewness value	Item-based recommender			User-based recommender		
		RMSE	ICoverage	UCoverage	RMSE	ICoverage	UCoverage
0.240	108.43	1.134	0.006	0.023	1.143	0.000	0.000
0.174	64.63	1.095	0.008	0.055	1.116	0.004	0.028
0.112	40.58	1.040	0.009	0.110	1.060	0.006	0.082
0.067	26.52	0.969	0.010	0.183	0.999	0.008	0.154
0.040	18.22	0.933	0.013	0.280	0.968	0.011	0.249
0.026	13.20	0.915	0.018	0.390	0.956	0.016	0.355
0.023	10.34	**0.907**	0.030	0.490	**0.951**	0.027	0.461
0.029	9.12	**0.902**	**0.060**	**0.561**	**0.951**	**0.052**	**0.534**
0.060	8.98	0.914	**0.176**	**0.610**	0.963	**0.162**	**0.584**

The characteristics of the examined data often affect the performance of recommender systems. It has been observed that propositions generated from denser data tend to be more accurate. However, there may also be less obvious relationships. When data is skewed towards popular items, the performance of algorithms is negatively affected, resulting in less accurate recommendations. This may seem counterintuitive since most consumers prefer popular products. However, the conclusions are the opposite. The long tail items are often more beneficial than the more popular ones. In addition, it is important to note that not all data characteristics are correlated with recommendation accuracy. For instance, the Shape value did not demonstrate any significant relationship with the indices.

6   Conclusions

The most important property of recommender systems is accuracy. To increase this factor, appropriate examination and preparation of data is necessary. While some data characteristics have an impact on the performance of a recommender system, others demonstrate no relationship. This paper presented experiments investigating 3 data characteristics: Shape, Skewness, and Density. The first of these showed no effect on the accuracy of the recommendation lists, whereas the following two statistics were strongly correlated with the efficiency of the algorithms.

The literature review shows that the correlated statistics depend on the data set. Examining data characteristics in recommendation tasks can improve the final algorithm performance. Identifying correlated features with the accuracy of recommendation lists enables the preparation and deployment of appropriate data improvement procedures.

Acknowledgments. The work was supported by a grant from the Bialystok University of Technology WZ/WI-IIT/3/2023 and funded with resources for research by the Ministry of Science and Higher Education in Poland.

References

1. Adomavicius, G., Zhang, J.: Impact of data characteristics on recommender systems performance. ACM Trans. Manag. Inf. Syst. **3**(1), 1–17 (2012)
2. Aggrawal, C.C.: Recommender Systems. The Textbook. Springer, Heidelberg (2016). https://doi.org/10.1007/978-3-319-29659-3
3. Alharbey, R., Ayub, M., Ghazanfar, M.A., Mehmood, Z., Saba, T.: Modeling user rating preference behavior to improve the performance of the collaborative filtering based recommender systems. PLOS ONE **14**(8) (2019)
4. Bellogin, A., Deldjoo, Y., Di Noia, T.: Explaining recommender systems fairness and accuracy through the lens of data characteristics. Inf. Process. Manage. **58**(5) (2021)
5. Bobadilla, J., Ortega, F., Hernando, A., Gutiérrez, A.: Recommender systems survey. Knowl.-Based Syst. **46**, 109–132 (2013)
6. Deldjoo, Y., Di Noia, T., Di Sciascio, E., Merra, F.A.: How dataset characteristics affect the robustness of collaborative recommendation models. In Proceedings of the 43rd International ACM SIGIR Conference on Research and Development in Information Retrieval (SIGIR 2020), pp. 951–960 (2020)
7. Gorgoglione, M., Pannielloa, U., Tuzhilin, A.: Recommendation strategies in personalization applications. Inf. Manage. **56**(6), 103143 (2019)
8. Gunawardana, A., Shani, G.: Evaluating recommendation systems. In: Ricci, F., Rokach, L., Shapira, B., Kantor, P. (eds.) Recommender Systems Handbook, pp. 257–297. Springer, Boston (2011). https://doi.org/10.1007/978-0-387-85820-3_8
9. Hsu, C.N., Chung, H.H., Huang, H.S.: Mining skewed and sparse transaction data for personalized shopping recommendation. Mach. Learn. **57**(1), 35–59 (2004)
10. Hug, N.: Surprise library. https://surprise.readthedocs.io/en/stable/. Accessed 12 Dec 2023
11. Jannach, D.: Recommender Systems: An Introduction. Cambridge University Press, Cambridge (2010)
12. Jarvelin, K., Kekalainen, J.: Cumulated gain-based evaluation of IR techniques. ACM Trans. Inf. Syst. (TOIS) **20**(4), 422–446 (2002)

13. Kaminskas, M., Bridge, D.: Diversity, serendipity, novelty, and coverage: a survey and empirical analysis of beyond-accuracy objectives in recommender systems. ACM Trans. Interact. Intell. Syst. (TiiS) **7**(1), 1–42 (2016)
14. Kużelewska, U.: Clustering algorithms for efficient neighbourhood identification in session-based recommender systems. In: Zamojski, W., Mazurkiewicz, J., Sugier, J., Walkowiak, T., Kacprzyk, J. (eds.) DepCoS-RELCOMEX 2022. LNNS, vol. 484, pp. 143–152. Springer, Cham (2022). https://doi.org/10.1007/978-3-031-06746-4_14
15. Ricci, F., Rokach, L., Shapira, B.: Recommender systems: introduction and challenges. In: Ricci, F., Rokach, L., Shapira, B. (eds.) Recommender Systems Handbook, pp. 1–34. Springer, Boston (2015). https://doi.org/10.1007/978-1-4899-7637-6_1
16. Schafer, J.B., Frankowski, D., Herlocker, J., Sen, S.: Collaborative filtering recommender systems, pp. 291–324. The Adaptive Web (2007)
17. Shaikh, S., Kagita, V.R., Kumar, V., Pujari, A.K.: Data augmentation and refinement for recommender system: a semi-supervised approach using maximum margin matrix factorization. Expert Syst. Appl. **238**(B) (2024)
18. Smyth, B., McClave, P.: Similarity vs. diversity. In: Aha, D.W., Watson, I. (eds.) ICCBR 2001. LNCS (LNAI), vol. 2080, pp. 347–361. Springer, Heidelberg (2001). https://doi.org/10.1007/3-540-44593-5_25
19. Yang, W., Fan, S., Wang, H.: An item-diversity-based collaborative filtering algorithm to improve the accuracy of recommender system. In: IEEE SmartWorld, Ubiquitous Intelligence & Computing, Advanced & Trusted Computing, Scalable Computing & Communications, Cloud & Big Data Computing, Internet of People and Smart City Innovation (SmartWorld/SCALCOM/UIC/ATC/CBDCom/IOP/SCI), pp. 106–110 (2018)

Randomly Initiated Cyclostationary Excitations for Dimensionality Reduction in Wiener System Identification

Gabriel Maik$^{(\boxtimes)}$ and Grzegorz Mzyk

Faculty of Information and Communication Technology, Wroclaw University of
Science and Technology, Wroclaw, Poland
{gabriel.maik,grzegorz.mzyk}@pwr.edu.pl

Abstract. The problem of nonparametric estimation of nonlinear characteristics in the Wiener system is considered. In this task, the traditional kernel algorithm suffers from dimensionality resulting from the memory length of the dynamic block. A special class of input sequences has been proposed that allows us to reduce the dimension and, consequently, improve the rate of convergence of the estimator to the true characteristics. Theoretical analysis of the proposed method is presented.

Keywords: System identification · Nonparametric methods · Wiener system · Dimension reduction · Cyclostationary processes · Regression estimation

1 Introduction

The article focuses on the common problem of input dimensionality in data modeling. In both system identification [5] and pattern recognition [10], the aim is to reduce the order of the model or the number of features considered. In particular, with a nonparametric approach to the modeling of nonlinear phenomena, the probability of finding input sequences in the vicinity of the reference point of interest decreases significantly as the input dimensionality increases. In the case of the kernel method, this results in a small number of selected cases and, consequently, a large variance of estimates. In turn, for approaches based on the nearest neighbors technique, a higher number of features results in a larger bias of the estimators. This phenomenon is often called the curse of dimensionality.

Much effort has been made over the last four decades to deal with this issue. Relevant examples in this topic would be the Akaike information criterion [1] method or L1 lasso regularization [13]. The use of small-size parametric models allows for the reduction of their variance with a limited number of observations, but asymptotically this is associated with a systematic non-zero approximation error (bias). Getting rid of the asymptotic approximation error is possible by increasing the complexity of the model as the number of observations increases to infinity. We call this approach nonparametric [3,4]. Examples of methods

© The Author(s), under exclusive license to Springer Nature Switzerland AG 2024
W. Zamojski et al. (Eds.): DepCoS-RELCOMEX 2024, LNNS 1026, pp. 143–151, 2024.
https://doi.org/10.1007/978-3-031-61857-4_14

include kernel algorithms or orthogonal series expansion algorithms. Unfortunately, due to the curse of dimensionality, the rate of increasing the model order (or the rate of kernel narrowing) to guarantee asymptotic consistency is usually very slow. Many different ideas to deal with this problem can be found in the literature. They are mainly based on the generation of special input sequences, e.g. multisinusoids [11], piecewise constants, periodic [8], or cyclostationary processes [9]. In general, the input is postulated to lie on on a manifold with a low internal dimension [7,12,14].

In this article, we consider a problem of nonparametric estimation of nonlinear characteristics in the Wiener system. This is one of the most popular and commonly used in practice so-called block-oriented structures [2]. It contains a linear dynamic object cascaded with a nonlinear static block. The dimensionality of the considered problem depends strictly on the length of the impulse response (memory) of the linear dynamic subsystem.

The original contribution made in this paper is based on the idea of using an additional autoregressive filter to generate a specific input signal. Identification of the Wiener system is carried out when the filter operates in a free state (when the generated input results only from the initial state of the filter). In such a situation, we show that it is possible to construct a kernel estimator of nonlinear characteristics, operating in a space of lower dimensions, corresponding to the order of autoregression of the filter used. A similar concept is presented in [6], but in this work, the class of excitations has been generalized to autoregressive processes of any order.

The paper is organized as follows. Section 2 formulates the problem in detail. Initial assumptions about the identified system and its input excitation are presented. Then, Sect. 3 describes the identification algorithm and formulates a theorem specifying the degree of reduction in the dimension of the considered problem. A brief discussion and concluding remarks can be found in the Sect. 4. The formal proof is located at the end in the Appendix in the Sect. A.

2   Problem Statement

The problem considered in the paper is the nonparametric identification of the nonlinear characteristics in the Wiener system using only input-output data $\{(u_k, y_k)\}_{k=1}^{N}$, where N is the number of the observations.

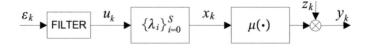

Fig. 1. Wiener system preceded by a filter generating excitations.

Assumption 1 (System). *The linear block of the Wiener system is an FIR filter with coefficients* $\lambda = [\lambda_0, \lambda_1, ..., \lambda_S]^T$. *The memory length, equal to S, is a priori known. The nonlinear block, however, is characterized by L-Lipschitz continuous function* $\mu(\cdot)$. *Summarising, hidden interconnection signal* $\{x_k\}$ *and noise-free output* $\{v_k\}$ *are as follows*

$$x_k = \sum_{j=0}^{S} \lambda_j u_{k-j}, \qquad v_k = \mu(x_k). \qquad (1)$$

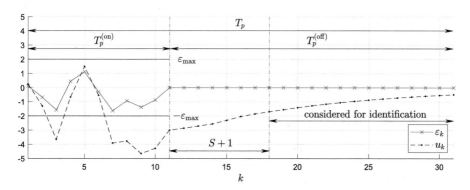

Fig. 2. Explanatory example to the Assumption 2, and introduced there symbols: T_p, $T_p^{(\mathrm{on})}$, $T_p^{(\mathrm{off})}$, $\varepsilon_{\max}$.

Assumption 2 (Input signal). *The input signal* $\{u_k\}$ *is generated with the help of unavailable for measurements sequence* $\{\varepsilon_k\}$ *and a stable[1] r'th order autoregressive input filter (see Fig. 1). A certain input filter order is lower than the order of the linear block in the Wiener system, i.e., $r < S + 1$.*

The input process consists of $\mathcal{P}$ intervals, each T_p-samples long, where $p = 0, 1, ..., \mathcal{P} - 1$. Every period starts with $T_p^{(on)}$ random ε_k samples, while for the rest $T_p^{(off)}$ samples within interval $\varepsilon_k = 0$. Random ε_k from all intervals form a bounded i.i.d. random variables sequence, such that $\forall_k P(|\varepsilon_k| \leq \varepsilon_{max} < \infty) = 1$. Additionally, $\forall_p (r \leq T_p^{(on)} \leq T_{max}^{(on)} < \infty \wedge S + 1 \leq T_p^{(off)} \leq T_{max}^{(off)} < \infty)$, where $T_p^{(on)}$ and $T_p^{(off)}$ are i.i.d. random variables sequences. As a subcase $T_p^{(on)}$ and $T_p^{(off)}$ can be const., which makes whole sequence $\{u_k\}$ cyclostationary in a strict-sense. Values T_p, $T_p^{(on)}$, $T_p^{(off)}$ are a priori known for $\forall p$.

Number of data, N, is the sum of all intervals lengths, i.e., $N = \sum_{p=0}^{\mathcal{P}-1} T_p = \sum_{p=0}^{\mathcal{P}-1} (T_p^{(on)} + T_p^{(off)})$. The initial state of the input filter can be any possible at the beginning of the interval $p \to \infty$.

---

[1] Proposed algorithm also works in the case of unstable input filter with the state being zeroed at the beginning of each interval.

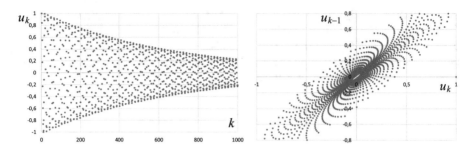

Fig. 3. Example input excitation $\{u_k\}$ for second order filter ($r = 2$) and the state vector trajectory $\{(u_k, u_{k-1})\}$.

To better understand Assumption 2, the symbols introduced therein are depicted in Fig. 2. Furthermore, an illustrative example with the state vector trajectory based on Assumption 2. is presented in Fig. 3. The dynamic input filter represents the environment preceding the Wiener system being examined, and the identification will be carried out when the filter operates in a free state.

Assumption 3 (Noise). *The additive noise $\{z_k\}$, which affects output measurements, $y_k = v_k + z_k$, is an i.i.d. random variables sequence, independent from input $\{u_k\}$. It has expected value equal to zero, $E\{z_k\} = 0$, and finite variance, $var\{z_k\} < \infty$.*

3   The Algorithm

Let us observe that the last r samples of the signal $\{u_k\}$ define the state of the input filter, and, in the case of proper initialization[2], also the state of the whole system. Thus, we introduce the state vector $\phi_k = [u_k, u_{k-1}, ..., u_{k-r+1}]^T$, which evolves in time obeying the following law

$$\phi_k = \underbrace{\begin{bmatrix} a^T \\ I_{r-1} \quad \mathbf{0} \end{bmatrix}}_{=A} \phi_{k-1} + \underbrace{\begin{bmatrix} a_0 \\ \mathbf{0} \end{bmatrix}}_{=B} \varepsilon_k, \tag{2}$$

where $a = [a_1, a_2, ..., a_r]^T$ is the vector of the input filter's coefficients, I_{r-1} is the $(r-1) \times (r-1)$ identity matrix, $\mathbf{0}$ is the $(r-1) \times 1$ vector of zeros, and a_0 is the instant gain of the excitation $\{\varepsilon_k\}$. For the clarity of the presentation, the initial state is defined as

$$\phi_0 = \alpha u, \tag{3}$$

where α is a unit direction $r \times 1$ vector in a state space, such that $\|\alpha\|_2 = 1$, and u plays a role of scale factor. When $\{\varepsilon_k\}$ is a sequence of zeros, it holds that

$$\phi_k = A^k \phi_0, \tag{4}$$

---

[2] Initialization such that $\varepsilon_k = 0$ for at least last $S + 1$ consecutive samples, i.e., for $k_1 : k - S \le k_1 \le k$.

$$u_k = [1, \mathbf{0}^T]\phi_k = \underbrace{[1, \mathbf{0}^T]A^k \alpha}_{=\beta_k(\alpha)} u, \tag{5}$$

and consequently

$$E\{y_S|\phi_0 = \alpha u\} = \mu\left(\sum_{i=0}^{S} \lambda_i u_{S-i} \bigg| \phi_0 = \alpha u\right) = \mu\left(\underbrace{\sum_{i=0}^{S} \lambda_i \beta_{S-i}(\alpha)}_{=c_S^{(\alpha)}} u\right) = \mu\left(c_S^{(\alpha)} u\right). \tag{6}$$

Thus, we propose the following estimator of the static nonlinearity

$$\widehat{\mu}_N^{(\alpha)}(x) = \frac{\sum_{k \in K_N} y_k \mathcal{K}\left(\frac{1}{h}\|\phi_k - \alpha x\|_\infty\right)}{\sum_{k \in K_N} \mathcal{K}\left(\frac{1}{h}\|\phi_k - \alpha x\|_\infty\right)}, \tag{7}$$

where set $K_N = \{k : [\varepsilon_k, \varepsilon_{k-1}, ..., \varepsilon_{k-S}]^T = \mathbf{0} \wedge 1 \le k \le N\}$, and h is bandwidth parameter dependent on N, i.e., $h = h(N)$. Kernel function $\mathcal{K}(\cdot)$ can be any function such that $\mathcal{K}(x) = 0$ for $|x| > 1$, and $0 < \mathcal{K}_{\min} \le \mathcal{K}(x) \le \mathcal{K}_{\max} < \infty$ when $|x| \le 1$. Function $\widehat{\mu}_N^{(\alpha)}(x)$ converges uniformly to its scaled counterpart $\mu^{(\alpha)}(x) = \mu(c^{(\alpha)}x)$. The domain of identifiability is defined as follows

$$\mathbb{D}^{(\alpha)} = \left\{x : \forall_{h>0, k \in K_{N,p(N)}, \phi^{(\text{old})} \in \Phi} \lim_{N \to \infty} P\left(\|\phi_k^{(\text{new})} + \phi_k^{(\text{old})} - \alpha x\|_\infty \le h\right) > 0\right\}, \tag{8}$$

where state $\phi_k = \phi_k^{(\text{old})} + \phi_k^{(\text{new})}$ depends on older input samples from previous intervals ($\phi_k^{(\text{old})}$) and newer ones from the current interval ($\phi_k^{(\text{new})}$), such that

$$\phi_k = \underbrace{A^k \phi_0 + \sum_{j=k_0(k)}^{k-1} A^j B \varepsilon_{k-j}}_{\overset{\text{def}}{=} A^{k_0(k)-1} \phi^{(\text{old})} \overset{\text{def}}{=} \phi_k^{(\text{old})}} + \underbrace{\sum_{j=0}^{k_0(k)-1} A^j B \varepsilon_{k-j}}_{\overset{\text{def}}{=} \phi_k^{(\text{new})}} \tag{9}$$

with $k_0(k)$ returning number of sample within period (from 1 to T_p), and $p(k)$ returning a period number, in which k'th sample is located. Furthermore, set $K_{N,p(N)}$ is such a subset of K_N, that all samples are within p'th interval, i.e.,

$$K_{N,p(N)} = \{k : p(k) = p(N) \wedge k \in K_N\}, \tag{10}$$

and Φ denotes a set of all possible $\phi_k^{(\text{old})}$ at the beginning of the p'th interval ($k_0(k) = 1$), asymptotically for $p \to \infty$, formally defined as follows

$$\Phi = \{\phi^{(\text{old})} : \forall_{h>0, k \in K_{N,p(N)}, k_0(k)=1} \lim_{N \to \infty} P(\|\phi_k^{(\text{old})} - \phi^{(\text{old})}\|_\infty \le h) > 0\}. \tag{11}$$

Remark 1. Nonlinearity can be also identified at some points outside the specified domain, but in general with a slower speed of convergence and more restrictive requirements. Additionally, in practice usually $\phi_k^{(\text{old})} \approx \mathbf{0}$, because it depends on non-zero old ε_k samples with indexes smaller by at least $r + 2S + 2$ than the current one, which makes the difference between the aforementioned domains of identifiability insignificant. Due to the aforementioned reasons, such a case is treated as out of the scope of the work.

Formally, we provide the following theorem with specified requirements for convergence.

Theorem 1. *If $h \to 0$, and $Nh^r \to \infty$, then $\widehat{\mu}_N^{(\alpha)}(x)$ converges to $\mu^{(\alpha)}(x)$ in the mean squared sense, i.e.,*

$$\mathrm{MSE}\left\{\widehat{\mu}_N^{(\alpha)}(x)\right\} = \mathrm{E}\left\{\left(\widehat{\mu}_N^{(\alpha)}(x) - \mu^{(\alpha)}(x)\right)^2\right\} \to 0 \tag{12}$$

as $N \to \infty$, in each point $x \in \mathbb{D}^{(\alpha)}$, for the given α.

Proof. See Appendix A.1.   □

4   Summary

The proposed approach seems to be very promising from a practical point of view. In reality, the system is usually stimulated by the output of another dynamic object. As soon as the object preceding the system with an autoregressive input filter is disabled, there is an opportunity to run the identification procedure with an accelerated rate of convergence.

Among possible generalizations of the proposed approach is to consider a system structure with broader approximation capabilities of dynamic-nonlinear phenomena, such as the Wiener-Hammerstein system. Furthermore, an interesting continuation of research will be the aggregation of estimators determined for different directions in a state space. Such a task is problematic because the aforementioned estimators model the true characteristics in different parts of its domain, which, simultaneously and in general, are not identically scaled.

A   Appendices

A.1   Proof of Theorem 1

Proof. First, for convenience, a set of selected samples' indexes for given α, h, and x is defined as $K_N^{(\alpha,h,x)} = \{k : \mathcal{K}(\frac{1}{h}\|\phi_k - \alpha x\|_\infty) > 0 \wedge k \in K_N\}$. In the considered domain of identifiability, $\mathbb{D}^{(\alpha)}$, the expected value of the selected samples' set size is, in the asymptotical sense for $N \to \infty$ and $h \to 0$, as follows

$$\mathrm{E}\left\{\#K_N^{(\alpha,h,x)}\right\} = \mathcal{P}\mathrm{E}\left\{\#\left(K_N^{(\alpha,h,x)} \cap K_{N,p}\right)\right\}$$

$$= \mathcal{P}\mathrm{E}\left\{\sum_{\tau=1}^{T_{\max}^{(\mathrm{off})} - S} \#\left\{k : k \in \left(K_N^{(\alpha,h,x)} \cap K_{N,p}\right) \wedge k = \underbrace{\sum_{q=0}^{p-1} T_q + T_p^{(\mathrm{on})} + S + \tau}_{\overset{\text{def}}{=}\kappa(p,\tau)}\right\}\right\}$$

$$= \underbrace{\mathcal{P}}_{=c_1 N} \sum_{\tau=1}^{T_{\max}^{(\text{off})}-S} \mathrm{P}\Big(T_p^{(\text{off})} - S \geq \tau \;\wedge\; \mathcal{K}\big(\tfrac{1}{h}\|\phi_{\kappa(p,\tau)} - \alpha x\|_\infty\big) > 0\Big)$$

$$= c_1 N \sum_{\tau=1}^{T_{\max}^{(\text{off})}-S} \underbrace{\mathrm{P}\Big(T_p^{(\text{off})} - S \geq \tau\Big)}_{=c_{2,1,\tau}} \underbrace{\mathrm{P}\Big(\mathcal{K}\big(\tfrac{1}{h}\|\phi_{\kappa(p,\tau)} - \alpha x\|_\infty\big) > 0 \Big| T_p^{(\text{off})} - S \geq \tau\Big)}_{=c_{2,2,\tau}h^r}$$

$$= c_1 N \sum_{\tau=1}^{T_{\max}^{(\text{off})}-S} c_{2,\tau} h^r = c_1 c_2 N h^r = c_3 N h^r, \tag{13}$$

where c_1, c_2, and c_3 are unknown constants. Thus, $\mathrm{E}\{\#K_N^{(\alpha,h,x)}\} \to \infty$, when $Nh^r \to \infty$. Consequently, asymptotically, it can be assumed that the realization of $K_N^{(\alpha,h,x)}$, defined as $\widetilde{K}_N^{(\alpha,h,x)}$, results in a non-empty set, and a positive number of selected observations, $\#\widetilde{K}_N^{(\alpha,h,x)} \geq 1$, which enables rewriting error as follows

$$\widehat{\mu}_N^{(\alpha)}(x) - \mu^{(\alpha)}(x) = \frac{\sum_{k\in K_N} \big(\overbrace{\mu(x_k) + z_k}^{=y_k} - \mu^{(\alpha)}(x)\big)\mathcal{K}\big(\tfrac{1}{h}\|\phi_k - \alpha x\|_\infty\big)}{\sum_{k\in K_N} \mathcal{K}\big(\tfrac{1}{h}\|\phi_k - \alpha x\|_\infty\big)} =$$

$$\underbrace{\frac{\sum_{k\in K_N} z_k \mathcal{K}\big(\tfrac{1}{h}\|\phi_k - \alpha x\|_\infty\big)}{\sum_{k\in K_N} \mathcal{K}\big(\tfrac{1}{h}\|\phi_k - \alpha x\|_\infty\big)}}_{\overset{\text{def}}{=} e_N^{(1)}} + \underbrace{\frac{\sum_{k\in K_N} \big(\mu(x_k) - \mu^{(\alpha)}(x)\big)\mathcal{K}\big(\tfrac{1}{h}\|\phi_k - \alpha x\|_\infty\big)}{\sum_{k\in K_N} \mathcal{K}\big(\tfrac{1}{h}\|\phi_k - \alpha x\|_\infty\big)}}_{\overset{\text{def}}{=} e_N^{(2)}}$$

$$\tag{14}$$

and, due to the independence of the noise from the input, it holds that

$$\mathrm{MSE}\Big\{\widehat{\mu}_N^{(\alpha)}(x)\Big\} = \mathrm{E}\Big\{e_N^{(1)} + e_N^{(2)}\Big\}^2 = \mathrm{E}\Big\{e_N^{(1)^2}\Big\} + 2\underbrace{\mathrm{E}\Big\{e_N^{(1)}\Big\}\mathrm{E}\Big\{e_N^{(2)}\Big\}}_{=0} + \mathrm{E}\Big\{e_N^{(2)^2}\Big\}.$$

$$\tag{15}$$

The component $\mathrm{E}\Big\{e_N^{(1)^2}\Big\}$ can be rewritten, in the asymptotical sense, as follows

$$\mathrm{E}\Big\{e_N^{(1)^2}\Big\} \leq \mathrm{E}\left\{\frac{\sum_{k\in K_N^{(\alpha,h,x)}} z_k^2 \mathcal{K}_{\max}^2}{\big(\sum_{k\in K_N^{(\alpha,h,x)}} \mathcal{K}_{\min}\big)^2}\right\} = \mathrm{var}\{z_k\}\frac{\mathcal{K}_{\max}}{\mathcal{K}_{\min}}\mathrm{E}\left\{\frac{1}{\#K_N^{(\alpha,h,x)}}\right\}.$$

$$\tag{16}$$

A new variable is defined as a set size, $\mathbb{K}_p \overset{\text{def}}{=} \#\{K_N^{(\alpha,h,x)} \cap K_{N,p}\}$, such that

$$\#K_N^{(\alpha,h,x)} = \sum_{p=0}^{\mathcal{P}-1} \mathbb{K}_p. \tag{17}$$

Hence $\frac{1}{\mathcal{P}h^r}\#K_N^{(\alpha,h,x)} = \frac{1}{\mathcal{P}}\sum_{p=0}^{\mathcal{P}-1}\frac{1}{h^r}\mathbb{K}_p$, where, based on the Eq. (13), asymptotically for $h \to 0$, expected value is as follows

$$\mathrm{E}\Big\{\tfrac{1}{h^r}\mathbb{K}_p\Big\} = \tfrac{1}{h^r}\underbrace{\mathrm{E}\Big\{\mathbb{K}_p\Big\}}_{=c_4 h^r} = c_4, \tag{18}$$

and variance is bounded as shown below

$$\operatorname{var}\{\tfrac{1}{h^r}\mathbb{K}_p\} \le \big(\max\{\tfrac{1}{h^r}\mathbb{K}_p\} - \mathrm{E}\{\mathbb{K}_p\}\big)^2 \mathrm{P}(\mathbb{K}_p \ge 1)$$
$$+ (0 - \mathrm{E}\{\mathbb{K}_p\})^2 \mathrm{P}(\mathbb{K}_p = 0)$$
$$\le \left(\frac{T_{\max}^{(\mathrm{off})} - S}{h^r} - c_4\right)^2 c_5 h^r + c_4^2(1 - c_5 h^r) \le \frac{c_6}{h^r}. \tag{19}$$

Thus, as $\{\mathbb{K}_p\}_{p=0}^{\mathcal{P}}$ is an i.i.d. sequence, and as

$$\sum_{p=0}^{\mathcal{P}-1} \frac{\operatorname{var}\{\mathbb{K}_p\}}{\mathcal{P}^2} \le \frac{\mathcal{P} c_6}{\mathcal{P}^2 h^r} = \frac{c_6}{\mathcal{P} h^r}, \tag{20}$$

by virtue of the strong law of large numbers, it holds that

$$\frac{1}{\mathcal{P} h^r} \# K_N^{(\alpha,h,x)} = \frac{1}{\mathcal{P}} \sum_{p=0}^{\mathcal{P}-1} \frac{1}{h^r} \mathbb{K}_p \to c_4 \tag{21}$$

with probability 1 as $\mathcal{P} h^r \to \infty$, which is obviously met as $N h^r \to \infty$ because asymptotically $\mathcal{P} = c_1 N$. Furthermore, as $\frac{1}{N h^r} \# K_N^{(\alpha,h,x)}$ converges to const. with probability 1, under the Slutsky's theorem, it holds that

$$c_7 \mathrm{E}\left\{\frac{1}{\# K_N^{(\alpha,h,x)}}\right\} = c_7 \mathrm{E}\left\{\frac{\frac{1}{N h^r}}{\frac{1}{N h^r} \# K_N^{(\alpha,h,x)}}\right\} = \frac{c_8}{N h^r}, \tag{22}$$

which results in the following convergence

$$\lim_{N\to\infty} \mathrm{E}\left\{e_N^{(1)^2}\right\} \le \lim_{N\to\infty} \frac{c_8}{N h^r} = 0 \iff N h^r \to \infty. \tag{23}$$

To analyze convergence of $\mathrm{E}\left\{e_N^{(2)^2}\right\}$, the Lipschitz condition can be used

$$\big|\mu(c_k^{(\alpha,h,x)} x_k) - \mu(c^{(\alpha)} x)\big| \le L\big|c_k^{(\alpha,h,x)} x_k - c^{(\alpha)} x\big|, \tag{24}$$

where $c_k^{(\alpha,h,x)}$ is an effective scale factor. When $k \in K_N^{(\alpha,h,x)} \wedge \# \widetilde{K}_N^{(\alpha,h,x)} \ge 1$, the absolute value can be further rewritten as below

$$\big|c_k^{(\alpha,h,x)} x_k - c^{(\alpha)} x\big| = \big|(c_k^{(\alpha,h,x)} - c^{(\alpha)}) x_k\big| + \big|c^{(\alpha)}(x_k - x)\big| \le \big|\Delta c_k^{(\alpha,h,x)} x_k\big| + c_9 h,$$

and $\big|\mu(x_k) - \mu^{(\alpha)}(x)\big| \le L\big(\big|\Delta c_k^{(\alpha,h,x)} x_k\big| + c_9 h\big) \stackrel{\text{def}}{=} b_k^{(\alpha,h,x)}$. Thus,

$$\mathrm{E}\left\{e_N^{(2)^2}\right\} \le \mathrm{E}\left\{\left|\frac{\sum_{k\in K_N} b_k^{(\alpha,h,x)} \mathcal{K}(\tfrac{1}{h}\|\phi_k - \alpha x\|_\infty)}{\sum_{k\in K_N} \mathcal{K}(\tfrac{1}{h}\|\phi_k - \alpha x\|_\infty)}\right|^2\right\} \le b_k^{(\alpha,h,x)^2}. \tag{25}$$

Difference $\Delta c_k^{(\alpha,h,x)}$, based on Eqs. (3–6), can be rewritten as follows

$$\Delta c_k^{(\alpha,h,x)} = c_k^{(\alpha,h,x)} - c^{(\alpha)} = C^T \alpha_{\text{eff}} + C^T \alpha = C^T(\alpha_{\text{eff}} - \alpha), \tag{26}$$

where C^T is a const. $1 \times r$ vector, and α_{eff} is an effective unit vector of direction of state in r-dimensional space similarly as in (3). When $x \neq 0$ and $h \to 0$, it holds that $\alpha_{\text{eff}} \to \alpha$, and consequently $\Delta c_k^{(\alpha,h,x)} \to 0$. Moreover, when $x = 0$ and $h \to 0$, selected values tend to zero $x_k \to 0$. Owing to the above, $b_k^{(\alpha,h,x)} \to 0$ as $h \to 0$, and the following convergence takes place

$$\lim_{N \to \infty} \mathrm{E}\left\{e_N^{(2)^2}\right\} \leq \lim_{N \to \infty} b_k^{(\alpha,h,x)^2} = 0 \iff h \to 0 \wedge \#\widetilde{K}_N^{(\alpha,h,x)} \geq 1, \qquad (27)$$

which is met when $Nh^r \to \infty$. Finally,

$$\lim_{N \to \infty} \mathrm{MSE}\left\{\widehat{\mu}_N^{(\alpha)}(x)\right\} = \lim_{N \to \infty} \mathrm{E}\left\{e_N^{(1)^2}\right\} + \lim_{N \to \infty} \mathrm{E}\left\{e_N^{(2)^2}\right\} = 0, \qquad (28)$$

as $h \to 0$, and $Nh^r \to \infty$, what ends the proof. $\qquad\square$

References

1. Akaike, H.: A new look at the statistical model identification. IEEE Trans. Autom. Control **19**(6), 716–723 (1974)
2. Giri, F., Bai, E.-W.: Block-Oriented Nonlinear System Identification, vol. 1. Springer, London (2010). https://doi.org/10.1007/978-1-84996-513-2
3. Greblicki, W., Pawlak, M.: Nonparametric System Identification, vol. 10. Cambridge University Press, Cambridge (2008)
4. Györfi, L., Kohler, M., Krzyżak, A., Walk, H., et al.: A Distribution-free Theory of Nonparametric Regression, vol. 1. Springer, New York (2002). https://doi.org/10.1007/b97848
5. Juditsky, A., et al.: Nonlinear black-box models in system identification: mathematical foundations. Automatica **31**(12), 1725–1750 (1995)
6. Maik, G., Mzyk, G., Wachel, P.: Exponential excitations for effective identification of Wiener system. Int. J. Control 1–11 (2023)
7. McInnes, L., Healy, J., Melville, J.: UMAP: uniform manifold approximation and projection for dimension reduction. arXiv preprint arXiv:1802.03426 (2018)
8. Mzyk, G., Hasiewicz, Z., Mielcarek, P.: Kernel identification of non-linear systems with general structure. Algorithms **13**(12), 328 (2020)
9. Mzyk, G., Maik, G.: Nonparametric identification of Wiener system with a subclass of wide-sense cyclostationary excitations. Int. J. Adapt. Control Signal Process. **38**(1), 323–341 (2024)
10. Peng, H., Long, F., Ding, C.: Feature selection based on mutual information criteria of max-dependency, max-relevance, and min-redundancy. IEEE Trans. Pattern Anal. Mach. Intell. **27**(8), 1226–1238 (2005)
11. Pintelon, R., Schoukens, J.: System Identification: A Frequency Domain Approach. Wiley, Hoboken (2012)
12. Roweis, S.T., Saul, L.K.: Nonlinear dimensionality reduction by locally linear embedding. Science **290**(5500), 2323–2326 (2000)
13. Tibshirani, R.: Regression shrinkage and selection via the lasso. J. R. Stat. Soc. Ser. B Stat. Methodol. **58**(1), 267–288 (1996)
14. Wachel, P., Śliwiński, P., Hasiewicz, Z.: Nonparametric identification of MISO Hammerstein system from structured data. J. Syst. Sci. Syst. Eng. **24**(1), 68–80 (2015)

Wireless Employee Safety Monitoring System with Measurement of Biomedical Parameters

Marcel Maj[✉]

Sieć Badawcza Łukasiewicz - EMAG, Katowice, Poland
marcel.maj@emag.lukasiewicz.gov.pl
https://emag.lukasiewicz.gov.pl/

Abstract. The article presents a safety system for monitoring an employee which serve as early warning systems for anticipating potential negative events. The system has two main functions: locating the worker and monitoring the worker parameters such as pulse, temperature, accelerometric measurement. The system is designed to help in emergency situations at large industrial facilities.

Keywords: Bioengineering · Biomedical measurement · Localization · Worker monitoring

1 Introduction

Large industrial facilities, including critical facilities like power plants, have a distributed field of operation. This means, that employees frequently work far from each other and, in addition, alone. Moreover, working conditions in such facilities are harsh [23] which can lead to poor worker well-being and, consequently, fainting [3,8,24]. Such working conditions create a space, in which a worker injured in an accident may not get the right help in a short period of time, or worse, may not get it at all in the absence of information about accident. In addition, at critical facilities, employees are not allowed to use cell phones for security reasons. It increases the worker's risk because it limits the ability to call for help in the mentioned facilities.

This problem creates demand for safety systems that monitor the worker and thus reduce the risk of not helping in the event of an accident [2,6,29]. The concept of the system, which is designed to assist the employer in protecting employees, consists of two main functions: locating the employee and measuring biomedical parameters of the employee. In this way, the employer or facility supervisor

Developed and published by Sieć Badawcza Łukasiewicz - EMAG on the basis of the results in the 6th stage of the multi-annual program "Rządowy Program Poprawy Bezpieczeństwa i Warunków Pracy", financed by Narodowe Centrum Badań i Rozwoju for project "Bezprzewodowy system monitorowania stanowiskowego pracowników wraz z pomiarem parametrów biometrycznych", number II.PN.07. Project commissioned by Centralny Instytut Ochrony Pracy - Państwowy Instytut Badawczy.

W. Zamojski et al. (Eds.): DepCoS-RELCOMEX 2024, LNNS 1026, pp. 152–162, 2024.
https://doi.org/10.1007/978-3-031-61857-4_15

will be able to assess the worker's condition through biomedical measurements and will additionally know the worker's location. This will translate into a quick response time when help is needed, even if the worker would be unconscious.

The biomedical measurement consist of recording 3 parameters: pulse, temperature and accelerometric measurement. The pulse is the most vital parameter which describe basic state of human. High level of pulse can show a stressfull situation which can potentially be dangerous. On the other hand low level of pulse can indicate the potential threat, for example fainting [5]. The temperature is also a good source of information, which can describe a state of worker. Higher temperature can also inform about stress full situation, which can correlate with pulse [26]. The last parameter, accelerometric measurement will present potentially dangerous situations, such as stillness or sudden acceleration. Stillness can be potentially dangerous, because it can be read as fainting, on the other hand sudden acceleration can be read as fall [10, 14]. All of mentioned measurements can describe wellness of the worker. At last, the localization of the worker is the vital point in the process of helping a worker in potentially dangerous situation. The site supervisor will be able to determine the location of each employee using the system at any time in case of emergency.

2   System Architecture

The system's architecture was developed based on the functions of the system which are: worker location and biomedical measurement. The system consists of 3 main areas which are represented by 3 devices respectively. Those are:

- personal devices,
- reference devices,
- master device.

More about each device in the following Sects. 3, 4 and 5. Each of these devices is responsible for a specific task created from the general system's tasks. Firstly, the personal device is responsible for biomedical measurements. It has modules which can measure pulse, temperature and acceleration. The data will be sent to the master device to save it in the database. Moreover, the personal device is responsible for localization itself with the help of the reference devices. The above is accomplished by receiving reference devices within range and transmitting information about the received reference devices to the master device. Than, the master device, compute the localization of the personal device with the information about reference devices.

2.1   Work Place of the System

It is expected that the system will be used in places with higher risks to the worker, such as factories. However, there is no contraindication to the system also being used in office buildings. The personal device will be worn by a worker. The reference devices will be dispersed around the user's building in such a way, to cover the area to be monitored. The master device will be a server, which process and store all information coming from the personal device. The master device will be the part of the system that binds the whole into one system. It will also be the system's interface for the custodian or facility manager. It is envisioned that the master device will be placed in the facility space according to the needs and policies of the facility.

3   Personal Device

In order for the system to meet the requirement about biomedical measurements, it was necessary to choose a device capable of doing so. What is more, it is important, that device does not affect the employee's work. In the world of biomedical measurements, the most desirable feature of measurements, for the comfort of the patient, is non-invasiveness [1,30]. For this reason, non-invasive methods related to pulse measurement were reviewed. The other measurements: temperature measurement and accelerometric measurement are inherently non-invasive. But, for the record, the epidermal temperature measurements will not be as accurate as measurements in the body's natural orifices. For the pulse measurement, the photoplethysmography (PPG) [11] method was chosen, due to the fact it is an non-invasive method [4].

Another issue related to the use of a personal device by an employee is where to wear it. Here an important factor is that the device should not affect the user's comfort and productivity. For this reason, a literature review was carried out, showing the optimal place for pulse measurement using the PPG method.

Pulse measurement must take place in specific places in order to obtain a signal of the best possible quality. The optimal places to measure heart rate using PPG are [7]:

- forehead,
- earlobe,
- chest,
- wrist,
- fingers,
- ankle area.

The effect of review possible places to were personal device is presented below.

1. *Continuous Blood Pressure Monitoring using Pulse Wave Transit Time* [12] - The article presents the optimal place for measuring the pulse using photoplethysmography: the wrist and fingers, with indications of possible movements that may negatively affect the measurement.
2. *Monitoring heart and respiratory rates at radial artery based on PPG* [28] - The article presents the optimal place for measuring the pulse using photoplethysmography: the radial artery, which runs through the wrist.
3. *Effective PPG Sensor Placement for Reflected Red and Green light, and Infrared Wristband-type Photoplethysmography* [13] - The article presents the optimal place for measuring the pulse using photoplethysmography: around the radius and ulna at the wrist.

4. *Development of a Wearable In-Ear PPG System for Continuous Monitoring* [20] - The article presents the optimal place for measuring the pulse using photoplethysmography on the wrist, which is the most frequently chosen place for measurements, due to the ulnar and radial arteries.
5. *Wrist-worn device combining PPG and ECG can be reliably used for atrial fibrillation detection in an outpatient setting* [21] - The article validated whether the optimal place for measuring the pulse using photoplethysmography was the wrist. The paper presents positive results for heart rate measurements using photoplethysmography on the wrist.
6. *Wrist band photoplethysmography in detection of individual pulses in atrial fibrillation and algorithm-based detection of atrial fibrillation* [27] - The article presents positive results of atrial fibrillation detection for pulse measurement using photoplethysmography on the wrist.
7. *Blood Pressure Estimation Using Custom Photoplethysmography Sensors Located on Radial Artery at Wrist* [19] - The article presents the problem of devices that are used for everyday use to measure the pulse using photoplethysmography. It was noted that small devices that fit on the wrist are the most convenient for the user.
8. *Guidelines for wrist-worn consumer wearable assessment of heart rate in biobehavioral research* [18] - The article presents devices that enable pulse measurement using photoplethysmography. Attention was made to the trend in devices intended for everyday use - the main place of wearing such devices is the wrist.

To summarize the cited articles, measuring the pulse on the wrist using PPG is a method used every day. It should be mentioned that the PPG method based on wrist pulse measurement does not carry any diagnostic information. However, for the needs of the current system it is sufficient. Another important point supporting the location of heart rate measurement on the wrist is user comfort. The measurement site on the wrist is not as burdensome as opposed to the measurement on the forearm or earlobe, which is clearly confirmed by the article [18]. The second aspect is the measurement method. The system is intended to help employees reduce dangers, which is why it was decided to use the PPG measurement method on the wrist to not limit employee's movements. An alternative to measuring the pulse could be measurement using electrodes. A simplified measurement method with a reduced number of electrodes could be used. In this measurement method, the electrodes must be attached to the skin all the time, which could cause discomfort or chafing to the employee during long-term use. For this reason, this option was rejected. Due to the fact that, in the system, the personal device should measure pulse by PPG. The measurement should be done on the wrist. In this case, the chosen device for the system is a band, which is located on the wrist.

With these assumptions in mind, a review of the market for available wristbands was conducted. For the project of the system it was considered various models of the personal device, but the best ones which were suitable for system are presented below:

- Vigomed band [25],
- Locon Life Plus band [17],
- T-wristband [16].

A comparison of the above bands was made which indicates, that the compared bands have similar functionality. Those are:

- pulse measurement,
- temperature measurement,
- accelerometric measurement,
- communication interface WIFI.

The T-wristband additionally has a Bluetooth communication interface and is fully programmable. For this reason, it was selected for the project as a personal device. The most important aspect is, T-wristband can be programmed according to own needs. This is important factor, due to the possibility of implementing mechanisms, that allow locating the employee.

The benefits of choosing a wrist band and especially the device T-wristband are:

- integration of all sensors in one place,
- using one microcontroller to operate all sensors,
- using one battery to operate all modules,
- preparing one communication channel from the personal device to the master device,
- wearing comfort for the employee, i.e. the size of the watch and its location, i.e. the wrist.

3.1   Algorithms and Libraries for Personal Device

In order to translate the IR signal that is used by the PPG into a pulse value, the penpheral beat amplitude (PBA) algorithm was used [22]. This algorithm is able to calculate the time between peeks of pulses to get a heart rate. The temperature is measured using the same sensor that also measures pulse rate. It is possible, because the sensor (MAX30102) has modules for measuring pulse and temperature. To operate the MAX30102 sensor, the library provided by Sparkfun [22] is used for both measurements. The acceleration measurement is done by the library presented by the developer of the T-wristband device [15]. The personal device sends the bool value if the fall was detected.

4   Reference Device

The reference devices in this system will be beacon devices. Beacon devices can work in two ways. Beacon can act as a "lighthouse", where it sends a Bluetooth signal with information about its data, e.g. ID. In this solution, the device that needs to be located finds the receivable beacons. In this way, it is possible to determine the location using receivable beacons by sending to the master device information about them. Another solution is a beacon that registers devices in its vicinity. Conceptually, this is the inverted solution given above. Beacons in the system register nearby devices and send this information to the central device. The advantage of this solution is that to the object, that needs to be located, needs only a tag registered by beacons. However, this solution places a burden on the network used by the location system, because each beacon reports a receivable device. On the other hand, a location system with "lighthouse" beacons requires a more complicated device that must have Bluetooth for locating.

In current system, it is opted for a solution in which the beacon works in the form of a "lighthouse". The personal device will actively locate itself by sending information about beacons receivable in the area. What is more, it will record the employee's parameters and send this information to the superior system. Additionally, this solution will improve and relieve network traffic because one packet of data sent by a personal device will contain all information including: found beacons and the employee's parameters. Due to the fact, that the chip present in the personal device has two transmission media: Bluetooth and WIFI, it was decided to monitor the location via Bluetooth transmission. Therefore, the selected beacons must transmit advertisement signal in the Bluetooth standard. For the purposes of the project, Bluetooth Eddystone beacon standard was chosen.

A company using the solution presented in this article must determine the monitoring area, so that beacons can be placed appropriately. The system works according to specific points where the beacons are placed. It is important to precisely determine the monitoring area and then place the beacons. Then, determine distance between them. In this way, the system will be able to correctly determine the location of the personal device.

5   Master Device

The master device, in the designed system, is the module of the system that connects all the modules into a whole. It is responsible for collecting data from personal devices as well as storing them in the database. The master device is a personal computer, which will be a server-type device that has a database, as well as a user interface, where user applications will be installed. An application present on the master device will inform the facility's caretaker of potential dangers threatening employees based on measured parameters, for example, a low pulse or a suspected fall.

The master device consists of the following solutions: operating system - Windows 11, database - PostgreSQL 16.1, web server - Apache 2.4.58 with PHP 8.3.0 for accepting, verifying the correctness and placing the received data in the database from personal devices. A PHP script was used for this purpose. The master device operates as an HTTP server accepting data in JSON format for receiving the data.

As part of the project, a communication protocol was also developed, including the design of the structure of transmitted data in the JSON format along with the appropriate database structure. Support for this protocol has been implemented both in the personal device and on the central server - master device. A sample of the data transferred by personal device is presented in the Table 1 and 2. The data pack form personal device is sent in average every 7 s, which ends with 8 measurements during a minute.

Table 1. A simplified view of the database with data from a personal device. Some columns were not displayed for readability reasons. (The data presented are actual data, but pulse and temperature were not measured in current example)

id	pulse	temp	fall	stime	uuid	rssi	rssi_avg
283102	0	226	false	2024-02-27 11:51:54	fe:2c:67:1f:d6:83	−96	−93
283102	0	226	false	2024-02-27 11:51:54	e1:4c:03:86:8c:6c	−97	−95
283102	0	226	false	2024-02-27 11:51:54	ce:98:a6:52:70:74	−89	−82
283102	0	226	false	2024-02-27 11:51:54	da:3a:7b:15:5b:ec	−91	−84
283102	0	225	false	2024-02-27 11:51:47	e1:4c:03:86:8c:6c	−94	−95
283102	0	225	false	2024-02-27 11:51:47	fe:2c:67:1f:d6:83	−96	−93
283102	0	225	false	2024-02-27 11:51:47	da:3a:7b:15:5b:ec	−85	−82
283102	0	225	false	2024-02-27 11:51:47	ce:98:a6:52:70:74	−83	−80

Table 2. A sample of data during pulse and temperature measurement

temp	pulse	stime
276	73	14.03.2024 11:27
280	73	14.03.2024 11:27
287	73	14.03.2024 11:27
294	73	14.03.2024 11:27
308	73	14.03.2024 11:27
321	73	14.03.2024 11:27
326	73	14.03.2024 11:27
326	73	14.03.2024 11:27
325	70	14.03.2024 11:26
325	66	14.03.2024 11:26
323	83	14.03.2024 11:26
324	87	14.03.2024 11:26

Biomedical measurements: pulse, temperature and accelerometer-based acceleration are placed respectively in the columns: "pulse", "temp", "fall". The first two are in integer form. In this case, the temperature values shown in the Table 1 and 2 must be divided by ten to get the actual temperature measurement. As the measurements show, the temperature is lower (e.g. 32,5) than accepted 36,6°, because measurement is done on the skin. Acceleration is shown in bool form. It takes the value true or false to indicate the fall of the person who wears the personal device.

The fall detection algorithm is based on a Eq. 1 [9]. The threshold of the fall detection is based on the article [9] - Table 1.

$$SV = \sqrt{(A_x)^2 + (A_y)^2 + (A_z)^2} \tag{1}$$

where: Ax, Ay, Az = acceleration (g) in x-, y-, and z-axes.

Localization mechanism is based on the measurements of the RSSI. Personal device sends the data in the one package. In the Table 1 there is a column "uuid" which indicate registered beacon. Next, there are two columns, which are RSSI values. The "rssi" column contains the raw measurement. On the other hand the column "rssi_avg" contains averaged value of the RSSI which is calculated using Eq. 2.

$$rssi\_avg = 0.3333 * rssi + 0.6666 * rssi\_avg \tag{2}$$

where:
rssi_avg = averaged RSSI value
rssi = current value of the RSSI.

Averaged RSSI and a raw RSSI value are used for calibrating the system in the localization module. The last two columns, "id" and "stime" represent respectively personal device and the time of the data to arrive to the server.

RSSI values from the local database are downloaded by the location application present in the master device. This application is responsible for calculating the location (coordinates) of the personal device. It uses triangulation technology specific to the BeaconFence library. The calculated location data goes back to the database. From there, they are downloaded by the visualization application. The application then allows to place personal tag icons on the map of the monitored area, in accordance with the calculated coordinates.

6   Connections and Workflow

The data flow in the system starts with reference devices (beacons). They transmit information about themselves to a personal device (T-wristband). The personal device receives beacons with Bluetooth interface. The wristband sends this data via the WIFI system along with the employee's current biomedical measurements to the master device. The master device is connected to the WIFI system via LAN connection. The whole system and its connections are represented in the Fig. 1.

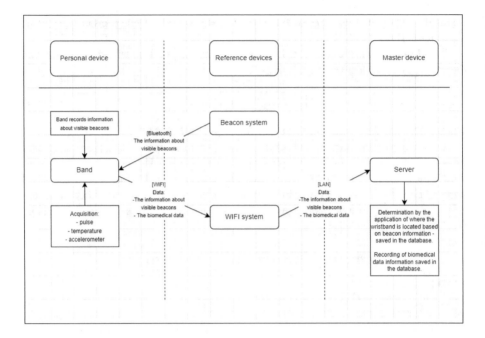

Fig. 1. System architecture

7 Summary

The system is still in the development state, where the main functionalities needs to be deeply tested for higher technological ready level. Nonetheless, the main goals of the system were met, such as: measurement of pulse, temperature and acceleration. The Table 1 and 2 presents that data about temperature is stored in the integer form. This is caused, due to the fact, that technically it is easier to send the whole number than decimal. Presented pulse values in Table 2 are nominal[1]. Developed system recognizes and measures the signal sent by beacons. The localization module is still development, where each of the personal devices will be presented on the map.

The system is a multidisciplinary platform combining many technological fields, such as biomedical measurements and beacon-based localization. The cited articles in this work indicate areas of interest related to the developed system. The presented personal device is not burdensome for the employee compared to other devices on the market that have electrodes glued to the body.

The system was designed to assist the employer in anticipating and preventing possible accidents at work. The system presented is mainly designed for establishments, where the danger to the employee is increased due to the nature of the facility, for example, a factory. However, the system can be successfully used in office buildings.

---

[1] Nominal pulse rages from 60 to 100 BMP during normal activities.

Acknowledgements. The author would like to thank Tomasz Molenda for his comments and whole team who participated in the development of the system and also the three anonymous reviewers for their insightful suggestions and careful reading of the manuscript.

References

1. Alotaibi, M., Honarvar Shakibaei Asli, B., Khan, M.: Non-invasive inspections: a review on methods and tools. Sensors **21**(24), 8474 (2021)
2. Bazaluk, O., et al.: Ergonomic risk management process for safety and health at work **11** (2023). https://doi.org/10.3389/fpubh.2023.1253141
3. Brown, G.: The global threats to workers' health and safety on the job. Soc. Just. **29**, 12–25 (2002)
4. Castaneda, D., Esparza, A., Ghamari, M., Soltanpur, C., Nazeran, H.: A review on wearable photoplethysmography sensors and their potential future applications in health care. Int. J. Biosens. Bioelectron. **4**(4), 195 (2018)
5. Charlton, P.H., Celka, P., Farukh, B., Chowienczyk, P., Alastruey, J.: Assessing mental stress from the photoplethysmogram: a numerical study. Physiol. Meas. **39**(5), 054001 (2018)
6. Emmerik, I.: For better and for worse: adverse working conditions and the beneficial effects of mentoring. Career Dev. Int. **9** (2004). https://doi.org/10.1108/13620430410526157
7. Ghamari, M., Nazeran, H., Soltanpur, C.: A review on wearable photoplethysmography sensors and their potential future applications in health care. **4**, 195–202 (2018). https://doi.org/10.15406/ijbsbe.2018.04.00125
8. Hasan, M.: Health and safety protection of employee in the workplace: a comparative analysis of labour law in between the UK and Bangladesh (2023)
9. Kangas, M., Konttila, A., Lindgren, P., Winblad, I., Jämsä, T.: Comparison of low-complexity fall detection algorithms for body attached accelerometers. Gait Posture **28**(2), 285–291 (2008). https://doi.org/10.1016/j.gaitpost.2008.01.003, https://www.sciencedirect.com/science/article/pii/S096663620800026X
10. Kim, S.H., Wang, C., Min, S.D., Lee, S.H.: Safety helmet wearing management system for construction workers using three-axis accelerometer sensor. Appl. Sci. **8**(12), 2400 (2018)
11. Kyriacou, P.A., Chatterjee, S.: The origin of photoplethysmography. In: Photoplethysmography, pp. 17–43. Elsevier (2022)
12. Lass, J., Meigas, K., Karai, D., Kattai, R., Kaik, J., Rossmann, M.: Continuous blood pressure monitoring during exercise using pulse wave transit time measurement. In: Conference Proceedings: Annual International Conference of the IEEE Engineering in Medicine and Biology Society. IEEE Engineering in Medicine and Biology Society. Conference, vol. 3, pp. 2239–2242 (2004). https://doi.org/10.1109/IEMBS.2004.1403652
13. Lee, S., Shin, H., Hahm, C.: Effective PPG sensor placement for reflected red and green light, and infrared wristband-type photoplethysmography, pp. 556–558 (2016). https://doi.org/10.1109/ICACT.2016.7423470
14. Lee, W., Lin, K.Y., Seto, E., Migliaccio, G.C.: Wearable sensors for monitoring on-duty and off-duty worker physiological status and activities in construction. Autom. Constr. **83**, 341–353 (2017)
15. LilyGO: LilyGO library and the projeckt of T-wristband. https://github.com/d-Raco/LilyGO_T-Wristband. Accessed 12 Mar 2024

16. LilyGO: T-wristband. https://github.com/Xinyuan-LilyGO/T-Wristband. Accessed 13 Jan 2024
17. Locon: Locon life plus band. https://bezpiecznarodzina.pl/p/opaska-locon-life-plus/. Accessed 13 Jan 2024
18. Nelson, B., Low, C., Jacobson, N., Arean, P., Torous, J., Allen, N.: Guidelines for wrist-worn wearable assessment of heart rate in biobehavioral research (2020). https://doi.org/10.31234/osf.io/3wk65
19. Nguyen, L., Chung, W.Y.: Blood pressure estimation using custom photoplethysmography sensors located on radial artery at wrist. In: ECS Meeting Abstracts, MA2020-01, p. 2009 (2020). https://doi.org/10.1149/MA2020-01272009mtgabs
20. Pedrana, A., Comotti, D., Re, V., Traversi, G.: Development of a wearable in-ear PPG system for continuous monitoring. IEEE Sens. J. 1 (2020). https://doi.org/10.1109/JSEN.2020.3008479
21. Saarinen, H., et al.: Wrist-worn device combining PPG and ECG can be reliably used for atrial fibrillation detection in an outpatient setting. Front. Cardiovasc. Med. **10**, 1100127 (2023). https://doi.org/10.3389/fcvm.2023.1100127
22. SparkFun: Max30105 particle and pulse ox sensor hookup guide. https://learn.sparkfun.com/tutorials/max30105-particle-and-pulse-ox-sensor-hookup-guide/all. Accessed 12 Mar 2024
23. Thomas, G.W., et al.: Low-cost, distributed environmental monitors for factory worker health. Sensors **18**(5), 1411 (2018)
24. Vandyck, E., Fianu, D., Papoe, M., Oppong, S.: Safety management systems, ergonomic features and accident causation among garment workers (2015)
25. VigoMed: VigoMed band. https://vigomed.pl/pl/p/ZAAWANSOWANY-WIELOFUNKCYJNY-SMARTWATCH-Opaska-SOS-dla-seniora-z-GPS%2C-pomiarem-pulsu-i-cisnienia%2C-czujnikiem-upadku-4G-WiFi-/406. Accessed 13 Jan 2024
26. Vinkers, C.H., et al.: The effect of stress on core and peripheral body temperature in humans. Stress **16**(5), 520–530 (2013)
27. Väliaho, E.S., et al.: Wrist band photoplethysmography in detection of individual pulses in atrial fibrillation and algorithm-based detection of atrial fibrillation. Europace **21**, 1031–1038 (2019). https://doi.org/10.1093/europace/euz060
28. Wang, C., Li, Z., Wei, X.: Monitoring heart and respiratory rates at radial artery based on PPG. Optik **124**, 3954–3956 (2013). https://doi.org/10.1016/j.ijleo.2012.11.044
29. Xu, X., Zhong, M., Wan, J., Yi, M., Gao, T.: Health monitoring and management for manufacturing workers in adverse working conditions. J. Med. Syst. **40** (2016). https://doi.org/10.1007/s10916-016-0584-4
30. Zemanova, M.A.: Towards more compassionate wildlife research through the 3Rs principles: moving from invasive to non-invasive methods. Wildl. Biol. **2020**(1), 1–17 (2020)

Artificial Intelligence Methods for Pet Emotions Recognition

Jacek Mazurkiewicz[✉] [ORCID]

Faculty of Information and Communication Technology, Wrocław University of Science and Technology, ul. Wybrzeże Wyspiańskiego 27, 50-370 Wrocław, Poland
jacek.mazurkiewicz@pwr.edu.pl

Abstract. Analyzing emotions in cats is a subject that comes with difficulties in interpretation. The subtle facial expressions of these animals make deciphering their emotional signals a challenge. The work focuses on the task of classifying feline emotions from images. It aims to test whether deep learning networks can recognize the primary emotional states of these animals on real (non-laboratory) image data. The results obtained in this work provide a basis for further research in this area. Future studies can focus on refining the parameters of existing models, developing advanced data analysis techniques, and exploring new classification models. Further research has the potential to contribute to a better understanding of emotional states in cats and improve the effectiveness of classification processes based on authentic images.

Keywords: artificial intelligence · emotions · cat

1 Introduction

According to recent studies, cats inhabit all continents and have been accompanying humans for about 10,000 years [10]. An estimated 371 million cats inhabit human homes worldwide [12]. Although people have so much contact with these animals, it can be problematic for them to recognize their pets' moods. One reason for this is the amount of facial muscles cats have - significantly fewer of them than in humans or dogs. In addition, the facial muscles in cats are less developed to convey emotion, which further reduces their facial expression capabilities. Unlike humans, the primary communication medium in cats is their entire bodies - posture, fur, and tails, as well as vocalization and smells. Such a divergent direction in the evolution of expression in cats is another barrier to communication between these animals and humans. Their ways are very unique, requiring exceptional human learning about their behavior. In addition, like humans, each cat has its character, will express itself slightly differently, and develop a unique language with its owner. Many myths about cat behavior mislead people trying to understand their pets. It is believed that they are loners when they are pack animals that do not like to be left alone for hours on end and that scratching various surfaces in the house is always for spite. That purring always means happiness, but it can mean significant pain and is intended to alleviate it. Anecdotes circulating in conversations are often misleading and

W. Zamojski et al. (Eds.): DepCoS-RELCOMEX 2024, LNNS 1026, pp. 163–176, 2024.
https://doi.org/10.1007/978-3-031-61857-4_16

lead to a lack of understanding of these animals [13]. Another mistake people make is to compare knowledge of dog behavior and attribute it to cats. For example, the tail wagging, a sign of contentment in most situations in dogs, is the opposite in cats. However, in some cases, it can be a sign of overstimulation in both animals. Such confusion often leads to a misreading of the signals the cat is sending and an unexpected reaction from the pet, which, in extreme cases, can be aggressive. As small predators, cats are in the middle of the food chain, so they try to hide injuries and pain, and their reactions are very subtle. Owners failing to recognize such subtleties may fail to react in time and cause their pets to deteriorate significantly before seeking veterinary help. Recognizing animal facial expressions is currently limited to manual decoding of facial muscles and observation, which is biased, time-consuming, and requires a lengthy training and certification process. It was decided to try to write an app that recognizes cat emotions to help owners better understand them and reduce the likelihood of exposing these animals to unpleasant situations. Deep neural networks and machine learning were chosen for the study. They are currently a prevalent topic. They were selected to see if they were suitable for such a task. Since humans have problems recognizing feline emotions, seeing how artificial intelligence will cope will be interesting. Especially since just recognizing human emotions, which are much more expressive, already causes problems for today's programs.

2   Cat Behaviorism Overview

Humans have difficulty understanding cat emotions when they read them from their mouths. This task is challenging because cats have fewer muscles on their snouts than humans and dogs. In addition, they need to be developed in terms of expressing emotions. Cats have only 32 muzzle muscles, less than at least dogs, which have 43. The task is so challenging that it poses difficulties even for veterinarians. For this reason, researchers at the University of Montreal in 2019 created the Feline Grimace Scale (FGS), [4] which provides guidelines describing how to tell if a cat is in pain by its mouth (Fig. 1).

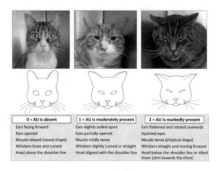

Fig. 1. Feline Grimace Scale

The scale uses "Action Unit," or AU for short, as the unit of measurement. Each of the five parameters, ear position, degree of eye-opening, muzzle tension, vibrissae angle, and head position relative to the back, is scored from zero to two AU. The cat is in significant pain if the final score is above four. It is necessary to define the primary emotions cats show to understand what these animals have to communicate beyond feeling pain. The paper [10] identifies five basic emotions. *Joy/playfulness* - is a positive emotional state manifested by behaviors, i.e., locomotor play, social play, or playing with objects. *Fear* - is a negative emotional state caused by perceiving an imminent threat or threat of danger. It is manifested by vigilance and attempts to withdraw. *Neutrality* is a positive emotional state caused by satisfying needs and desires. It is manifested by rest, calmness, and affiliative behavior. *Interest* - is a positive emotional state caused by the presence of a new stimulus. It manifests itself as attention and orientation to the stimulus. *Anger* - a negative emotional state caused by a frustrated desire to act or compete for resources. It manifests as aggression or the threat of aggression. A significant indicator of a cat's emotional state is its body position. If a cat exposes its belly, it signals that it is relaxed and trusts its surroundings, as it is revealing the most sensitive spot on its body. The opposite is an animal that arches its back and fluffs its fur - this is a sign of fear, as the cat is trying to appear larger than it is to scare off a potential adversary. However, if the fur is not fluffed, the cat may invite stroking - in the evaluation, pay attention to various indicators. If a cat is lying down with all its limbs tucked underneath it, it is usually not wholly calm at that point and remains alert. Cats' signals are often very subtle, so check several body parts. A good indicator of the mood of these animals is their tails. A tail set high, which is not moody, indicates that the cat is comfortable; however, if it is gloomy, the animal is nervous. A tail parallel to the back is a sign that the cat is curious. A tail tucked close to the torso means the cat is unsure of the situation. Another indicator of a cat's mood is their ears. They can be divided into three basic positions. *Neutral position* - when a cat is relaxed, its ears point forward in what is known as a neutral position. This position of the ears indicates the cat's calmness and happiness. *Straight up and forward position* - the cat moves its ears to an alert position when it wants to pay more attention to what is happening around it. Sometimes, the ears are turned in different directions. *Low and lateral position* - if a cat's ears are flattened against its head, it means it is frightened or nervous, which can lead to aggressive behavior. When a cat's ears are in this position, the cat feels uncomfortable. Another part of the cat's body you can pay attention to is their eyes. Not only do they dilate and contract with the change of light, but they also depend on their mood. A cat's pupils dilate when the cat is in a playful mood. If they look like they are covering most of the iris, it could mean that the cat is angry or scared. If the pupils are constricted, it may be a sign that the cat is calm or relaxed. As was mentioned in the cat grimace scale, vibrissae are an indicator of the cat's condition. If the cat's whiskers are facing forward and spread out, the cat is interested in something. If the cat's whiskers are pressed firmly against the face, the cat is most likely afraid and trying to look smaller. Pay attention to the cat's teeth - specifically, whether it exposes them. If it is not yawning, it usually means the cat is frightened and trying to scare off an opponent. To understand a cat's mood, it is necessary to pay attention to all these mentioned signs [2].

3   Assumptions and Data Preprocessing

There is a database on the website kaggle.com [1] containing more than nine thousand image data with cats, where each image marked nine landmarks on the cat's muzzle: three for the left ear, three for the right ear, two for the centers of the eyes and one for the center of the muzzle. The photos are in different sizes and resolutions. The selected base will not contain lab photos; only natural ones will represent the photographs the potential user took. Data were prepared depending on the tested classifier, including size adjustment or color scale change. An expert determines the criteria by which the photos of cats would be evaluated in terms of their emotional state depicted in the photo. The following criteria were set. *Cat's mouth* - position of the cat's ears - whether neutral or backward, whether the cat reveals its teeth, whether the cat's mouth is tight, cat pupil size, the position of the cat's vibrissae - whether they are lowered. *The entire body of the cat* - the position of the tail - whether it is up or teeny, whether the cat exposes its belly, whether the cat has a visibly rounded spine, whether the cat's head is below the shoulder line, whether the cat is in a hunched over position, whether the cat is in a position to attack. Based on these criteria, cat emotions were determined: satisfaction, neutrality, sadness, and anger. Unfortunately, the database from kaggle.com [1] has unbalanced data regarding emotions shown by cats - this may result from users' preference for sharing photos of cats in a contented or relaxed mood. It was hard to find photos of cats experiencing pain. A search was made for photos of such cats on the Internet, but few were there. Therefore, the database comprises photos divided into designated categories, totaling 120 photos per category. Another obstacle is the quality of photos for the sad category in terms of analysis by the classifier. These are often photos of cats in a collar or cage, hiding, or being held by a human. A sub-baseline for testing was also done, where each category was divided into long-haired and short-haired cats to see if there would be a difference in recognizing their emotions. In the papers describing the recognition of cats in a photo, long-haired cats are, on average, recognized worse by the algorithms. The next step was to prepare a program that cut out the cats' mouths from photos for part of the tests and to do so; it relied on Haar's classifier and the OpenCV library [11]. To create a Haar classifier, you need a "positive" and "negative" database, i.e., having the object in question or not. In the database, you need to label this object in the image. This is a time-consuming task, so a ready-made classifier from the official OpenCV library [13] was used for the project. It has many object classifiers, including those for cat snouts that can be seen from the front. Two other classifiers were tested. However, the one from the official OpenCV distribution achieved the best results - correctly recognizing the most cat snouts. Unfortunately, the classifier did not cope with long-haired and black cats.

4   Deep Learning Classifier

4.1   Transfer Learning

The simplest solution was tested first to determine whether deep neural networks with or without transfer learning would suit this task. CNN, or convolutional neural network models, were used [14]. These models are a type of deep neural network and are considered particularly effective for image processing tasks. The layer that distinguishes them from other models is the convolutional layer. It extracts features from an image by moving a filter over it and calculating convolution values [9]. The result of this process is a feature map that contains information about the detected patterns. Transfer learning involves using a previously trained model on an extensive data set to solve a similar task. It uses the weights and features learned as a starting point for a new problem. This study used transfer learning with the ResNet50 model and architecture [3] (Fig. 2).

Layer (type)	Output Shape	Param #
resnet50 (Functional)	(None, 7, 7, 2048)	23587712
dropout (Dropout)	(None, 7, 7, 2048)	0
flatten (Flatten)	(None, 100352)	0
batch_normalization (BatchN ormalization)	(None, 100352)	401408
dense (Dense)	(None, 512)	51380736
batch_normalization_1 (Batc hNormalization)	(None, 512)	2048
activation (Activation)	(None, 512)	0
dropout_1 (Dropout)	(None, 512)	0
dense_1 (Dense)	(None, 256)	131328
batch_normalization_2 (Batc hNormalization)	(None, 256)	1024
activation_1 (Activation)	(None, 256)	0
dropout_2 (Dropout)	(None, 256)	0
dense_2 (Dense)	(None, 128)	32896
batch_normalization_3 (Batc hNormalization)	(None, 128)	512
activation_2 (Activation)	(None, 128)	0
dropout_3 (Dropout)	(None, 128)	0
dense_3 (Dense)	(None, 64)	8256
batch_normalization_4 (Batc hNormalization)	(None, 64)	256
activation_3 (Activation)	(None, 64)	0
dropout_4 (Dropout)	(None, 64)	0
dense_4 (Dense)	(None, 2)	130

Fig. 2. ResNet50 model architecture for four emotions recognition

It was checked whether using methods from this work would give the desired results with non-laboratory images and four categories. The size of the images was set at 224 × 224 pixels, as this size provides the best results for ResNet50. Weights from ImageNet [7] were used. This gives each category 240 photos. The results are presented in Table 1.

Table 1. Results of emotion recognition in the ResNet50 model for four classes

Class	Global	Short-haired cats	Long-haired cats
	Correct Answers [%]	Correct Answers [%]	Correct Answers [%]
Sad	72	79	69
Satisfied	83	94	86
Neutral	75	80	73
Bad	81	85	78

4.2 Traditional Learning

In this approach, a previously untrained model is given an isolated database for its training. The LeNet model [14] was used because it performed best in the studies found. A 9:1 ratio of training to test data was established, as in the source paper. The size of the images was set to 50 × 50, and the images were converted to grayscale. The architecture used was reproduced from the description in the source paper (Fig. 3).

```
Layer (type)                 Output Shape          Param #
=================================================================
conv2d_6 (Conv2D)            (None, 50, 50, 20)    520

activation_14 (Activation)   (None, 50, 50, 20)    0

max_pooling2d_6 (MaxPooling  (None, 25, 25, 20)    0
2D)

conv2d_7 (Conv2D)            (None, 13, 13, 50)    25050

activation_15 (Activation)   (None, 13, 13, 50)    0

max_pooling2d_7 (MaxPooling  (None, 6, 6, 50)      0
2D)

flatten_5 (Flatten)          (None, 1800)          0

dense_16 (Dense)             (None, 500)           900500

dense_17 (Dense)             (None, 4)             2004
```

Fig. 3. Architecture of the LeNet model

The results are presented in Table 2.

Table 2. Results of emotion recognition in the LeNet model for four classes

Class	Global	Short-haired cats	Long-haired cats
	Correct Answers [%]	Correct Answers [%]	Correct Answers [%]
Sad	67	77	68
Satisfied	79	91	86
Neutral	71	83	79
Bad	76	88	89

5 MLP Classifier

Using a multilayer perceptron for categorization was chosen as the second path. This model type was used because MLP is used for photos with designated characteristics [3]. The model was written to divide the images into four categories. To get the desired results, the architecture was modified from the one described in the article due to a different approach to photo recognition. The photos were set to a size of 128×128 pixels. The model's architecture is shown below (Fig. 4). Results are available in Table 3.

```
Layer (type)                 Output Shape         Param #
=================================================================
conv2d_24 (Conv2D)           (None, 50, 50, 20)   520

activation_2 (Activation)    (None, 50, 50, 20)   0

max_pooling2d_24 (MaxPooli   (None, 25, 25, 20)   0
ng2D)

conv2d_25 (Conv2D)           (None, 13, 13, 50)   25050

activation_3 (Activation)    (None, 13, 13, 50)   0

max_pooling2d_25 (MaxPooli   (None, 6, 6, 50)     0
ng2D)

flatten_9 (Flatten)          (None, 1800)         0

dense_28 (Dense)             (None, 500)          900500

dense_29 (Dense)             (None, 4)            2004

=================================================================
```

Fig. 4. MLP model architecture

Table 3. Results of emotion recognition for four classes using MLP classifier

Class	Global	Short-haired cats	Long-haired cats
	Correct Answers [%]	Correct Answers [%]	Correct Answers [%]
Sad	67	72	68
Satisfied	69	73	76
Neutral	71	68	69
Bad	76	81	83

6   Classification Based on Feature Point Prediction

Another attempt was to use feature points and a decision tree [14]. The VGG19 model and a database from kaggle.com [1] were used to generate the feature points (Fig. 5).

```
Model: "sequential_1"
_____
 Layer (type)                Output Shape              Param #
=================================================================
 vgg19 (Functional)          (None, 7, 7, 512)         20024384

 flatten_1 (Flatten)         (None, 25088)             0

 dense_3 (Dense)             (None, 512)               12845568

 dropout_1 (Dropout)         (None, 512)               0

 dense_4 (Dense)             (None, 256)               131328

 dense_5 (Dense)             (None, 1)                 257

=================================================================
```

Fig. 5. Architecture of the VGG19 model for feature point prediction

Weights from the ImageNet collection [7] were used to train the model on the data. This large-scale labeled image dataset is widely used in computer vision research and machine learning. The photos from the "Cat Dataset" [1] that were used to train the model were set to a resolution of 224 × 224 pixels. The images were used in color mode. One thousand photos were selected for training - not all photos in the database are correctly labeled. Additional characteristics were added to the images, based on the [5], which describes the cat's morphometrics: the set of 48 points on the cat's mouth and the directions they move when the animal is in pain. The additional characteristic points are intended to help determine the relationship between the distinct points and the emotion the cat shows in the photo. The points are given: on the upper and lower eyelids to determine whether the cat is closing its eyes, on the lower lip to determine whether the cat opens its muzzle. In addition, photos of 1000 cats from the database showing their tails and backs were selected to add feature points that define these body parts. The results for feature prediction on the muzzle (Fig. 6) and whole-body (Fig. 7) images are presented in the contingency matrices and Table 4. Having a working model for predicting characteristic points, an algorithm was written to determine what mathematical relations between the points describe a given emotion. The exact numerical values were determined experimentally, describing the measure of angles between different parts of the animal body [10].

Table 4. Comparison of quality measure results for feature point prediction

Points of characteristic	Recall	Accuracy	F1	Precision
Muzzle	0.907	0.911	0.923	0.917
Whole body	0.801	0.809	0.818	0.812

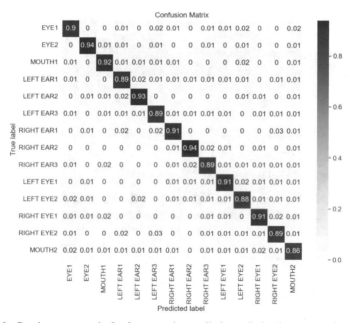

Fig. 6. Contingency matrix for feature point prediction - distinctive points of muzzle

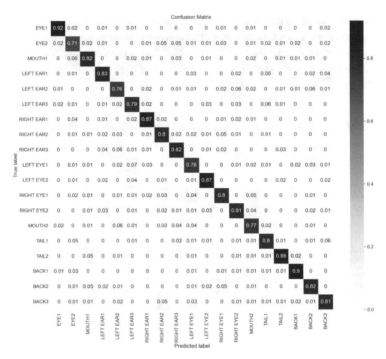

Fig. 7. Contingency matrix for feature point prediction - whole body landmarks

Description of the parameters used in the algorithm.

Key Points: $p1$ - the center of the left eye, $p2$ - the center of the right eye, $p3$ - the center of the mouth, $p4$ - the outer corner of the left ear, $p5$ - top of left ear, $p6$ - the inner corner of the left ear, $p7$ - the inner corner of the right ear, $p8$ - the top of the right ear, $p9$ - the outer corner of the right ear, $p10$ - upper eyelid of the left eye, $p11$ - lower eyelid of the left ear, $p12$ - upper eyelid of the right ear, $p13$ - lower eyelid of the right ear, $p14$ - lower part of the muzzle, $p15$ - beginning of tail, $p16$ - tail end, $p17$ - the start of the ridge, $p18$ - center back, $p19$ - end of ridge.

Lines: $l1$ - between points $p1$ and $p2$, between the centers of the eyes; l2 - between points $p5$ and $p6$, between the peak and the inner corner of the left ear; l3 - between points $p8$ and $p7$, between the peak and the inner corner of the right ear, l4 - between points $p3$ and $p13$, between the tips of the muzzle, l5 - between points $p15$ and $p16$, between the ends of the tail, l6 - between points $p17$ and $p19$, between the beginning and end of the ridge, l7 - between points $p17$ and $p18$, between the beginning and center of the ridge, l8 - between points $p18$ and $p19$, between center and end of ridge.

Distances: $|A|$ - between points $p5$ and $p6$, between the peak and the inner corner of the left ear, $|B|$ - between points $p4$ and $p6$, between the corners of the left ear, $|C|$ - between points $p8$ and $p7$, between the peak and the inner corner of the right ear, $|D|$ - between points $p7$ and $p9$, between the corners of the right ear, $|E|$ - between points $p10$ and $p11$, between the eyelids of the left eye, $|F|$ - between points $p12$ and $p13$, between the eyelids of the right eye, $|G|$ - between points $p1$ and $p2$, between the centers of the eyes, $|H|$ - between points $p3$ and $p14$, between parts of the muzzle, $|J|$ - between the point $p3$ and the point determined by the lines $l1$ and $l4$, that is, the center of the muzzle, and the midpoint between the eyes.

Angles: $k1$ - between the lines $l1$ and $l2$, between the line passing through the centers of the eyes, and the line passing through the top and inner corner of the left ear; $k2$ - between the lines $l1$ and $l4$, between the line passing through the centers of the eyes, and the line passing through the top and inner corner of the right ear, $k3$ - between the $l5$ and $l6$ straights, between the lines passing through the ends of the tail and back, $k4$ - between lines $l7$ and $l8$, between the line passing through the beginning and middle of the ridge and the line passing through the middle and end of the ridge.

Assumptions: $Z1$ - if $k1 \leq 38°$ and $k2 \leq 38°$ - tilted ears. $Z2$ - if $38° < k1 < 68.5°$ and $38° < k2 < 68.5°$ - ears in neutral position, $Z3$ - if $k1 \geq 68.5°$ and $k2 \geq 68.5°$ - ears in an upright position, $Z4$ - if $k1 < 38°$ and $k2 \leq 42.5°$ - inclined left ear, $Z5$ - if $k1 \leq 42.5°$ and $k2 < 38°$ - inclined right ear, $Z6$ - if $15° > k3 > 285°$ - tail in horizontal position, $Z7$ - if $85° \leq k3 < 93°$ - tail upright, $Z8$ - if $180° < k3 \leq 270°$ - tail curled, $Z9$ - if $270° < k3 \leq 285°$ - tail tucked, $Z10$ - if $15° \leq k3 < 85°$ or $93° \leq k3 \leq 180°$ - tail in neutral position, $Z11$ - if $k4 > 45°$ - bent back, $Z12$ - if $k4 \leq 45°$ - back in neutral position, $Z13$ - if $|A| > |B|$-1.4 and $|C| > |D|$-1.4 - flattened ears, $Z14$ - if $|A| > |B|$-1.6 and $|C| > |D|$-1.4 - flattened ears, $Z15$ - if $|A| \leq |B|$-1.6 and $|C| \leq |D|$-1.4 - ears erect, $Z16$ - if $|A| > |B|$-1.4 and $|C| > |D|$-1.6 - flattened ears, $Z17$ - if $1/3|G| \geq |E| > 1/5|G|$ and $1/3|G| \geq |F| > 1/5|G|$ - open eyes, $Z18$ - if $1/5|G| \geq |E|$ and $1/5|G| \geq |F|$ - wide open eyes, $Z19$ - if $1/3|G| < |E|$ and $1/3|G| < |F|$ - closed eyes, $Z20$ - if $|E| = 0$ and $|J| = 0$ - closed eyes, $Z21$ - if $|H| \geq 1.5$-$|J|$ - mouth open, $Z22$ - if $|H| < 1.5$-$|J|$ - muzzle closed, $Z23$ - if $y[p1] < y[p18]$ or $y[p2] < y[p18]$ - head tilted.

The algorithm checks two cases: the whole body of the cat or the cat's mouth alone. For both alternatives, the algorithm separately checks sets of assumptions to assign the photo to one of four categories - angry cat, sad cat, happy cat, neutral cat.

PSEUDOCODE - Categorization of cat emotions

```
 1: function CategorizeEmotions(pointsCharacteristic, mode)
 2:  if mode == "all cat" then
 3:      if ((Z2 ∪ Z3) ∧ Z15) ∧ Z7 ∧ Z12 ∧ Z17 then emotion = cheerful
 4:          else if (Z3 ∧ Z15) ∧ Z6 ∧ Z12 ∧ Z18 then emotion = cheerful
 5:          else if Z3 ∧ Z15) ∧ (Z6 ∪ Z7) ∧ Z12 ∧ (Z19 ∪ Z20) then emotion = cheerful
 6:          else if Z2 ∧ Z15) ∧ Z8 ∧ Z12 ∧ Z20 ∧ Z23 then emotion = neutral
 7:          else if Z2 ∧ Z15) ∧ Z8 ∧ Z12 ∧ (Z17 ∪ Z19) then emotion = neutral
 8:          else if Z2 ∧ Z15) ∧ Z10 ∧ Z12 ∧ Z17 then emotion = neutral
 9:          else if ((Z1 ∪ Z4 ∪ Z5) ∧ (Z13 ∪ Z14 ∪ Z16)) ∧ Z7 ∧ Z11 ∧ Z19 ∧ (Z21 ∪ Z22)
10:              then emotion = angry
11:          else if ((Z1 ∪ Z4 ∪ Z5) ∧ (Z13 ∪ Z14 ∪ Z16)) ∧ Z9 ∧ Z11 ∧ Z18 ∧ Z22
12:              then emotion = angry
13:          else if ((Z1 ∪ Z4 ∪ Z5) ∧ (Z13 ∪ Z14 ∪ Z16)) ∧ Z9 ∧ Z19 ∧ Z21 then emotion = bad
14:          else if ((Z1 ∪ Z4 ∪ Z5) ∧ (Z13 ∪ Z14 ∪ Z16)) ∧ Z8 ∧ (Z19 ∪ Z20) ∧ Z23
15:              then emotion = sad
16:          else
17:              return "emotion detection error"
18:      end if
19: else if mode == "muzzle alone" then
20:      if (Z3 ∧ Z15) ∧ Z18 ∧ (Z21 ∪ Z22) then emotion = cheerful
21:          else if (Z2 ∧ Z15) ∧ (Z19 ∪ Z20) ∧ Z22 then emotion = cheerful
22:          else if (Z2 ∧ Z15) ∧ Z20 ∧ Z21 then emotion = neutral
23:          else if (Z2 ∧ Z15) ∧ Z17 ∧ Z22 then emotion = neutral
24:          else if ((Z1 ∪ Z4 ∪ Z5) ∧ (Z13 ∪ Z14 ∪ Z16)) ∧ Z17 ∧ (Z19 ∪ Z20) ∧ Z22
25:              then emotion = angry
26:          else if ((Z1 ∪ Z4 ∪ Z5) ∧ (Z13 ∪ Z14 ∪ Z16)) ∧ Z19 ∧ Z21 then emotion = bad
27:          else if ((Z1 ∪ Z4 ∪ Z5) ∧ (Z13 ∪ Z14 ∪ Z16)) ∧ (Z19 ∪ Z20) ∧ Z22
28:              then emotion = neutral
29:      else
30:          return "emotion detection error"
31:      end if
32: else
33:      return to "unknown mode"
34: end if
35: end function
```

7 Conclusions

The paper examined how different artificial intelligence methods were used to recognize feline emotions from non-laboratory photos. The following approaches were used: deep learning – convolutional - -transfer method, traditional deep learning – convolution - method, multilayer perceptron method, and the feature prediction method of the deep learning network, and then sorting them into appropriate categories by the expert-system type algorithm [8]. The transferable deep learning network - the ResNet50 model with weights from ImageNet, used to identify whether a cat was in pain, based on lab photos, was in action. Four different datasets were used for this test. The model architecture was modified to recognize four categories. The tested database contained all of the photos, including photos of long-haired and short-haired cats, to eliminate the suggestion that short-haired cats were correctly recognized in a higher percentage. The accuracy was at 70–80%. Traditional deep learning network - the LeNet architecture previously used for human emotion recognition based on black-and-white lab photos [6]. The color of the images in the database was changed to grayscale, and the architecture was modified to recognize four emotions. The accuracy of this model was almost 90% in the tested database. The result was higher than with transfer learning but still much lower than for human faces. Models used to recognize emotions on human faces may need to be improved for feline emotions [8]. For the next step of the experiment, a multilayer perceptron was used to classify the images. The accuracy is at most 80%. The results are comparable to the ResNet50 network: feature prediction and categorization by algorithm. The study bypassed a small sample of labeled data to categorize emotions in cat photos correctly. A model was constructed to predict characteristics. Transfer learning with the VGG19 model and weights from ImageNet was used. It was used for two datasets with characteristics. The first dataset contained images of one thousand cats' mouths. The accuracy of the model on this dataset was 91%. The second part of the photos includes a thousand whole cats - the accuracy was 88%. This number of photos was chosen because it was the smallest number of pictures at which the algorithm correctly categorized them. Characteristics were manually added to the pictures. Then, an algorithm was used to check various relationships between feature points and categorize the photos. A sample of 200 photos was selected to test the program's performance. The correctness of the categorization of emotions in the images was estimated at 98%. Improvements could be achieved by collecting more images for each category in the database. Deep learning models could achieve higher results. A layered perceptron model would likely achieve higher results in categorizing photos with a larger sample of data. Creating a database of labeled photos with characteristics would allow the algorithm to be replaced by a decision tree. Adding more feature points would make it possible to distinguish more subtle emotions, e.g., distinguish the category of happy cat into excited and relaxed in the algorithm. Constructing a model that would recognize different amounts of characteristics depending on the parts of the cat visible in the photo would help streamline the program. Two separate models would not be needed, and the algorithm would be more transparent. A mobile application to collect images for the database could help improve the final neural network model.

References

1. Cat Dataset | Kaggle (2024). https://www.kaggle.com/datasets/crawford/cat-dataset
2. Ellis, S.L.: Recognising and assessing feline emotions during the consultation: history, body language and behaviour. J. Feline Med. Surg. **20**(5), 445–456 (2018). https://doi.org/10.1177/1098612X18771206
3. Feighelstein, M., Shimshoni, I., Finka, L.R., et al.: Automated recognition of pain in cats. Sci. Rep. **12**, 9575 (2022). https://doi.org/10.1038/s41598-022-13348-1

4. Feline Grimace Scale. Université de Montréal (2019). https://www.felinegrimacescale.com/
5. Finka, L.R., Luna, S.P., Brondani, J.T., et al.: Geometric morphometrics for the study of facial expressions in non-human animals, using the domestic cat as an exemplar. Sci. Rep. **9**, 9883 (2019). https://doi.org/10.1038/s41598-019-46330-5
6. Gaddam, D.K.R., Ansari, M.D., Vauppala, S., Gunjan V.K., Sati, M.M.: Human facial emotion detection using deep learning. In: ICDSMLA 2020, pp. 1417–1427 (2022). https://doi.org/10.1007/978-981-16-3690-5_136
7. Imagenet (2021). https://www.image-net.org/
8. Karn-Buehler, J., Kuhne, F.: Perception of stress in cats by german cat owners and influencing factors regarding veterinary care. J. Feline Med. Surg. **24**(8), 700–708 (2022). https://doi.org/10.1177/1098612X211041307
9. Khan, A.R.: Facial emotion recognition using conventional machine learning and deep learning methods: current achievements, analysis and remaining challenges. Comput. Vis. Secur. Appl. Inf. **13**(6), 268 (2022). https://doi.org/10.3390/info13060268
10. Nicholson, S.L., O'Carroll, R.Á.: Development of an ethogram/guide for identifying feline emotions: a new approach to feline interactions and welfare assessment in practice. Ir. Vet. J. **74**, 8 (2021). https://doi.org/10.1186/s13620-021-00189-z
11. OpenCV Library (2022). https://docs.opencv.org/4.7.0/
12. Schötz, S.: The Secret Language of Cats: How to Understand Your Cat for a Better, Happier Relationship. Hanover Square Press (2018). ISBN-13: 978–1335013897 ISBN-10: 133501389X
13. Twelve Common Cat Myths Debunked (2023). https://www.bluecross.org.uk/advice/cat/wellbeing-and-care/12-common-cat-myths-debunked
14. Zimmermann, H.G., Neuneier, R.: Neural network architectures for the modeling of dynamical systems. In: A Field Guide to Dynamical Recurrent Networks, pp. 311–350. IEEE Press, Los Alamitos (2001)

Parallel Swarm Intelligence: Efficiency Study with Fast Range Search in Euclidean Space

Łukasz Michalski⬤, Andrzej Sołtysik⬤, and Marek Woda$^{(\boxtimes)}$⬤

Department of Computer Engineering, Wroclaw University of Technology,
Janiszewskiego 11-17, 50-372 Wrocław, Poland
marek.woda@pwr.edu.pl

Abstract. Swarm intelligence algorithms are recognised for their effectiveness in solving complex optimisation problems. However, scalability can be a significant challenge, especially for large problem instances. This study focuses on examining the time performance of swarm intelligence algorithms when used with parallel computing on both central processing units (CPUs) and graphics processing units (GPUs). The investigation focuses on algorithms designed for range search in Euclidean space. The research optimizes these algorithms for GPU execution and assesses their effectiveness in handling large-scale instances. The inquiry also explores swarm-inspired solutions tailored for GPU implementation, with an emphasis on enhancing efficiency in video rendering and computer simulations. The research findings show potential for advancing the field by addressing challenges related to large-scale optimization through innovative GPU-accelerated swarm intelligence solutions.

Keywords: swarm intelligence · parallel computing · range search · CUDA

1 Introduction

Numerous organisms in natural settings exhibit collective movement patterns, with examples including fish forming schools, birds flying in flocks, sheep organizing into herds, and insects assembling in swarms. In particular, ants display a coordinated distribution strategy when foraging for food, wherein they collectively follow a designated path. The emulation of such aggregate motion holds significant relevance within the realms of artificial life and computer animation, finding application in diverse domains such as gaming and cinematography. Furthermore, these behavioral dynamics are harnessed to address optimization challenges, exemplified by ant colony optimization and particle swarm optimization methodologies [8]. The conceptualization of a computational model to simulate group animal motion was pioneered by Craig Reynolds in 1987, who introduced the *boids* model [10]. The term "boids" denotes simulated creatures

W. Zamojski et al. (Eds.): DepCoS-RELCOMEX 2024, LNNS 1026, pp. 177–186, 2024.
https://doi.org/10.1007/978-3-031-61857-4_17

within a generic flock. The collective motion of boids emerges from the interplay of uncomplicated behaviors exhibited by individual entities. The boids model encompasses three fundamental rules governing the behavior of each simulated boid: aversion to crowding with neighbors, synchronization and coordination of movements with neighboring entities, and convergence towards group cohesion. Subsequent refinements to the model incorporated additional rules, such as navigating around obstacles and pursuing predefined objectives, within the broader framework of steering behavior. Noteworthy adaptations of the boids model include extensions by Delgado et al. [1], who introduced fear effects by simulating emotional transmission between entities using smell, represented as pheromones modeled as particles in a free expansion gas. Additionally, Hartman et al. [4] introduced a supplementary force termed the "change of leadership," influencing a boid's likelihood to assume a leadership role and attempt an escape. Since its inception, the boids model has been extensively employed in computer graphics to generate lifelike representations of collective motion in groups. The Valve Video Game Company, for instance, utilized the boids model in the 1998 video game Half-Life to simulate bird-like creatures in flight. This marked a significant departure from traditional techniques in computer animation for motion pictures. The inaugural application of the boids model in animation occurred in the short film *Stanley and Stella in Breaking the Ice* (1987), with subsequent integration into the introduction of feature film *Batman Returns* (1992). Over the years, the boids model has found widespread adoption in various gaming and cinematic contexts, underscoring its versatility and enduring significance.

The study implements a swarm intelligence algorithm on both multithreaded Central Processing Units and Graphics Processing Units using CUDA (*Compute Unified Device Architecture*) technology to accelerate the algorithm runtime. The contribution to the existing body of knowledge is clear. The research involves a detailed examination of performance measurements and comparative analyses between two distinct computational architectures. The investigation explores visualizing swarm behavior in constrained three-dimensional Euclidean space problems. This aspect of our work aims to improve the interpretability and comprehensibility of the algorithmic results, especially in situations with limited space. Additionally, our research includes experiments to enhance the implementation of the swarm intelligence algorithm on GPU platforms. This optimization process involves adjusting block sizes strategically and using profiling tools judiciously. The main goal is to improve the computational efficiency of the algorithm on GPU architectures.

2   Swarm Intelligence

The boids swarm model represents each rule with a vector that adapts to the environment through its magnitude and direction. The boid's movement vector is a linear combination of all behavior rule vectors. As the number of behavior rules increases, determining and optimizing the coefficients for the moving vector becomes more challenging. It is crucial to establish realistic moving behavior by setting these coefficients appropriately. Flock behaviour is the result of the motion and interaction of boids. Each boid follows three simple rules of steering behaviour that describe how it moves based on the positions and velocities of its flock mates (social reaction) (Fig. 1).

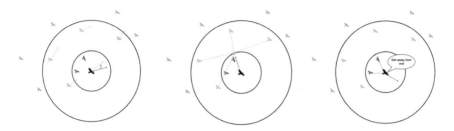

Fig. 1. Three main rules describing flocking behavior. Alignment rule (most left) - align with an average velocity of boids in the visible range. Cohesion rule (center image) - move toward the center of mass of boids in the perception range. Separation rule (most right) - move away from boids in the protected range. Source: *Cornell University ECE 4760 - Obsolete Designing with Microcontrollers.*

Cohesion rule defines the steering force to move toward the average position of local flock mates (as in the original Reynolds' model). Applying the cohesion rule keeps the boids together. This rule acts as the complement of the separation. If only the cohesion rule is applied, all the boids in the flock will merge into one position. Cohesion $\left(\overrightarrow{Coh_i}\right)$ of the boid (b_i) is calculated in two steps. First, the center $\left(\overrightarrow{Fc_i}\right)$ of the flock (f) that has this boid is calculated as in Eq. 1. Then the tendency of the boid to navigate toward the center of the density of the flock is calculated as the cohesion displacement vector as in Eq. 2. Where p_j is the position of boid j and N is the total number of boids in f.

$$\overrightarrow{Fc_i} = \sum_{\forall b_j \in f} \frac{\overrightarrow{p_j}}{N} \tag{1}$$

$$\overrightarrow{Coh_i} = \overrightarrow{Fc_i} - \overrightarrow{p_i} \tag{2}$$

Alignment steer to match the heading and the speed of its neighbors. This rule tries to make the boids mimic each other's course and speed. Boids tend to align with the velocity of their flock mates. The alignment (Ali_i) is calculated in two steps. First, the average velocity vector of the flock mates $\left(\overrightarrow{Fv_i}\right)$ is calculated by Eq. 3. Then Ali_i is calculated as the displacement vector in Eq. 4. Where $\overrightarrow{v_i}$ is the velocity vector of boid i. If this rule were not used, the boids would bounce around a lot and not form the beautiful flocking behavior that can be seen in nature.

$$\overrightarrow{Fv_i} = \sum_{\forall b_j \in f} \frac{\overrightarrow{v_j}}{N} \tag{3}$$

$$\overrightarrow{Ali_i} = \overrightarrow{Fv_i} - \overrightarrow{v_i} \tag{4}$$

Separation rule avoids collisions and overcrowding with other flock mates. There are many ways to implement this rule. An efficient solution to calculate the separation (Sep_i) is by applying Eq. 5. Vectors defined by the position of the boid b_i and each visible boid b_j are summed, then separation steer $\left(\overrightarrow{Sep_i}\right)$ is calculated as the negative sum of these vectors.

$$\overrightarrow{Sep_i} = -\sum_{\forall b_j \in f} (\overrightarrow{p_i} - \overrightarrow{p_j}) \tag{5}$$

Overall boid model moving vector V_i for boid b_i is calculated by combining all the steering behavior vectors as in Eq. 6. Where w_i are the coefficients describing the influences of each steering rule and used to balance the rules.

$$\overrightarrow{V_i} = w_1 \cdot \overrightarrow{Coh_i} + w_2 \cdot \overrightarrow{Ali_i} + w_3 \cdot \overrightarrow{Sep_i} \tag{6}$$

The foundational operation of the algorithm involves the iterative refinement of both the motion vector and position grouping for every particle p (??). This meticulous process not only governs the trajectory of individual particles but also facilitates the cohesive grouping of moving objects into well-defined clusters.

Range Search. Within the realm of the algorithm, a crucial computational undertaking commands attention - the judicious allocation of resources to discern and isolate local particles residing within the perceptual purview of a given entity. Conceptually, one can envision each simulated particle as an object on a virtual stage, that identifies and interacts with neighboring objects. The essence of the algorithmic framework lies in the fast filtering of neighboring particles within the immediate perceptual radius of a focal entity, thereby laying the groundwork for subsequent decision-making processes. Selective filtration of particles within this localized sphere of influence serves as a fundamental aspect for optimizing computational efficiency while concurrently affording the algorithm an insightful comprehension of the microcosmic environment surrounding each simulated entity. Not merely a computational optimization strategy, this targeted filtration process becomes the basis for the emergent dynamics defining the collective behavior of the simulated entities. By discerning the presence and relative positioning of proximate particles, the algorithm affords the simulated entities the capacity for responsive actions, encapsulating the principles of cohesion, alignment, and separation emblematic of natural flocking phenomena.

Linear Search is a naive approach for computing the new positions and velocities for the boid b_i. Each particle performs an exhausting search of the entire flock f and filters out boids in their perception range. Then accumulate forces for each boid b_j in the neighborhood and apply the update on boid b_i. This is extremely slow with a runtime of $O(n^2)$. Where n is the number of boids in the flock f.

Faster Range Search can be achieved using tree data structures like k-d tree or octree. In computer science and computational geometry, k-d trees have become a popular data structure used to organize points in k-dimensional space. This is because these structures allow for very efficient searches of points in multidimensional space, including nearest-neighbor searches and range searches (Fig. 2). In a k-d tree, the space is recursively divided into half-spaces along a specific dimension, alternating at each level of the tree - this process is referred to as a spatial partitioning technique [2,3,6]. The selection of the dimension along which to split the space is made based on various criteria, typically maximizing data distribution or minimizing the variance along a particular dimension.

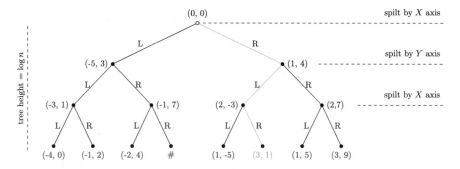

Fig. 2. k-d tree data structure, with a specific value $k = 2$. The steps for searching the nearest neighbor to the point $(3, 3)$ are highlighted in green. This is done in logarithmic time [5].

Octrees similar to k-d trees divide space but into octants (sub-cubes), where each node in the tree corresponds to an octant (Fig. 3). The term *octree* is derived from the fact that each node has eight children, corresponding to the eight octants resulting from the subdivision of a cube. Octrees are particularly well-suited for representing 3D spatial data and volumes. They provide an efficient way to organize and query spatial information hierarchically. One of the strengths of octrees is their ability to adapt to varying point densities in space. They can efficiently represent both dense and sparse regions by adjusting the level of detail in different parts of the space. On the other hand, octree may not be as efficient for uniformly distributed data compared to kd-trees.

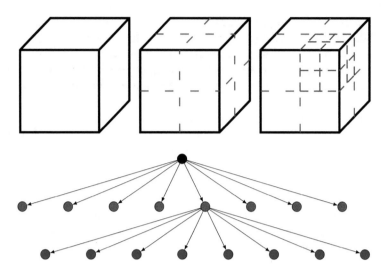

Fig. 3. Octree Data Structure: recursive subdivision of a cube into octants. The resulting octree showcases the hierarchical organization of Euclidean space into octants.

3 Implementation

While the proposed solution reduces time complexity, implementing it on the GPU side using trees is challenging [9]. The tree is a 'pointer machine' data structure with possible gaps in computer memory. To implement this in CUDA, one thread can be assigned to each boid. Each thread would loop over the entire position and velocity buffers (skipping itself) and compute a new velocity for that boid. Another CUDA kernel would then apply the position update using the new velocity data.

Proper Data Representation is a crucial concept in designing parallel algorithms to reduce memory throughput. When organizing an agent's diverse properties in GPU memory, it is important to balance the trade-off between fast access and easy maintenance. Storing data in the struct-of-array format is an effective approach to enhance coalesced access to global memory. In this configuration, values for a common property of all agents are stored in a separate array, reducing the number of cache misses. It is worth noting that agents may have varying properties, and not all properties are necessarily shared by every agent. Therefore, creating a distinct array for each property in the struct-of-array format can introduce programming complexities. Conversely, the array-of-struct format is more favourable in terms of programmability. However, the adoption of this approach does not optimize coalesced access, as the data for a specific agent property becomes interleaved in global memory. Therefore, in our approach, we choose to represent agents using the struct-of-array format to strike a balance between facilitating coalesced access and maintaining programmatic simplicity.

Efficient Range Search on GPU. It is clear that implementing a spatial data structure can significantly improve the performance of the algorithm. By reducing the number of boids that each individual boid must evaluate, we can effectively minimise the computational load per thread. Since the three governing rules apply within a certain radius, organising the boids into a uniformly spaced grid proves to be crucial for performing an efficient neighbour search [7]. A uniform grid consists of cells that are at least as wide as the neighborhood distance and cover the entire simulation domain. In a preprocessing step, we assign the boids to the grid before computing their new velocities. This approach reduces the number of boids that need to be considered. If the cell width is double the neighborhood distance, each boid only needs to be checked against other boids in 8 cells, or 4 in the 2D case (see Fig. 4).

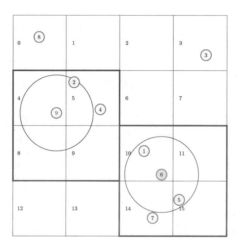

Fig. 4. Range search in uniform scattered grid particles space

To build a uniform grid on the CPU, one must iterate over the boids, determine their enclosing cell, and keep a pointer to the boid in a resizable array that represents the cell. However, this approach is not suitable for the GPU due to unresizable arrays and the potential for race conditions when naively parallelizing the iteration. The construction phase of the uniform grid is achieved through sorting. Each boid is labelled with an index representing its enclosing cell, and the list of boids is sorted by these indices to ensure that pointers to boids in the same cells are contiguous in memory. The array of sorted uniform grid indices is then traversed, and every pair of values is examined. If the values differ, it indicates that we are at the border of the representation of two different cells. Representing the uniform grid can be achieved by storing the locations in a table with an entry for each cell. This table can be an array with space for all cells. The process is data-parallel and can be parallelized.

4   Results

The comparative analysis involved assessing the performance of the algorithm across various implementations, specifically in CPU single-threaded (ST), multithreaded (MT), and GPU parallel versions (LS - linear search & U-SG - uniform scattered grid). The metric utilized for evaluation was frames per second (FPS), and the particle number N served as a key parameter. For the CPU-based implementation on an AMD Ryzen 5 7600 processor, the simulation encountered limitations, allowing for the successful execution of up to 50k and 100k particles for single-threaded (ST) and multithreaded (MT) scenarios, respectively. In contrast, the GPU implementation running on NVIDIA GeForce RTX 4070Ti using the CUDA framework demonstrated greater scalability, running simulations with $N = 1M$ particles in naive implementation and 5M using fast range search. However, it is noteworthy that the achieved FPS rate did not meet the desired level of responsiveness in this configuration. Results are presented in Table 1 metric was rounded to two significant digits (Fig. 5).

Table 1. Performance Results - frames per second/particle number N. The ✗ marker indicates simulation launch failure.

N	AMD Ryzen 5 7600		NVIDIA GeForce RTX 4070Ti	
	ST [FPS]	MT [FPS]	GPU LS [FPS]	GPU U-SG [FPS]
1 000	182	450	1700	2300
2 500	38	120	1000	2300
5 000	9.9	42	590	1700
10 000	2.5	13	280	1800
25 000	0.43	2.6	120	1500
50 000	0.97	0.68	36	1100
100 000	✗	0.16	9.8	460
250 000	✗	✗	2.1	200
500 000	✗	✗	0.53	55
1 000 000	✗	✗	0.13	14
2 500 000	✗	✗	✗	0.58
5 000 000	✗	✗	✗	0.1

Furthermore, the assessment of algorithmic performance on the GPU was conducted by varying the threads per block configuration in simulations comprising $N = 25 \cdot 10^3$ particles. In the pursuit of determining the optimal thread block size, a systematic approach involving experimentation and profiling was employed. Through profiling of the thread block size, warp occupancy, and other pertinent parameters, we successfully identified the configuration that maximizes performance for our distinct workloads and hardware specifications. Ultimately, the most effective choice for a specific N was found to be 128 threads

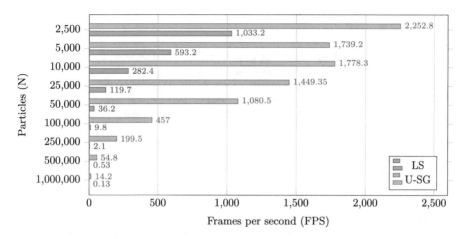

Fig. 5. Comparison of GPU linear search (LS) and fast range search using a uniform grid (U-SG) based on FPS/N - particles

per block. It is noteworthy to highlight the characteristics of warps, particularly in instances where the block size $T < 32$. In such cases, a block consists of a single warp, and despite the presence of unused threads, each thread is executed individually without contributing to any algorithmic advantages. The utilization of smaller blocks necessitates a larger number of blocks. Allocating memory for each of these numerous blocks, each containing a single warp, results in a loss of the performance benefits associated with shared memory within a block. The utilization of the GPU for parallel calculations demonstrates an average speedup of 50 times compared to employing a CPU multithreaded solution when $N \geq 25 \cdot 10^3$. Notably, the bottleneck associated with data transfer between the host and device via the PCIe bus becomes negligible. This is attributed to the substantial advantages derived from highly parallel calculations, particularly when dealing with simulation sizes numbering in the thousands.

5   Summary

The author's parallel implementation of the algorithm demonstrates significant performance gains achieved by implementing the parallel computation model of CUDA technology on GPU compared to traditional CPU approaches. The use of GPU parallelism, particularly in CUDA, leads to significant improvements in the computational efficiency of the multivariate cooperative algorithm analysed. It is noteworthy to mention the potential for further improvements by building on the shared memory concept introduced in the reference paper [7]. This suggests that improvements in shared memory utilisation could contribute to even greater performance improvements in GPU-based parallel computing.

References

1. Delgado-Mata, C., Ibáñez-Martínez, J., Bee, S., Ruiz-Rodarte, R., Aylett, R.: On the use of virtual animals with artificial fear in virtual environments. New Gener. Comput. **25**, 145–169 (2007). https://doi.org/10.1007/s00354-007-0009-5
2. Eldawy, A., Alarabi, L., Mokbel, M.F.: Spatial partitioning techniques in spatial-hadoop. Proc. VLDB Endow. **8**(12), 1602–1605 (2015).https://doi.org/10.14778/2824032.2824057
3. Gomes, A.J.P., Voiculescu, I., Jorge, J., Wyvill, B., Galbraith, C. (eds.): Spatial Partitioning Methods, pp. 187–225. Springer, London (2009).https://doi.org/10.1007/978-1-84882-406-5_7
4. Hartman, C., Benes, B.: Autonomous boids. J. Vis. Comput. Animat. **17**, 199–206 (2006)
5. Karger, D.R.: Advanced Algorithms. In: Advanced Algorithms MIT Course No.6.5210/18.415. MIT OpenCourseWare (2022). https://6.5210.csail.mit.edu/
6. Li, B.: A Comparative Analysis of Spatial Partitioning Methods for Large-scale, Real-time Crowd Simulation (2014). https://api.semanticscholar.org/CorpusID:54175377
7. Li, X., Cai, W., Turner, S.J.: Efficient Neighbor Searching for Agent-Based Simulation on GPU, pp. 87–96. DS-RT '14, IEEE Computer Society, USA (2014). https://doi.org/10.1109/DS-RT.2014.19
8. Michelakos, I., Mallios, N., Papageorgiou, E., Vassilakopoulos, M.: Ant Colony Optimization and Data Mining, pp. 31–60. Springer, Berlin, Heidelberg (2011). https://doi.org/10.1007/978-3-642-20344-2_2
9. NVIDIA Corporation: NVIDIA CUDA C Programming Guide (2023). https://docs.nvidia.com/cuda/pdf/CUDA_C_Programming_Guide.pdf
10. Reynolds, C.W.: Flocks, herds and schools: a distributed behavioral model. In: Proceedings of the 14th Annual Conference on Computer Graphics and Interactive Techniques, pp. 25–34 (1987)

Digital Transformation Impacts on Industry 4.0 Evolution

Issam A. R. Moghrabi(⊠)

Kuwait Technical College, Information Systems and Technology Department, Kuwait City, Kuwait
i.moghrabi@ktech.edu.kw

Abstract. This paper explores digital metamorphosis and its contribution to smart logistics. Technological developments have disintegrated business processes and conditioning by taking enterprises to utilize digital technologies in all operations. The world is currently undergoing Industry 4.0, a new industrial revolution characterized by the utilization of novel technologies such as the Internet of Things (IoT), artificial intelligence (AI), and big data in manufacturing processes. This has led to significant changes in the way industries operate, and it has also impacted other sectors of the economy. In this paper, we will focus on two aspects of Industry 4.0 – Smart Logistics and Smart Economy and highlight their impact on the industrial sector and the economy as a whole.

Keywords: smart economy · digital transformation · Industry 4.0 · smart logistics

1 Introduction

To investigate and incorporate digital technology into their systems and procedures, businesses have recently implemented a variety of tactics. Thus, to improve performance and efficiency, digital metamorphosis has become a process that ignites organizational changes by combining information, computional, and communication advances. Digital transformation, driven by Industry 4.0 technologies, has emerged as a significant development in recent decades, impacting various aspects of personal, social, and economic life [1]. It is considered an inevitable trend and a matter of survival for countries, organizations, businesses, and consumers worldwide [2]. The implementation of Industry 4.0 and digital transformation has advantages for manufacturing industries, while also affecting the employment structure and social life. The fast growth of technology during the Industry 4.0 era has led to the formation of new or evolved organizations, with changes in the skills and capacities of individuals and organizations [3]. The impact of Industry 4.0 on the digital transformation of industries has been studied, highlighting the need for transformation at multiple levels, including industrial equipment, IT systems, and internal business processes [4]. The digital transformation of manufacturing enterprises in Ukraine, for example, under the influence of Industry 4.0 involves the optimization of equipment operation modes, improvement in productivity and labor safety, logistic optimization, and improvement in product quality, among other factors.

The implementation of Industry 4.0 technologies in production has the potential to optimize equipment operation, increase productivity and labor safety, improve product quality, and shorten time-to-market, among other benefits.

2 Problem Statement and Research Objectives

In recent years, digital transformation has emerged as a pivotal force reshaping industries across the globe. This transformation is characterized by the integration of digital technologies into all aspects of business operations, leading to profound changes in how companies operate, compete, and create value. Concurrently, Industry 4.0, the fourth industrial revolution, is driving the convergence of physical and digital technologies in manufacturing and production environments. However, the precise influence and implications of digital transformation on the evolution of Industry 4.0 remain a topic of significant interest and exploration. Thus, the research objectives of this paper can be stated as:

i. Examine the key drivers and catalysts of digital transformation in various industries, including manufacturing, healthcare, finance, and retail.
ii. Explore the impacts of digital transformation on traditional manufacturing processes, supply chain management, product development, and customer engagement within the context of Industry 4.0.
iii. Analyze the challenges and barriers faced by organizations in adopting and implementing digital transformation initiatives, including issues related to infrastructure, cybersecurity, talent acquisition, and cultural change.
iv. Identify best practices, strategies, and frameworks for effectively managing and leveraging digital transformation efforts to maximize benefits and minimize risks in the Industry 4.0 era.
v. Provide insights and recommendations for policymakers, industry leaders, and stakeholders to navigate and harness the opportunities presented by digital transformation in the context of Industry 4.0.

Expected Outcomes:

1. A comprehensive understanding of the interplay between digital transformation and Industry 4.0, including the synergies, challenges, and transformative potential.
2. Insights into the role of digital technologies in driving innovation, agility, and competitiveness across industries undergoing digital transformation.
3. Guidelines and recommendations for organizations seeking to embark on or accelerate their digital transformation journey within the Industry 4.0 paradigm.
4. Implications for policy formulation, investment strategies, and talent development initiatives to support and capitalize on the digital revolution and Industry 4.0 advancements.

3 Impact of Industry 4.0 on the Logistics Industry

The impact of Industry 4.0 on the logistics industry is profound, ushering in a brand-new period of creativity and efficiency. By combining cutting-edge technology like data analytics, artificial intelligence, and the Internet of Things (IoT), logistics operations are becoming increasingly streamlined and intelligent. Real-time tracking of shipments, coupled with predictive analytics, enables precise demand forecasting and inventory management. Autonomous vehicles and drones are transforming last-mile delivery, enhancing speed and reducing costs. Furthermore, the use of smart warehouses equipped with automated systems optimizes storage, retrieval processes, and order fulfillment. The result is a logistics landscape characterized by improved responsiveness, reduced operational errors, and enhanced overall supply chain visibility, illustrating the transformative power of Industry 4.0 in reshaping the logistics industry [5].

The logistics industry is responsible for commodities transportation and storage from the source to the final destination. It is a complex and dynamic industry, with various stakeholders involved, such as manufacturers, suppliers, distributors, and retailers. Industry 4.0 technologies are transforming the logistics industry in the following ways:

- Real-Time Tracking and Visibility: With the use of IoT sensors and devices, Industry 4.0 enables real-time tracking and visibility of goods throughout the entire supply chain. This allows logistics companies to monitor the location, condition, and status of shipments, leading to improved transparency and efficiency.
- Predictive Maintenance: Smart logistics Systems gather data in real-time using IoT-enabled sensors. From vehicles and equipment, allowing for predictive maintenance. This helps to prevent breakdowns and delays, ensuring timely delivery of goods.
- Automated Warehousing: Industry 4.0 technologies such as robotics and automation are being used to optimize warehouse operations. Automated storage and retrieval systems, as well as autonomous robots, are being used to move and store goods, reducing the need for manual labor and increasing efficiency.
- Optimization of Routes and Delivery: Big data analytics and Delivery routes are optimized by AI algorithms that consider variables like traffic, weather, and fuel costs. This results in faster and more efficient deliveries, reducing costs and improving customer satisfaction.
- Demand Forecasting: Industry 4.0 technologies enable logistics companies to gather and examine enormous volumes of data, such as market trends and consumer behavior. This data can be used to forecast demand accurately, allowing logistics companies to plan and optimize their operations accordingly.

4   Challenges and Future Outlook

Despite the numerous benefits of Industry 4.0 and smart logistics, there are still some challenges that need to be addressed. The high implementation costs of these technologies represent one of the primary challenges, which may be a barrier for smaller logistics companies. Additionally, there is a need for skilled workers who can operate and maintain these advanced systems.

However, the future outlook for smart logistics is promising. As technology continues to advance and becomes more affordable, more logistics companies will be able to adopt Industry 4.0 technologies. This will result in increased competition and further innovation in the industry, leading to even more efficient and sustainable logistics operations [6].

Industry 4.0 and smart logistics are transforming the traditional logistics industry, leading to increased efficiency, cost savings, and sustainability. Utilizing the right cutting-edge technologies can optimize the efficiency of their operations, supply a better customer experience, and contribute to a more sustainable future. As we continue to move towards a more connected and digital world, the impact of Industry 4.0 on the logistics industry will only continue to grow [7].

Among the main advantages of smart logistics is the reduction of manual labor. With the use of automation and robotics, tasks such as inventory management, order fulfillment, and transportation can be carried out more accurately and quickly, reducing the need for human labor. This also allows for a more flexible and agile supply chain, as changes in demand can be quickly accommodated without the need for manual adjustments. Smart Logistics also enables better visibility and transparency in the flow of the supply chain, which is critical for effective supply chain management. Using IoT sensors, real-time data on the location and status of goods can be collected and analyzed, allowing for better decision-making and risk management. This also facilitates the implementation of sustainable methods, so businesses may monitor and lower their carbon footprint by optimizing transportation routes and reducing waste.

5   Impact on the Industrial Sector

The integration of Smart Logistics into the industrial sector has brought about significant changes and has had a favorable effect on the sector at large. Making advantage of advanced technologies, companies can now produce goods more efficiently and at a lower cost, which leads to increased competitiveness in the global market. This has also enabled companies to offer personalized and customized products to meet the changing needs and preferences of consumers.

Additionally, Smart Logistics has enabled companies to improve their inventory management, reducing the risk of overstocking or stockouts. This reduces expenses and raises consumer satisfaction by guaranteeing that the products are available when and where they are needed [8]. The use of real-time data and analytics has also enabled predictive maintenance, reducing downtime and maintenance costs for machinery and equipment.

In their work, Warke et al. [6] highlight four dimensions for digital transformation: financial considerations, alterations to the structure, technological use, and shifts in the value-creation process. The feature related to the use of technology involves an organization's stance toward new manufacturing technologies and its ability to leverage these technologies for changes and improvements. According to [8], a successful digital business transformation necessitates the effective integration of new technologies across various business domains, fundamentally altering how a company operates. In the context of sustainable manufacturing, companies need to identify and incorporate technologies such as equipment and processes to influence the manufacturing of products. The choice of technologies depends on the organization's operations, skills, and technical requirements, resulting in varying technological needs across different industries. Companies must choose between developing new technologies to become market leaders and adapting current ones to fulfill requirements in this aspect of the digital transformation plan.

The research outcomes suggest that significant changes in value creation result from the use of digital technologies in sustainable manufacturing, impacting value chains.

Machado et al. [9] stress the critical role of value-creation networks in sustainable manufacturing, transforming corporate relationships both within and across companies. For instance, when a company embraces another's technologies, like cloud computing, the value chains calibrate to adjust to these collaborations. Additionally, Frosch et al. [10] note that certain technologies may lead to deviations from a company's core business, presenting expansion opportunities. However, such deviations may need new capabilities in digital to ensure a successful transformation. This supports the claims that reengineering and optimizing business processes in the light of new strategies and needs is necessary for a successful digital transformation. Digital change in sustainable marketing encompasses elements including circumstances, organizational culture, policies, and processes, emphasizing the necessity for a flexible value-creation network.

The dimension of structural alterations pertains to modifications made to a business structure concerning the location of newly implemented digital initiatives. According to Frosch et al. [10], digital transformation can impact processes, skills, and products, necessitating structural changes within the firm. For instance, the increased societal awareness of the environmental impact of industrial practices, driven by the adoption of ICT technologies, has influenced the market demand for sustainable products. Machado et al. [9] emphasize the global impacts of manufacturing, highlighting digital technology's necessity to enable a successful shift to sustainable practices. The authors in [11, 12] identify major digital aspects in the manufacturing sector, such as cloud computing, mobile technology, the Internet of Things (IoT), big data, and data analysis, which may require substantial investments. While these investments can result in increased performance, efficiency, and productivity that raise sales and revenues, achieving these advantages requires significant funding. Moghrabi et al. [13] acknowledge the uncertainty of the economics of digital transformation, highlighting the need for additional study. Therefore, insufficient financial resources significantly undermine the adoption of sustainable manufacturing technologies.

Throughout the literature search, the circular economy and Industry 4.0 emerged as recurrent themes. Industry 4.0 is defined as the use of state-of-the-art technology to digitally transform production and manufacturing processes to improve value creation and real-time optimization. Machado et al. [9] identify Industry 4.0 as the creation of smart commodities or services, smart factories, and extended value networks that can be created by fusing the Internet of Things and manufacturing processes with the Cyber-Physical System (CPS). It entails technologies that are affordable, versatile, adaptable, agile, and reconfigurable. Shahbazi et al. [14] reveal how Industry 4.0 integration has increased manufacturing flexibility and enabled customized mass production. Sustainability is a key component of smart manufacturing, with smart factories that use renewable energy to increase their level of independence. Modern technologies that reinforce organizational data infrastructures include Industrial IoT and integrated systems, which enable embedded computing communications and real-time interactions. Shahbazi et al. [15] explain how Industry 4.0 technology can integrate and transparently monitor a product's lifecycle throughout the manufacturing process, contributing to sustainable manufacturing dimensions—social, environmental, and economic. Table 1 presents a quick summary of the influence of Industry 4.0 technologies on sustainable industry.

Table 1. Effects of Industry 4.0 on Sustainable Industry

Technology 4.0	Opportunities Associated with Sustainable Manufacturing
collaborative and autonomous robots	These autonomously functioning machines can operate alongside humans in production environments. Manufacturing facilities combine them with human talents to boost productivity and decrease workload. These qualities play a major role in mass production They improve the production process's flexibility
Industrial IoT	Utilizing technologies, industrial processes' data is gathered and analyzed to prevent waste and pointless manufacturing steps The monitoring capabilities of the technologies improve equipment and operator safety by offering real-time solutions and hazard warnings. Maintenance issues can be resolved by integrating industrial IoT with sustainable manufacturing technologies
Cloud computing	Energy is saved because virtualized data centers take the place of physical resources Sharing computer and storage infrastructures amongst organizations lowers expenses and improves access to virtual data from any location
Systems Integration	Manufacturing facilities use vertical and horizontal system integration to improve visibility and enable close observation of operations Manufacturing waste and energy consumption can be decreased with the use of integrated data from all organizational domains, including administrative and manufacturing

6   Smart Economy and Its Impact

The concept of a Smart Economy is closely related to Industry 4.0, as it involves the integration of advanced technologies into the overall economy. It encompasses various sectors such as manufacturing, transportation, healthcare, energy, and agriculture. The use of IoT, AI, and big data in these sectors has led to increased efficiency, productivity, and innovation, ultimately contributing to economic growth. One of the key impacts of the Smart Economy is the creation of new job opportunities. While it is true that automation and robotics may replace some manual labor, the integration of advanced technologies also develops a demand for skilled workers capable of developing, operating, and maintaining these systems. This leads to a shift in the job market, with a higher demand for workers with technical and digital skills [12].

Smart Economy also leads to the creation of new business models and opportunities. With the use of advanced technologies, companies can now offer new products and services, such as personalized and on-demand manufacturing, which were not possible before. This creates new revenue streams and increases competitiveness in the global market [8].

The adoption of Industry 4.0 technologies has had a considerable impact on the industrial sector. The following are some of the major ways in which this revolution has changed the industrial landscape:

i. Increased Productivity and Efficiency

One of the most significant impacts of Industry 4.0 on the industrial sector is the increased efficiency and productivity of manufacturing processes. The integration of advanced technologies such as robotics, automation, and IoT has enabled industries to streamline their operations and improve their overall efficiency. With the help of real-time data, manufacturers can monitor their processes and identify areas for improvement, leading to increased productivity.

ii. Cost Reduction

Industry 4.0 technologies have also helped industries to reduce their costs significantly. By automating various tasks, manufacturers can save on labor costs and increase their overall efficiency. Predictive maintenance enabled by IoT and big data analytics has also reduced downtime and maintenance costs, leading to cost savings for industrial companies.

iii. Improved Quality and Customization

The use of advanced technologies in the industrial sector has also led to improved product quality and customization. Data analytics can be used by manufacturers to identify flaws in real time and make the required corrections to raise the standard of their products. The integration of IoT and big data also allows industries to gather customer data and personalize their products to meet the specific needs and preferences of their customers.

iv. Job Displacement and Changes in Skills Requirements

While Industry 4.0 has brought numerous benefits to the industrial sector, Furthermore, it has resulted in the loss of jobs and modifications to the skills needed for employment. With the adoption of automation and robotics, many manual jobs have been replaced, leading to a decrease in the demand for labor in some industries.

v. Supply Chain Optimization

Industry 4.0 has also had a profound influence on supply chain management. With the integration of IoT and real-time data analytics, manufacturers can monitor their supply chain in real time, leading to improved efficiency and reduced costs. This has also enabled industries to respond quickly to changes in demand and supply, leading to better customer satisfaction.

To sum up, the impact of Industry 4.0 on innovative organizations includes changes in business models, improved career satisfaction and well-being for professionals, and the ability to adapt quickly to client requests.

7   Challenges of Industry 4.0 for the Industrial Sector

While Industry 4.0 has brought numerous benefits to the industrial sector, it has also brought some challenges. The following are some of the key challenges faced by industries in this revolution:

1. High Initial Investment

    The adoption of Industry 4.0 technology comes with a hefty upfront cost, which might be prohibitive for small and medium-sized businesses (SMEs). The high costs associated with implementing advanced technologies such as IoT, robotics, and data analytics can make it difficult for smaller companies to keep up with larger competitors.

2. Data Security and Privacy Concerns

    Modern technology's incorporation into the industrial sector has raised concerns about data security and privacy. With the increased use of connected devices and the collection of vast amounts of data, industries need to ensure that their systems are secure from cyber threats. They also need to comply with data privacy regulations to protect their customers' data.

3. Workforce Adaptation and Reskilling

    As mentioned earlier, Industry 4.0 technologies resulted in the loss of jobs and modifications to the skills needed for employment. Industries need to invest in training and reskilling their employees to adapt to the changing technological landscape. Some businesses may find this to be a major difficulty, particularly those with a sizable workforce.

The advent of Industry 4.0 has brought forth a wave of transformative technologies, but its integration into the industrial sector is not without challenges. One significant obstacle is the substantial financial investment required for the usage of innovative technologies like the Internet of Things (IoT), smart automation, and artificial intelligence. Another formidable challenge lies in the heightened cybersecurity risks associated with Industry 4.0 implementation. As industrial systems become more interconnected through IoT devices and cloud computing, they become susceptible to cyber threats that can interfere with operations, compromise private information, and even pose safety risks. Securing the industrial infrastructure against cyberattacks demands robust measures, advanced cybersecurity protocols, and continuous monitoring.

8   Conclusion

In conclusion, Industry 4.0 has had an impactful influence on the industrial sector, bringing numerous benefits such as increased efficiency, productivity, and cost reduction. However, it has also brought challenges such as job displacement, changes in skills requirements, and data security concerns. To thrive in this revolution, industries need to embrace these technologies and invest in their workforce to adapt to the changing landscape. With proper planning and implementation, the industrial sector can continue to reap the benefits of Industry 4.0 and stay competitive in the global market. The integration of Industry 4.0 technologies into the supply chain and the overall economy has brought about significant changes and benefits. Smart Logistics has improved the efficiency and cost-effectiveness of the supply chain, while also enabling better sustainability practices. The industrial sector has also seen improvements in productivity, competitiveness, and job creation.

References

1. Sartal, A., Bellas, R., Mejías, A.M., García-Collado, A.: The sustainable manufacturing concept, evolution and opportunities within Industry 4.0: a literature review. Adv. Mech. Eng. **12**(5), 23–37 (2020)
2. Hương, T.P., Pham, T.G., Pham, T.M., Ta, L.A.: Impact of digital transformation on knowledge-sharing activities of accounting students at universities in VietNam. Int. J. Sci. Res. Manag. **11**(03), 4689–4704 (2023). https://doi.org/10.18535/ijsrm/v11i03.em4
3. Karakuş, G.: The Impact of Digital Transformation on the Quality of Work Life of Female Professionals in the Industry 4.0 Environment, pp. 123–148 (2023). https://doi.org/10.4018/978-1-6684-6118-1.ch008
4. Kryshtal, H., Zgalat-Lozynska, L., Denysiuk, O., Skyba, H., Panin, Y.: The impact of Industry 4.0 on the digital transformation of manufacturing enterprises in Ukraine. Naukovyi Visnyk Natsionalnoho Hirnychoho Universytetu **2**, 149–153 (2023). https://doi.org/10.33271/nvngu/2023-2/149
5. Kamel Boulos, M.N., Zhang, P.: Digital twins: from personalised medicine to precision public health. J. Pers. Med. **11**(8), 745 (2021)
6. Warke, V., Kumar, S., Bongale, A., Kotecha, K.: Sustainable development of smart manufacturing driven by the digital twin framework: a statistical analysis. Sustainability (Switzerland) **13**(18) (2021). https://doi.org/10.3390/su131810139
7. Žídek, K., Pitel', J., Adámek, M., Lazorík, P., Hošovský, A.: Digital twin of experimental smart manufacturing assembly system for industry 4.0 concept. Sustainability (Switzerland) **12**(9), 1–16 (2020). https://doi.org/10.3390/su12093658
8. Zuo, Y.: Making smart manufacturing smarter–a survey on blockchain technology in Industry 4.0. Enterp. Inf. Syst. 1–31 (2020)
9. Machado, C.G., Winroth, M.P., Ribeiro da Silva, E.H.D.: Sustainable manufacturing in Industry 4.0: an emerging research agenda. Int. J. Prod. Res. **58**(5), 1462–1484 (2020)
10. Frosch, R.A., Gallopoulos, N.E.: Strategies for manufacturing. Sci. Am. **261**(3), 144–153 (1989)
11. Meindl, B., Ayala, N.F., Mendonça, J., Frank, A.G.: The four smarts of Industry 4.0: evolution of ten years of research and future perspectives. Technol. Forecast. Soc. Change **168**, 120784 (2021). https://doi.org/10.1016/j.techfore.2021.120784
12. Soori, M., Arezoo, B., Dastres, R.: Virtual manufacturing in industry 4.0: a review. Data Sci. Manag. **7**(1), 47–63 (2024). https://doi.org/10.1016/j.dsm.2023.10.006
13. Moghrabi, I.A.R., Bhat, S.A., Szczuko, P., AlKhaled, R.A., Dar, M.A.: Digital transformation and its influence on sustainable manufacturing and business practices. Sustainability (Switzerland) **15**(4), 34–61 (2023)
14. Shahbazi, Z., Byun, Y.-C.: Smart manufacturing real-time analysis based on blockchain and machine learning approaches. Appl. Sci. **11**(8), 3535 (2021)
15. Shahbazi, Z., Byun, Y.-C.: Integration of blockchain, IoT and machine learning for multistage quality control and enhancing security in smart manufacturing. Sensors **21**(4), 1467 (2021)

Optimization of Procurement Strategy Supported by Simulated Annealing and Genetic Algorithm

Szymon Niewiadomski[(✉)] and Grzegorz Mzyk

Wroclaw University of Science and Technology, Wroclaw, Poland
{szymon.niewiadomski,grzegorz.mzyk}@pwr.edu.pl
http://staff.iiar.pwr.wroc.pl/grzegorz.mzyk/

Abstract. The problem of optimizing purchasing plans in a large enterprise, taking into account non-linear constraints, was analyzed and formalized. A critical review of available methods was made in the context of their potential application. Experimental results using the simulated annealing method and the genetic algorithm are presented and compared.

Keywords: Non-convex optimization · Non-linear constraints · Simulated annealing · Genetic algorithm · Forecasting

1 Introduction

This article is the result of work on the strategy for purchasing IT resources in a company from the energy sector. A characteristic feature of IT purchases is the continuously growing demand for IT resources, the lifespan of which is limited. Resources must be phased out from the infrastructure due to the end of support or degradation. The price of resources fluctuates over time and depends on the order quantity [14].

On one hand, we have a procurement strategy involving a single large purchase that ensures resources for many years. On the other hand, we can conduct procurement procedures as frequently as possible, ensuring maximum utilization. The first of these strategies will be optimal with a constant resource price over time and significant discounts based on order quantity, while the second will be optimal with decreasing commodity prices and a volume-independent price [15].

Another factor to consider is the cost of the procurement process. The company for which the strategy was developed operates in the public sector and is subject to public procurement law. This law defines threshold amounts, exceeding which imposes an additional regime in conducting the procurement process. In summary, the larger the purchase, the more expensive the procedure, with

This work is supported by the Polish Minister of Education and Science as part of an implementation doctorate, grant No. DWD/5/0286/2021.

significant cost increases when exceeding the thresholds, affecting the differentiability of the function we will be seeking the optimum for.

The proposed algorithm was primarily developed with a focus on disk space resources but can be applied, for example, to telecommunication connections, products sold in an increasingly popular subscription model, and even to support decisions regarding the duration of contracts for telecommunication services.

In the problem, we omitted the impact of the time value of money, inflation, and the uncertainty associated with forecasting resource demand. Also, we assume that the resource is available immediately after purchase (no delivery time). In Sect. 8 we describe the possibilities for algorithm development and the addition of these elements to the definition of the problem.

The work focuses on solving a specific optimization problem of great economic importance. Large corporations face a constant dilemma related to developing a purchasing strategy. On the one hand, there is price variability over time, and secondly, resource consumption is probabilistic/uncertain. Thirdly, the costs of the purchasing procedure depend non-linearly on the size of the order.

The problem posed in this way is very general, i.e. there is no guarantee of convexity of the criterion, there are strongly non-linear constraints, and there is also a random factor related to the fact that the demand function is a risky forecast.

2   Statement of the Problem

The problem is finding the optimal purchasing strategy. We assume that for the considered time interval $t \in [0, T]$ we have the following given (or estimated) a priori: (i) price function, $p(t)$, for 1 unit of the resource (e.g. 1TB of storage), and (ii) nondecreasing demand function, $u(t)$. At equidistant discrete time points $\{t_i\}_{i=1}^n$, $t_i = \frac{i-1}{n}T$, we can purchase any quantities of resources $\{m_i\}_{i=1}^n$. The period the resource can be available in the infrastructure (lifetime) is assumed to be identical for all orders and denoted by d. Hence, the resource (memory) availability function related to the ith order has the following form

$$s_i(t) = m_i \{1(t - t_i) - 1(t - t_i - d)\}. \tag{1}$$

The boundary condition (constraint) is to ensure resources in a quantity not less than the demand, i.e.,

$$\sum_{i=1}^n s_i(t) \geq u(t) \tag{2}$$

fot each $t \in [0, T]$. Moreover, we consider the cost ξ of the ith purchasing procedure, which depends non-linearly on the order quantity, i.e., $\xi = \xi(m_i)$. Owing to above, the total expense/cost for the i-th order is

$$c_i(m_i) = m_i p(t_i) + \xi(m_i). \tag{3}$$

The decision variable is therefore a vector

$$m = (m_1, m_2, .., m_i, ..., m_n) \in \mathcal{R}^n \tag{4}$$

and the optimization criterion is the total price of all purchases and costs of procurement procedures

$$Q\left(m\right) = \sum_{i=1}^{n} c_i\left(m_i\right) \to \min_m. \tag{5}$$

3   Method Selection

The first idea to solve the above problem was, of course, a complete inspection. Unfortunately, even with relatively radical quantization of m_i variables, it leads to unacceptable computational complexity. Some effort has been made to try to apply Bellman dynamic programming [3] to the problem at hand. Nevertheless, it is difficult to formalize the concept of state here, as optimal future paths and decisions are strictly dependent on previous decisions. It also entails large memory requirements. Since the convexity of the criterion function is not guaranteed, convex optimization techniques ([5,12]) are also rather excluded. The fundamental problem is that the optimized function is not even differentiable, due to the discontinuity of the $\xi\left(\cdot\right)$ function (see (3)). Therefore, the use of neural networks based on stochastic gradient (e.g. Adam algorithm [9]) becomes problematic. The situation is further complicated by non-linear constraints (2) and a large number of hyperparameters that need to be set. The considered problem is difficult to represent as a graph which makes it difficult to apply the ant search algorithm [6]. The tabu-search technique was also potentially considered, but it would have involved a relatively large number of tunable parameters ([8,13]). Due to the above limitations, it was decided to use and compare the genetic algorithm ([1,2]) and the simulated annealing method ([10]). The results obtained are presented in the following sections. The presented methods require tuning, i.e., setting the appropriate mutation and crossover operators or the cooling rate to avoid the risk of convergence to local minimum. To ensure that the constraints are met, the concept of coordination using the penalty function, presented in [7], was applied. In the future, an attempt to solve the problem under consideration using the particle swarm optimization technique ([11]) is also pronounced, as well as the use of branch and bound ([11]) methodology. The developed pre-implementation system will also be equipped with a tool to forecast the $u(t)$ function (and possibly a confidence interval) based on historical observations ([4]).

4   Generation of Initial Solutions

Let $\Delta \triangleq t_i - t_{i-1} = \frac{T}{n}$ be the time interval between consecutive purchases (constant for all i's). Assuming that $d \geq \Delta$, purchase m_i affects resources in $t_i, t_{i+1}, ..., t_{i+h}$, i.e., in the horizon

$$h = \left\lfloor \frac{d}{\Delta} \right\rfloor. \tag{6}$$

The components m_i of each new individual (candidate solution) (4) are generated successively $(i = 1, 2, ..., n)$, from a uniformly distributed random generator

$$m_i \sim \mathcal{U}\left[m_{i,\min}, m_{i,\max}\right],\tag{7}$$

where $m_{i,\min}$ stands for the minimal order, which guarantees sufficient resources, only in the horizon until the next, $(i + 1)$ th, purchasing procedure

$$m_{i,\min} = u(t_{i+1}) - \sum_{j<i} s_j\left(t_{i+1}\right),\tag{8}$$

whereas $m_{i,\max}$ guarantees sufficient resources until time t_{i+h}, i.e.,

$$m_{i,\max} = u(t_{i+h}) - \sum_{j<i} s_j\left(t_{i+h}\right).\tag{9}$$

5   Application of Genetic Algorithm (GA)

5.1   Chromosome, Gene, Allele

In a genetic algorithm, a chromosome (individual) represents a single purchasing strategy that satisfies the assumptions of the problem. A gene corresponds to a point in time at which a purchase can occur, and the allele is the decision at that point within a given strategy:

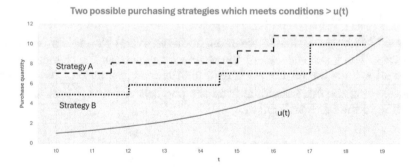

Fig. 1. Two purchasing strategies.

- $m = (m_1, ..., m_n)$ - chromosome (individual)
- i - gene
- m_i - allele

5.2   Initial Population

Our crossover method tends to move into average and therefore it is important to generate a significant portion of individuals close to the boundaries of possible solutions. We decided to use the following rule: 20% individuals which represents one big order in the first possible order, 20% representing as many small purchases as possible and the remaining 60% fulfill the area randomly. An example of initial population is shown in Fig. 2.

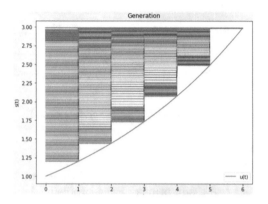

Fig. 2. Initial population.

5.3   Parent Selection

We used fitness proportionate selection (roulette wheel) to select a population of parents. The lower the cost of the strategy, the greater the chance that the individual will qualify for the set generating the next generation. The probability of selecting the kth individual from the population $\left\{m^{(\kappa)}\right\}_{\kappa=1}^{N}$ is calculated according to the following equation

$$P\left(m^{(k)} \text{ is selected}\right) = \frac{q\left(m^{(k)}\right)}{\sum_{\kappa=1}^{N} q\left(m^{(\kappa)}\right)}, \tag{10}$$

where $q\left(\cdot\right)$ represents the adaptation function, e.g., $q\left(m^{(\kappa)}\right) = Q_{\max} - Q\left(m^{(\kappa)}\right)$ with $Q_{\max} \triangleq \max_{\kappa=1,2,\ldots,N} Q(m^{(\kappa)})$.

5.4   Crossover

Our approach to combining genetic features must take into account boundary conditions and ensure that the subsequent generation adheres to the assumptions of the problem. The permissible strategy space after exchanging features between strategies A and B is illustrated in Fig. 1. Our proposed method for generating offspring involves randomly selecting a point in time to transition from one strategy, as shown in Fig. 3, to strategy 2 and further utilizing the features of the second individual. Feature blending can be performed in two directions, and in Fig. 4, you can observe the offspring of two merged individuals (strategies). The second method for a crossover is an average for each m.

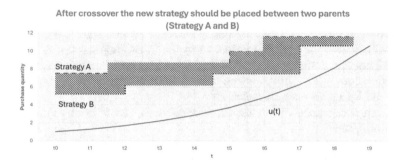

Fig. 3. Valid crossover area.

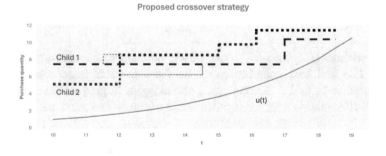

Fig. 4. Valid crossover - proposed solition.

5.5 Mutation

In the algorithm, we introduced mutations that occur each time a new population is generated. The mutation involves shifting an order from a random time point to the previous or next one (as long as the strategy remains a valid one).

5.6 Termination Condition

The average cost for the whole population is changing less than ϵ during the consequent 3 generations. This parameter depends on the total cost of the strategies and, therefore must be considered every time a new resource is analyzed.

6   Simulated Annealing (SA)

In the simulated annealing algorithm, we have assumed that searching for a solution may involve exploring infeasible solutions that incur a penalty. Infeasible solutions include purchasing a negative quantity of resources and failing to provide the minimum quantity required by the demand. We used following algorithm parameters:

- Initial Seed - To ensure comparability with the genetic algorithm, we decided to run a search for each individual from the initial population of the genetic algorithm and compare the best $Q(m)$ found by both algorithms.
- Temperature - The initial temperature for the algorithm is set to 1 and reduced by 1% at each iteration step.
- Solution acceptance - To accept a new solution, the following condition has been adopted:

```
NeighborQ(m) < CurrentQ(m)
or
rand() < exp((CurrentQ(m)-NeighborQ(m))/Temperature)
```

- Interations - we decided to use 1000 iterations per each initial m

7   Results

Both algorithms (GA and SA) were tested on edge cases where solutions are obvious. The first such case involves a constant product price over time, zero procurement cost, and a discount for purchasing in large quantities - here, the most advantageous strategy is to make the largest one-time purchase. The second case involves no discount, zero procurement cost, and a decreasing product price over time - in this scenario, the preference is for the strategy of purchasing the resource as late as possible. The cost function in these cases is monotonic, and both algorithms find the optimal solution. The population behavior for the genetic algorithm is depicted in Fig. 5.

However, for more complex cases, when problem parameters were set considering product life cycles and abrupt procurement procedure thresholds, the behavior of the genetic algorithm proved to be more stable. For the majority of the examined parameter sets, both algorithms were able to find the same solution. However, there were instances where the solution obtained by the genetic algorithm was superior, and the simulated annealing algorithm found a local minimum that was not optimal. No case was identified where simulated annealing found a solution better than the genetic algorithm.

The analysis of the problem during the operation of the SA algorithm indicated that it is possible to configure the algorithm parameters, especially the penalties for forbidden solutions, in a way that allows the algorithm to start finding the global minimum. However, this improvement came at the expense of

Preference for one big order   Preference for the expense delay

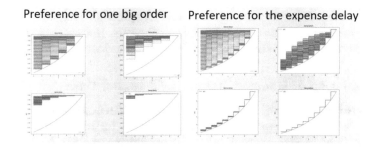

Fig. 5. Searching for solutions using a genetic algorithm for simplified problems - significant discount and decreasing resource costs. Generations (1, 5, 10, 15).

deteriorating performance in other cases. It was not possible to find a method for dynamically adjusting the parameters of the SA algorithm based on the problem parameters.

An example of a solution where GA outperformed SA is shown in Fig. 6.

The case with limited lifetime of the resource

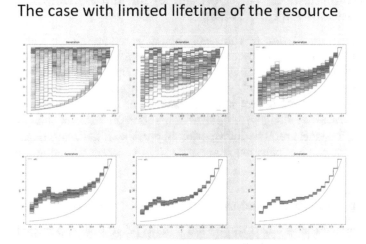

Fig. 6. Searching for a genetic algorithm solution to a problem with limited time availability of resources. (Generations: 1, 5, 10, 15, 20, 25).

8   Open Problems

8.1   New Ideas for Crossover

An extension of this algorithm could involve the development of alternative methods for blending features. The region of valid solutions presented in the article can be expanded, which may mitigate the algorithm's tendency to average successive generations and search for niche solutions. However, it is crucial to ensure the correctness of strategies in subsequent generations.

8.2 Estimation of Demand Based on Historical Data

A natural extension of the problem is to introduce forecasting of demand functions. Forecasting comes with inherent risks, and the further we project into the future, the larger the margin of safety we should incorporate in procurement planning. A single upfront purchasing strategy may pose higher risks, and the necessity of adopting an appropriate reserve could impact the feasibility of alternative strategies. An example proposal may be the use of a simple exponential model

$$\overline{u}\left(t\right) = \alpha_1 e^{\alpha_2 t} + \alpha_3 \tag{11}$$

in which parameters $\alpha_1, \alpha_2, \alpha_3$ are calculated based on historical data. To reduce the risk associated with model uncertainty, it is possible to periodically recalculate it and initiate the entire optimization procedure anew.

8.3 Introduction of Delay Between Order and Delivery. Initial Conditions for the Problem

The practical application of the algorithm may require the introduction of a time delay between the resource ordering time and its delivery, along with its implementation. The time delay, along with the risks associated with delivery delays, could be an additional, very interesting extension of the presented problem. Another practical aspect is the possibility of introducing the initial state of existing resources along with information about the dates of their withdrawal from the infrastructure. The information about the initial state does not affect the algorithm but can offer much greater utility to teams planning purchases in the enterprise.

8.4 Introduction of Resource Degradation in Time

Some resources that we aim to provide in enterprises may experience a decline in performance, leading to degradation of their capacity and usability. In the algorithm, a binary degree of resource availability was adopted, but the algorithm can be extended to a solution that takes into account functions defining the manner of resource degradation. A good example could be photovoltaic panels, and the analysis of purchasing strategies may consider the decline in the efficiency of devices.

8.5 Dynamic Parameterization of SA

All parameters of the SA algorithm can be expressed as a function of the problem parameters. If such dependencies could be identified, the algorithm could potentially prove to be more computationally attractive than GA.

9   Conclusion

To sum up, the created tool will make planning purchases in a large corporation easier. Given the multitude of constraints and non-linear dependencies between variables, the optimal strategy is difficult to develop intuitively. The results of experimental studies carried out on simulated data with a specificity similar to real problems show a certain advantage of the genetic algorithm over simulated annealing. With a comparable computational effort and effort put into tuning the methods, in practice the genetic algorithm was less likely to get stuck in the local minimum.

References

1. Ashlock, D.: Evolutionary Computation for Modeling and Optimization, vol. 571. Springer, New York (2006). https://doi.org/10.1007/0-387-31909-3
2. Bauer, R.J.: Genetic Algorithms and Investment Strategies, vol. 19. John Wiley & Sons, Hoboken (1994)
3. Bellman, R.: Dynamic programming. Science **153**(3731), 34–37 (1966)
4. Box, G.E., Jenkins, G.M., Reinsel, G.C., Ljung, G.M.: Time Series Analysis: Forecasting and Control. John Wiley & Sons, Hoboken (2015)
5. Boyd, S.P., Vandenberghe, L.: Convex Optimization. Cambridge University Press, Cambridge (2004)
6. Dorigo, M., Stützle, T.: Ant colony optimization: overview and recent advances. In: Gendreau, M., Potvin, J.Y. (eds.) Handbook of Metaheuristics. International Series in Operations Research and Management Science, LNCS, vol. 272, pp. 311–351. Springer, Cham (2019). https://doi.org/10.1007/978-3-319-91086-4_10
7. Findeisen, W., Bailey, F.N., Brdyś, M., Malinowski, K., Tatjewski, P., Wozniak, A.: Control and Coordination in Hierarchical Systems. John Wiley & Sons, Hoboken (1980)
8. Glover, F., Laguna, M.: Tabu Search. In: Du, D.Z., Pardalos, P.M. (eds.) Handbook of Combinatorial Optimization, pp. 2093–2229. Springer, Boston, MA (1998). https://doi.org/10.1007/978-1-4613-0303-9_33
9. Kingma, D.P., Ba, J.: Adam: a method for stochastic optimization (2014). arXiv preprint arXiv:1412.6980
10. Kirkpatrick, S., Gelatt, C.D., Jr., Vecchi, M.P.: Optimization by simulated annealing. Science **220**(4598), 671–680 (1983)
11. Lawler, E.L., Wood, D.E.: Branch-and-bound methods: a survey. Oper. Res. **14**(4), 699–719 (1966)
12. Moré, J.J.: The Levenberg-Marquardt algorithm: implementation and theory. In: Watson, G.A. (ed.) Numerical Analysis. LNM, vol. 630, pp. 105–116. Springer, Heidelberg (1978). https://doi.org/10.1007/BFb0067700
13. Nowicki, E., Smutnicki, C.: A fast taboo search algorithm for the job shop problem. Manag. Sci. **42**(6), 797–813 (1996)
14. Prince, M., Smith, J.C., Geunes, J.: A three-stage procurement optimization problem under uncertainty. Nav. Res. Logist. (NRL) **60**(5), 395–412 (2013)
15. Sundarraj, R., Talluri, S.: A multi-period optimization model for the procurement of component-based enterprise information technologies. Eur. J. Oper. Res. **146**(2), 339–351 (2003)
16. Wang, D., Tan, D., Liu, L.: Particle swarm optimization algorithm: an overview. Soft Comput. **22**, 387–408 (2018)

Tumor Volume Measurements in Animal Experiments: Current Approaches and Their Limitations

Melánia Puskás[1,2]($\boxtimes$) (ID), Borbála Gergics[1,2](ID), Levente Kovács[1](ID), and Dániel András Drexler[1](ID)

[1] Physiological Controls Research Center, University Research and Innovation Center, Óbuda University, Budapest, Hungary
puskas.melania@uni-obuda.hu
[2] Applied Informatics and Applied Mathematics Doctoral School, Óbuda University, Budapest, Hungary
https://physcon.uni-obuda.hu/

Abstract. Animal experiments are often used in cancer research as an alternative to clinical trials to determine the effect of various drugs and other biological conditions on tumor behavior. The tumor volume measurement is necessary in these experiments. The operation of imaging devices is expensive, and it is necessary to anesthetize the animals during the measurement. Anesthesiology induces stress and deviation from homeostatic functioning, which greatly affects the results of the experiment. To address these problems, the digital caliper method, which is safe and cost-effective, is applied to measure tumor volume during in vitro studies. However, it is only rarely possible to accurately measure the volume of the tumor with a digital caliper. It is, therefore, important to examine how we can estimate the three-dimensional volume of the tumor from measurements carried out with a digital caliper. This study summarizes the approximation formulas used in the literature and their scope of use and analyzes the advantages and disadvantages of digital caliper measurements. We aim to highlight the importance of creating a noise model in cancer research that generally describes the difference between real tumor volumes and those measured with calipers.

Keywords: tumor measurements · animal experiments · digital caliper · measurements noise

Project no. 2019-1.3.1-KK-2019-00007. has been implemented with the support provided from the National Research, Development and Innovation Fund of Hungary, financed under the 2019-1.3.1-KK funding scheme. This research was supported partially by Horizon2020-2017-RISE-777911 project. This project has been supported by the Hungarian National Research, Development and Innovation Fund of Hungary, financed under the TKP2021-NKTA-36 funding scheme. The work of Dr. Dániel András Drexler was supported by the Starting Excellence Researcher Program of Óbuda University, Budapest, Hungary. Melánia Puskás was supported by the ÚNKP-23-3 New National Excellence Program of the Ministry for Culture and Innovation from the source of the National Research, Development and Innovation Fund.

W. Zamojski et al. (Eds.): DepCoS-RELCOMEX 2024, LNNS 1026, pp. 206–217, 2024.
https://doi.org/10.1007/978-3-031-61857-4_20

1   Introduction

Preclinical studies offer valuable information for designing optimal therapies. However, measuring the tumor volume change during therapy can cause obstacles. Medical imaging of experimental animals is challenging due to several reasons. During imaging such as MRI, CT, PET, or SPECT, the immobility of the animal is required. This is difficult to do without anesthesia; however, anesthesia carries many risks and side effects. Creating a high-quality dataset requires frequent sampling, although the frequent anesthesia causes stress to the animal, which leads to different reactions to the therapy in the later stages of the experiment (e.g., loss of appetite, weight loss, dehydration) [12,31]. Imaging devices are also expensive to operate [46].

Considering these issues, tumor volume measurements are taken with a digital caliper during the preclinical experiments. Since the caliper can only measure the tumor length and width under the skin, the measurement may yield inaccurate values due to the connective tissue and skin between the tumor and the caliper. In addition, with this method, we do not have information about the entire three-dimensional structure of the tumor, but as an approximation, we can use different formulas (detailed in Sect. 3) to estimate the tumor volume.

The motivation of our work is enhancing personalized tumor therapy research, which is often based on animal experiments in the preclinical phase. In our previous works, we treated cancerous mice with chemotherapy drugs [16,19,30]. Tumors were measured with a digital caliper and an approximation formula was used to estimate the volume of the tumor [46]. Based on the estimated volumes, we fitted a mathematical model [17,18] to obtain the parameters of the tumors of mice [29]. The measurement was quite noisy since we did not get accurate values with the caliper. We did not have an imaging device available, so the fitting of the model was imprecise as well. Our goal in the future is to determine the noise on measurements taken with a digital caliper. Based on our preliminary work, we have already experimented with a simpler noise model [40], but in order to clarify this, it is important to summarize the specifics of measuring with a digital caliper.

The use of approximate formulas incorporates an error in the measurements since we have to measure a soft amorphous tumor. The person performing the measurement also affects the result, and the skin of the animal may also thicken during the treatment. Our article serves to highlight the importance of such a noise model in cancer research to optimize therapy [15] and summarizes the methods used in the current literature.

2   Measurement of Tumor Volume During Preclinical Studies

Various tools and techniques are used for tumor volume measurement in preclinical studies. Many imaging techniques are utilized based on the type of cancer under consideration. The choice of method depends on factors such as the experimental design, the nature of the tumor, and the desired level of precision. Several volume measurement procedures are known to require the sacrifice of the experimental animal or invasive surgery of the tumor. The researchers measure the volume of the tumor removed in this way, for example, using a caliper [48] or applying Archimedes' principle on an operated tumor [42,52]. Longitudinal studies require several tumor volume measurements per week. In order to meet the requirements of the 3Rs (reduction, replacement, and refinement of animal experiments) and reduce the number of experimental animals used, we have to avoid invasive tumor volume measurement methods during longitudinal studies [26]. Imaging techniques such as magnetic resonance imaging (MRI), computed tomography (CT), positron emission tomography (PET), single-photon emission computed tomography (SPECT), and ultrasound (US) can provide detailed and non-invasive assessments of tumor volume [28].

Ultrasound is the predominant imaging modality among the commonly used imaging techniques. Internal organs of the organism are examined in real-time, including imaging of the reproductive system, evaluation of cardiac valve function, and identification of abdominal tumors through the assessment of liver perfusion. US is cost-effective, many-sided, multifunctional, and portable compared to CT and MRI [21,22,51]. In the last decades, US modalities have been re-engineered for researchers who use animal models in disease research in longitudinal studies of disease models and normal development [47]. Literature often uses the term 'ultrasound biomicroscopy' (UBM). UBM was first researched in three important clinical applications: ophthalmology [36], dermatology [25], and intravascular ultrasound [33]. Within the realm of preclinical imaging, UBM is commonly denoted as 'micro-ultrasound' [51].

X-ray and CT are widely used in cancer diagnosis in clinical practice. Applying this medical imaging technique to preclinical studies is promising for detecting and monitoring tumor growth, as well as for assessing the efficacy of anticancer therapies. CT, applied in small animal models called micro-CT, is used to study various types of cancer such as pheochromocytoma [34], respiratory cancer [8], sarcoma [37], brain tumor [38] and many more. With the help of micro-CT, tumor-specific angiogenesis [4] and metastasis [35] can also be investigated. A main obstacle of X-ray and micro-CT is that the sum of low-dose radiation might impact animal welfare and physiological parameters of laboratory animals [5].

MRI is employed on the same in vivo animal tumor models as CT. However, compared to CT, MRI does not apply radiation exposure and is usually non-invasive. Imaging does not affect the tissues of the body. MRI offers advantages due to its higher resolution, operating within a micrometer range. MRI has the capability to concurrently extract physiological, molecular, and anatomical information [7,39].

The disadvantages of the tumor imaging methods mentioned above are that the animal needs to be anesthetized, and the handling might cause undesirable stress for the animal. Repeated anesthesiology and stress can influence the examined chemotherapeutic effect during longitudinal studies, which leads to distortion in the outcomes. Moreover, these imaging devices are not the most cost-effective methods [5].

In preclinical tumor research, digital calipers are commonly used to measure tumor dimensions in experimental animals, such as mice. Caliper measurements offer a non-invasive, cost-effective, simple method for tumor volume measurement. Tumor volume measurement with a digital caliper does not require an expensive, complicated device or imaging specialists. Apart from the small stress caused by animal handling, measuring with a caliper is completely harmless, and it can be performed without anesthesiology [32].

Usually, a nude mouse model is used for this purpose; thus, it is avoided that animal hair interferes with the measurement. The caliper is placed on the skin over the tumor, and the two longest perpendicular axes in the x/y plane of each orthotopic tumor are measured. The third dimension is assumed from the two measured dimensions using an approximation formula. In Sect. 3, we collected the commonly used formulas for this purpose found in the literature [6].

However, the unreliability of established tumor measurement methods reduces the accuracy and reproducibility of the outcome. In orthotopic mouse models, the usage of caliper measurements to estimate tumor volumes faces imprecision due to geometrical challenges. Moreover, the mechanical characteristics of calipers introduce additional variability by enabling the compression of the tumor, thereby affecting its shape and the recorded dimensions [10].

3 Estimation of Tumor Volumes from Tumor Length and Width Measured by Caliper

Manual methods (such as digital calipers) used to measure tumors in mice are often inaccurate and time-consuming. The inaccuracy comes from the fact that we are only able to measure in two dimensions. Since we can only measure the width (the shorter diameter part of the tumor) and length (the longer diameter part of the tumor) of the tumor with the digital caliper, we can only estimate the volume of the tumor. In the literature, many approximate formulas estimate the volume of the 3D structure from 2D values measured with a digital caliper.

In the literature, the solution of assuming an ellipsoidal geometric shape for the tumor is often used [20, 50]. The volume bounded by the ellipsoid is

$$V = \frac{4}{3}\pi \cdot l \cdot w \cdot h, \tag{1}$$

where l is length and w is width and h is height. Most of the approximate formulas used to calculate tumor volumes are based on formula (1).

Court et al. [10] work with a head and neck cancer mouse model. In their research, the tumor volume was approximated with a caliper, and these measurements were used to compare groups. The geometrical challenges also make this procedure imprecise because the tumor shape is amorphous and difficult to estimate. The approximate formula that they used is determined based on the Xenograft tumor model protocol [1], where the tumor volume should be calculated using the following simpler formula:

$$V_{cal} = \frac{1}{2}l \cdot w^2,$$ (2)

where l is the length and w is the width of the tumor, so the longer or shorter diameter of the tumor from two roughly perpendicular measurements, respectively. This formula estimates that tumor volume is equivalent to half the volume of the quadrangular prism [20], and this is the most accurate volume calculation that was obtained using caliper measurements in the literature. The formula has a distortion due to the fact that the shorter dimension is squared; thus, the effect of the change in the shorter dimension is second order, while the effect of the change in the longer dimension is only linear.

Court et al. also proved that, in the case of small tumor volumes, the volume estimated with the help of a caliper only weakly correlates with the true volume value. Measurements taken with an imaging device were considered to be the real values (ground truth) because in addition to the caliper, CBCT (Cone Beam Computed Tomography) [2,9] measurements were also performed on the tumors of the mice. They observed that tumor volume values that were smaller than $200\,\mathrm{mm}^3$ measured with a caliper were almost always larger than the corresponding CT volume measurements. However, the caliper measurements that were larger than $200\,\mathrm{mm}^3$ were almost always smaller than the corresponding CT volume. Using linear regression, they obtained that caliper measurements significantly underestimate changes in tumor volume.

Jensen et al. [27] also used an approximate formula (2) to estimate the tumor volume in the case of values measured with a digital caliper. In their study, they demonstrated that the caliper was inaccurate and encumbered with a significant size-dependent bias. They compared the caliper measurements with the measurements carried out with microCT [13,24] and microPET. The results showed that microCT was more accurate than both external caliper and 18F FDG-microPET [11,43] for in vivo measurement of subcutaneous tumors in mice, and 18F FDG-microPET was declared unsuitable for tumor size determination.

In another study, Harris et al. [23] investigated the chemopreventive effects of the specific COX-2 blocker celecoxib [44] against the development of chemically induced breast cancer. At the end of their experiment, each tumor diameter was measured by a micrometer caliper, and the tumor volume was calculated using the formula

$$V_{cal} = \frac{4}{3}\pi \cdot r^3,$$ (3)

where r is half of the average tumor diameter. This formula (3) assumes that tumors grow to form spheres [20,23]. Harris et al. found that administering a COX-2 inhibitor reduced the incidence, frequency, and size of DMBA-induced mammary tumors [3] in female Sprague Dawley rats.

In our works, we use the following formula to approximate the volume of the tumor:

$$V_{Cal} = \frac{\pi}{2}(w \cdot l)^{3/2}.$$ (4)

This approximate formula was determined in a study by Sápi et al. [45,46], based on MRI measurements. Currently, MRI is the most widely used non-invasive method for tumor volume measurement. With this formula, the tumor volume can also be approximated from the length (l) and width (w) values, but it resulted in a much more accurate approximation of the tumor volumes measured by MRI than the protocol-based calculation. Compared to (2), both the width and length are taken into consideration with an exponent of $3/2$, thus the distortion present in (2) is not present here.

Overall, most research found in the literature adapts the approximation formula of the tumor volume to the experimental circumstances. The volume of the tumor is approximated according to the assumed shape. Many articles face challenges because the shape of the tumor is not comparable to any fundamental geometric shape since it has an amorphous shape. Also, it is often difficult to access because it can appear in any part of the animal (e.g., in Fig. 1.).

Fig. 1. The shape of the tumor is amorphous and can appear in any part of the animal.

Due to its soft texture, it is difficult to measure, and the measurement results are significantly influenced by the person performing the measurement. The results can be improved if the values are not measured through the skin, but the animal is anesthetized and the tumor is removed, but this is expensive and does not show much more accurate results.

Figure 2 shows a real experimental case [30]. The upper part of the figure shows two tumor volume curves. The blue curve was determined using the approximation formula (2), while the red curve shows tumor volumes obtained using the approximation formula (4), which was validated based on MRI measurements [46]. The middle part of the figure shows the weight of the mouse, while the bottom part shows the doses administered during the experiment. The figure indicates that using different approximation formulas results in different tumor volumes. Since the MRI measurements are considered more accurate, we used the approximation formula (4) to model the measurement noise during our previous work [40]. The figure shows that the estimations using formula (4) yielded significantly larger volumes compared to the ones calculated with (2) when the volumes are large.

In our previous work, we created a noise model that can describe the measurement noise. During the measurements, we estimated the tumor volumes based on (4) [40]. Our goal was to generate virtual noise measurements as training

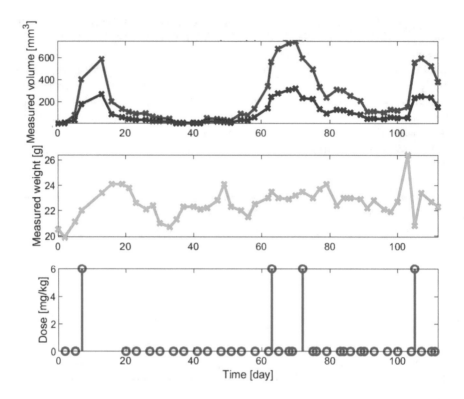

Fig. 2. Results from an in vivo experiment [30]. Top: Tumor volume approximation based on (2) (blue curve) and tumor volume approximation based on (4) (red
curve). Middle: Weight measurements of the mouse. Bottom: The injected doses of the
chemotherapeutic drug. (Color figure online)

data for neural networks [41] and create virtual patients for in silico algorithm
development. We applied these networks to estimate tumor model parameters
[16], which can be used to generate therapy [14,49]. In the first step, we had
to reproduce the measurement error in the virtual measurements so we could
find the appropriate distribution to generate the realistic measurement noise. We
smoothed the original noisy measurements and then performed noise whitening
on the error set. We found that the noise does not follow a uniform or normal distribution, so we smoothed it out with a transformation to make the distribution
of the errors be close to normal.

We also found that the measurement noise depends on the current volume
of the tumor. The smaller the tumor, the more difficult it is to measure with
the digital caliper, thus the measurement noise is larger. By whitening the noise,
we made the error independent of the tumor volume. We denote this whitening
transformation by Φ and define it as the following function:

$$\Phi(\mathbf{y}) = \sqrt[3]{\mathbf{y}}, \tag{5}$$

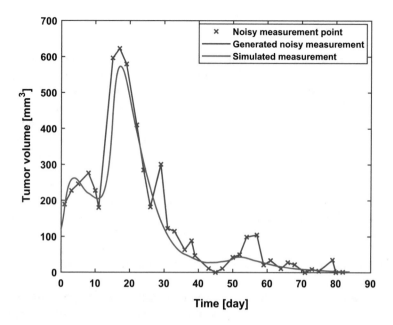

Fig. 3. Simulated tumor volume measurement (red curve) and noisy measurement (blue curve) of a virtual patient with the generated noisy measurement points (blue crosses). (Color figure online)

and use it to transform the noise as

$$\tilde{e} = \Phi(y)e, \tag{6}$$

where $\tilde{e}$ is the transformed error, y is the tumor volume, and $\Phi(\mathbf{y})$ is the transformation used for whitening. We fitted different probability distributions to this transformed error and evaluated the best ones. As a result, the Weibull distribution was the best fit for the noise with a random variable x:

$$C(x, y) = \begin{cases} \frac{B}{A} \left(\frac{x}{A}\right)^{(B-1)} \exp\left(-\frac{x}{A}\right)^B \cdot \frac{1}{\sqrt[3]{y}}, & \text{if } x \geq 0, \\ 0, & \text{if } x = 0. \end{cases} \tag{7}$$

This distribution (denoted by C) describes the noise. Parameters A and B are the scale and shape parameters of the Weibull distribution. The total tumor volume is denoted by y.

Using the Weibull distribution in the simulation of virtual patients, realistic measurement noise was generated for the simulated measurements with the help of (5), the results are shown in Fig. 3. The simulated tumor volume is the red curve, and the blue curve shows the noisy measurement. The blue crosses show the generated noisy measurement points. Based on the data from [46] and the analysis carried out in [40], this figure shows the estimation error when the tumor volume is approximated using (4).

4   Conclusion

In summary, although we can only approximate tumor volume from the width and length of the tumor, the digital caliper is still the most common non-invasive tumor measurement tool in animal experiments. In the literature, many articles report the difficulties of measurements with a digital caliper, yet there are few initiatives to clarify the approximate formulas. Several analyses in the literature support that imaging modalities measure the volume of the tumor more accurately, but the digital caliper is more cost-effective and causes less stress to the animals than frequent anesthesia.

In this work, we have summarized the most commonly used tumor measurement methods and their drawbacks. We have collected the currently applied approximation formulas, all of which are based on assumptions. The most common method is when the shape of the tumor is compared to an ellipsoid, but the shape of the tumor is not predictable at all; it takes on an amorphous shape. Finally, we additionally described a noise model that we created based on preclinical mice measurements.

The measurements could be made more precise if parallel measurements were made using an imaging modality and a digital caliper, and then the difference between the two would be described with a model. The measurement noise would be understood from the model, which would greatly contribute to current research using animal experiments.

We have already attempted this in our research, and the modeling is constantly being refined [40]. Our goal is to use a noise model to accurately take measurements in the future, which can lead to optimal therapy generation.

References

1. Protocol Online (2005): Xenograft tumor model protocol. https://www.protocol-online.org/prot/Protocols/Xenograft-Tumor-Model-Protocol-3810.html. Accessed 19 Jan 2024
2. Chapter 1 - image-guided radiation therapy of tumors in preclinical models. In: Sato, A., Kraynak, J., Marciscano, A.E., Galluzzi, L. (eds.) Radiation Oncology and Radiotherapy Part C, Methods in Cell Biology, vol. 180, pp. 1–13. Academic Press (2023). https://doi.org/10.1016/bs.mcb.2023.02.008
3. Abba, M.C., et al.: Dmba induced mouse mammary tumors display high incidence of activating pik3cah1047 and loss of function pten mutations. Oncotarget **7**(39), 64289–64299 (2016). https://doi.org/10.18632/oncotarget.11733, https://www.oncotarget.com/article/11733/
4. Ayala-Domínguez, L., Brandan, M.: Quantification of tumor angiogenesis with contrast-enhanced x-ray imaging in preclinical studies: a review. Biomed. Phys. Eng. Express **4**(6), 062001 (2018)
5. Baier, J., et al.: Repeated contrast-enhanced micro-CT examinations decrease animal welfare and influence tumor physiology. Invest. Radiol. **58**(5), 327–336 (2023)
6. Baris, M.M., et al.: Xenograft tumor volume measurement in nude mice: estimation of 3d ultrasound volume measurements based on manual caliper measurements. J. Basic Clin. Health Sci. **4**(2), 90–95 (2020)

7. Beckmann, N., et al.: Macrophage labeling by SPIO as an early marker of allograft chronic rejection in a rat model of kidney transplantation. Magn. Reson. Med. Off. J. Int. Soc. Magn. Reson. Med. **49**(3), 459–467 (2003)
8. Camara, J.A., Pujol, A., Jimenez, J.J., Donate, J., Ferrer, M., Vande Velde, G.: Lung volume calculation in preclinical MicroCT: a fast geometrical approach. J. Imaging **8**(8), 204 (2022)
9. Chiu, T.D., Arai, T.J., Campbell III, J., Jiang, S.B., Mason, R.P., Stojadinovic, S.: MR-CBCT image-guided system for radiotherapy of orthotopic rat prostate tumors. PLOS ONE **13**(5), 1–19 (2018). https://doi.org/10.1371/journal.pone.0198065
10. Court, B.V., Neupert, B., Nguyen, D., Ross, R.B., Knitz, M.W., Karam, S.D.: Measurement of mouse head and neck tumors by automated analysis of CBCT images. Sci. Rep. **13** (2023). https://api.semanticscholar.org/CorpusID:260164003
11. Dandekar, M., Tseng, J.R., Gambhir, S.S.: Reproducibility of 18F-FDG MicroPET studies in mouse tumor xenografts. J. Nuclear Med. **48**(4), 602–607 (2007). https://doi.org/10.2967/jnumed.106.036608, https://jnm.snmjournals.org/content/48/4/602
12. Dholakia, U., Clark-Price, S.C., Keating, S.C.J., Stern, A.W.: Anesthetic effects and body weight changes associated with ketamine-xylazine-lidocaine administered to CD-1 mice. PLOS ONE **12**, 1–11 (2017). https://doi.org/10.1371/journal.pone.0184911
13. Dizbay Sak, S., Sevim, S., Buyuksungur, A., Kayı Cangır, A., Orhan, K.: The value of Micro-CT in the diagnosis of lung carcinoma: a radio-histopathological perspective. Diagnostics **13**(20) (2023). https://doi.org/10.3390/diagnostics13203262, https://www.mdpi.com/2075-4418/13/20/3262
14. Dömény, M.F., Puskás, M., Kovács, L., Drexler, D.A.: In silico chemotherapy optimization with genetic algorithm. In: 2023 IEEE 17th International Symposium on Applied Computational Intelligence and Informatics (SACI), pp. 97–102. IEEE, May 2023. https://doi.org/10.1109/saci58269.2023.10158619
15. Dömény, M.F., Puskás, M., Kovács, L., Drexler, D.A.: Population-based chemotherapy optimization using genetic algorithm. In: 2023 IEEE 21st International Symposium on Intelligent Systems and Informatics (SISY), pp. 23–28. IEEE, September 2023
16. Drexler, D.A., Ferenci, T., Füredi, A., Szakács, G., Kovács, L.: Experimental data-driven tumor modeling for chemotherapy. In: Proceedings of the 21st IFAC World Congress, pp. 16466–16471 (2020)
17. Drexler, D.A., Sápi, J., Kovács, L.: Modeling of tumor growth incorporating the effects of necrosis and the effect of bevacizumab. Complexity **2017**, 1–10 (2017). https://doi.org/10.1155/2017/5985031
18. Drexler, D.A., Ferenci, T., Füredi, A., Szakács, G., Kovács, L.: Experimental data-driven tumor modeling for chemotherapy. IFAC-PapersOnLine **53**(2), 16245–16250 (2020). https://doi.org/10.1016/j.ifacol.2020.12.619, 21st IFAC World Congress
19. Drexler, D.A., Ferenci, T., Lovrics, A., Kovács, L.: Tumor Dynamics Modeling based on Formal Reaction Kinetics. Acta Polytech. Hung. **16**, 31–44 (2019). https://doi.org/10.12700/APH.16.10.2019.10.3
20. Faustino-Rocha, A.I., et al.: Estimation of rat mammary tumor volume using caliper and ultrasonography measurements. Lab Anim. **42**, 217–224 (2013). https://api.semanticscholar.org/CorpusID:295046
21. Foster, F.S., Pavlin, C.J., Harasiewicz, K.A., Christopher, D.A., Turnbull, D.H.: Advances in ultrasound biomicroscopy. Ultrasound Med. Biol. **26**(1), 1–27 (2000)

22. Foster, F., et al.: A new ultrasound instrument for in vivo microimaging of mice. Ultrasound Med. Biol. **28**(9), 1165–1172 (2002)

23. Harris, R., Alshafie, G.A., Abou-Issa, H., Seibert, K.: Chemoprevention of breast cancer in rats by celecoxib, a cyclooxygenase 2 inhibitor. Cancer Res. **60 8**, 2101–3 (2000). https://api.semanticscholar.org/CorpusID:2928733

24. Hegab, A.E., et al.: Using micro-computed tomography for the assessment of tumor development and follow-up of response to treatment in a mouse model of lung cancer. J. Vis. Exp. JoVE (111) (2016). https://doi.org/10.3791/53904, https://europepmc.org/articles/PMC4927707

25. Hoffmann, K., el Gammal, S., Altmeyer, P.: B-scan ultrasound in dermatology. Der Hautarzt; Zeitschrift fur Dermatologie, Venerologie, und verwandte Gebiete **41**(9), W7–W16 (1990)

26. Jean-Quartier, C., Jeanquartier, F., Jurisica, I., Holzinger, A.: In silico cancer research towards 3R. BMC Cancer **18**(1), 1–12 (2018)

27. Jensen, M., Jørgensen, J.T., Binderup, T., Kjær, A.: Tumor volume in subcutaneous mouse xenografts measured by microCT is more accurate and reproducible than determined by 18F-FDG-microPET or external caliper. BMC Med. Imaging **8**, 16 (2008). https://api.semanticscholar.org/CorpusID:264775535

28. Jose, J.V., Maurya, R.P.P.: Biomedical engineering in cancer diagnosis and therapy. In: Dwivedi, A., Tripathi, A., Ray, R.S., Singh, A.K. (eds.) Skin Cancer: Pathogenesis and Diagnosis, pp. 173–191. Springer, Singapore (2021). https://doi.org/10.1007/978-981-16-0364-8_10

29. Kisbenedek, L., Puskás, M., Kovács, L., Drexler, D.A.: Indirect supervised fine-tuning of a tumor model parameter estimator neural network. In: 2023 IEEE 17th International Symposium on Applied Computational Intelligence and Informatics (SACI), pp. 000109–000116 (2023). https://doi.org/10.1109/SACI58269.2023.10158651

30. Kovács, L., et al.: Positive impulsive control of tumor therapy—a cyber-medical approach. IEEE Transactions on Systems, Man, and Cybernetics: Systems, pp. 1–12 (2023). https://doi.org/10.1109/tsmc.2023.3315637

31. Layton, R., Layton, D., Beggs, D., Fisher, A., Mansell, P., Stanger, K.J.: The impact of stress and anesthesia on animal models of infectious disease. Front. Vet. Sci. **10** (2023). https://doi.org/10.3389/fvets.2023.1086003, https://www.frontiersin.org/articles/10.3389/fvets.2023.1086003

32. Murkin, J.T., Amos, H.E., Brough, D.W., Turley, K.D.: In silico modeling demonstrates that user variability during tumor measurement can affect in vivo therapeutic efficacy outcomes. Cancer Inform. **21**, 11769351221139256 (2022)

33. Nissen, S.E., et al.: Application of a new phased-array ultrasound imaging catheter in the assessment of vascular dimensions. In vivo comparison to cineangiography. Circulation **81**(2), 660–666 (1990)

34. Ohta, S., et al.: MicroCT for high-resolution imaging of ectopic pheochromocytoma tumors in the liver of nude mice. Int. J. Cancer **119**(9), 2236–2241 (2006)

35. Pandit, P., Johnston, S.M., Qi, Y., Story, J., Nelson, R., Johnson, G.A.: The utility of Micro-CT and MRI in the assessment of longitudinal growth of liver metastases in a preclinical model of colon carcinoma. Acad. Radiol. **20**(4), 430–439 (2013)

36. Pavlin, C.J., Harasiewicz, K., Sherar, M.D., Foster, F.S.: Clinical use of ultrasound biomicroscopy. Ophthalmology **98**(3), 287–295 (1991)

37. Popova, E., Tkachev, S., Reshetov, I., Timashev, P., Ulasov, I.: Imaging hallmarks of sarcoma progression via x-ray computed tomography: Beholding the flower of evil. Cancers **14**(20), 5112 (2022)

38. Prajapati, S.I., et al.: Microct-based virtual histology evaluation of preclinical medulloblastoma. Mol. Imaging Biol. **13**, 493–499 (2011)
39. Preda, A., et al.: MRI monitoring of Avastin^{TM} antiangiogenesis therapy using b22956/1, a new blood pool contrast agent, in an experimental model of human cancer. J. Magn. Reson. Imaging Off. J. Int. Soc. Magn. Reson. Med. **20**(5), 865–873 (2004)
40. Puskás, M., et al.: Noise modeling of tumor size measurements from animal experiments for virtual patient generation. In: 2023 IEEE 27th International Conference on Intelligent Engineering Systems (INES). IEEE, July 2023. https://doi.org/10.1109/ines59282.2023.10297747
41. Puskás, M., Gergics, B., Ládi, A., Drexler, D.A.: Parameter estimation from realistic experiment scenario using artificial neural networks. In: 2022 IEEE 16th International Symposium on Applied Computational Intelligence and Informatics (SACI), pp. 000161–000168. IEEE (2022)
42. Rajuddin, R., Oscar, R., Dewi, T.P.: Serum vascular endothelial growth factor levels and uterine fibroid volume. Indones. J. Obstet. Gynecol. 107–111 (2020)
43. Rezaei Aghdam, H., et al.: 18F-FDG micropet and MRI targeting breast cancer mouse model with designed synthesis nanoparticles, vol. 2022, January 2022. https://doi.org/10.1155/2022/5737835
44. Saini, S.S., Gessell-Lee, D.L., Peterson, J.W.: The cox-2-specific inhibitor celecoxib inhibits adenylyl cyclase. Inflammation **27**, 79–88 (2003). https://api.semanticscholar.org/CorpusID:23087722
45. Sápi, J., Drexler, D.A., Sápi, Z., Kovács, L.: Identification of C38 colon adenocarcinoma growth under bevacizumab therapy and without therapy. In: CINTI 2014 15th IEEE International Symposium on Computational Intelligence and Informatics, pp. 443–448 (2014)
46. Sápi, J., Kovács, L., Drexler, D.A., Kocsis, P., Gajári, D., Sápi, Z.: Tumor volume estimation and quasi-continuous administration for most effective bevacizumab therapy. PLoS ONE **10**(11), e0142190 (2015). https://doi.org/10.1371/journal.pone.0142190
47. Sherar, M., Noss, M., Foster, F.: Ultrasound backscatter microscopy images the internal structure of living tumour spheroids. Nature **330**(6147), 493–495 (1987)
48. Shirvalilou, S., Khoei, S., Khoee, S.: In vivo 3t magnetic resonance imaging (MRI) of rat brain glioma-bearing tumor: a comparison with digital caliper measurement and histology. Front. Biomed. Technol. **6**(2), 73–78 (2019)
49. Szűcs, T.D., Puskás, M., Drexler, D.A., Kovács, L.: Model predictive fuzzy control in chemotherapy optimization. In: 2023 IEEE 17th International Symposium on Applied Computational Intelligence and Informatics (SACI), pp. 103–108 (2023). https://doi.org/10.1109/SACI58269.2023.10158569
50. Tomayko, M.M., Reynolds, C.P.: Determination of subcutaneous tumor size in athymic (nude) mice. Cancer Chemother. Pharmacol. **24**, 148–154 (2004). https://api.semanticscholar.org/CorpusID:36523909
51. Turnbull, D.H., Bloomfield, T.S., Baldwin, H.S., Foster, F.S., Joyner, A.L.: Ultrasound backscatter microscope analysis of early mouse embryonic brain development. Proc. Natl. Acad. Sci. **92**(6), 2239–2243 (1995)
52. Zivković, N., Zivković, K., Despot, A., Paić, J., Zelić, A.: Measuring the volume of uterine fibroids using 2-and 3-dimensional ultrasound and comparison with histopathology. Acta Clin. Croat **51**(4), 579–589 (2012)

Utilizing CNN Architectures for Non-invasive Diagnosis of Speech Disorders

Filip Ratajczak$^{(\boxtimes)}$, Mikołaj Najda, and Kamil Szyc

Wrocław University of Science and Technology, Wrocław, Poland
fratajczak124@gmail.com, najdamikolaj@protonmail.com,
kamil.szyc@pwr.edu.pl

Abstract. This study explored the potential of convolutional neural networks (CNNs), like VGG and ResNet, in diagnosing diseases affecting the voice and speech apparatus through non-invasive analysis of vowel sound recordings. Utilizing the Saarbruecken Voice Database, voice recordings were transformed into spectrograms for model training, focusing on the vowels /a/, /u/, and /i/. The research employed Explainable Artificial Intelligence (XAI) techniques to highlight critical features for disease detection within these spectrograms, aiming to provide medical professionals with deeper insights into disease manifestations in voice patterns. Based on the F1-score, performance evaluation revealed that ResNet18 models achieved a score of 71.51 ± 2.49, outperforming VGG19 models, which scored 68.51 ± 1.78. The study's findings suggested that vowel selection and data augmentation strategies did not significantly enhance model performance. Additionally, it highlighted the inefficacy of using the 'U' and 'I' vowels as standalone indicators for disease diagnosis, recommending the aggregation of multichannel data to improve feature extraction by the models. (All results are fully reproducible, the source code is available at https://github.com/Tesla2000/DepCoS2024).

Keyword: Convolutional Neural Networks (CNNs), Voice Disorder Diagnosis, Explainable Artificial Intelligence (XAI), Vowel Sound Analysis

1 Introduction

In medical diagnostics, the quest for non-invasive, efficient, and accessible testing methods has become a paramount concern. The traditional approach to diagnosing many diseases often involves specialized, invasive, and sometimes costly procedures. However, the advent of machine learning (ML) technologies offers promising possibilities for revolutionizing this aspect [1,2]. Early detection and diagnosis of diseases is a pivotal trend in contemporary medicine and can significantly benefit from the analysis based on thousands of patients processed by deep learning (DL) algorithms. In this study, we focused on the analysis of speech

W. Zamojski et al. (Eds.): DepCoS-RELCOMEX 2024, LNNS 1026, pp. 218–226, 2024.
https://doi.org/10.1007/978-3-031-61857-4_21

recordings. This approach is grounded in the premise that initial symptoms of various diseases can be discerned through changes in voice or speech patterns, thereby enabling early intervention.

This study focuses on harnessing deep learning methodologies to detect diseases affecting the voice and speech apparatus. By analyzing simple voice recordings of vowel sounds such as /a/, /u/, and /i/ our models aim to determine the health status of patients whose samples were recorded.

In practice, we have utilized the Saarbruecken Voice Database as the foundation of our research. Our methodology involved converting voice recordings into visual representations using spectrograms combined with image or audio augmentation techniques. We trained convolutional neural network (CNN) models, specifically VGG and ResNet architectures, assessing their performance across various configurations, including the choice of vowel sound or augmentation strategies, with the F1-score serving as our primary evaluation metric. Furthermore, our study leverages Explainable Artificial Intelligence (XAI) techniques to elucidate the distinguishing features of spectrograms between healthy and diseased samples. This approach not only aids in the disease detection process but also provides valuable insights for medical professionals, enhancing their understanding of disease manifestations within voice patterns.

The paper is organized as follows: Sect. 2 delves into related work such as voice analysis, speech-to-spectrogram conversion, CNN models, XAI, and studies of classifying pathological voice recordings. Section 3 covers data preparation, model evaluation, and presenting Results. Finally, Sect. 4 provides an overview of the paper's findings.

2 Related Works

The methodologies employed in voice analysis can be broadly categorized into three main streams [3,4]: static analysis [5], time-series analysis [6,7], and transforming the signal to image and then using image models [8,9]. Static methods focus on extracting and analyzing specific features from voice samples providing voice characteristics such as pitch, volume, timbre, formant frequencies, jitter (frequency variation), and shimmer (amplitude variation). Dynamic approaches treat the voice as a time-series input, employing models like Long Short-Term Memory (LSTM) networks to capture temporal variations and patterns over time. An exciting approach, where the voice signal is transformed into an image, such as a spectrogram, has also gained traction, allowing the application of image analysis techniques, notably Convolutional Neural Networks (CNNs).

Our research focuses on the last approach, recognizing the potential of spectrograms to encapsulate both the frequency and temporal information of voice signals in a visual format. Alternatives to spectrograms [10] include short-time Fourier transform (STFT) with linear and Mel scales, constant-Q transform (CQT) and continuous Wavelet transform (CWT), each offering different perspectives on the signal's characteristics. However, spectrograms stand out for their ability to represent the signal's energy distribution across time and frequency in a manner intuitively accessible for CNN analysis.

Convolutional Neural Networks (CNNs) [11] have been pivotal in advancing image recognition tasks, with architectures like VGG [12], ResNet [13], or the more recent EfficientNet [14] setting benchmarks in performance. These architectures provide a solid foundation for analyzing the complex features within spectrograms of voice samples.

In the context of disease detection through voice analysis, our study is aligned with the direction of utilizing simple voice recordings, specifically vowel sounds, as diagnostic inputs. This approach is grounded in the observation that certain diseases manifest through subtle changes in vocal characteristics, detectable in the articulation of vowels. Research leveraging the Saarbruecken Voice Database [15] exemplifies the potential of voice recordings as a rich dataset for training ML models to identify and classify pathological conditions. This database, encompassing a wide range of voice disorders, offers a comprehensive resource for developing and testing algorithms aimed at disease detection.

Several groups of researchers have explored the use of CNNs for classifying pathological voice recordings using Saarbruecken Voice Database. The authors of [16] investigate the application of deep learning, specifically through adapting the VGG. Their methodology focuses on organic dysphonia disorders and uses ensemble models of networks trained on different vowel subsets. They report achieving an 82% accuracy in detecting pathological speech. This work highlights the effectiveness of using model ensembles and transfer learning to address the challenge of limited dataset sizes in pathological speech recognition. The [17] study investigated using a novel algorithm, OSELM, to identify and classify voice disorders. The researchers used a combination of vowel sounds and continuous sentences spoken at different pitches to achieve high accuracy in detecting voice pathologies. This method achieved promising results, reaching over 87% accuracy in various metrics, suggesting its potential for real-time clinical applications. The authors of [18] proposed CS-PVC, a system for classifying pathological voices. It analyzes speech features to distinguish healthy from unhealthy voices, achieving 81.6% accuracy on the Saarbruecken Voice Database. CS-PVC extracts Mel-frequency cepstral coefficients (MFSC) features and feeds them into a DCA-ResNet architecture. This network incorporates attention modules that focus on relevant features and communicate with each other, improving the model's effectiveness. It's important to note that different studies, including ours, utilize diverse voice sample subsets for training and evaluating their voice pathology classification models.

Using CNNs in voice-based disease detection allows for applying Explainable Artificial Intelligence (XAI) techniques [19]. These techniques, like Grad-CAM [20], provide visual insights into how the CNN models make their predictions, highlighting essential areas in the spectrograms. This can offer clinicians interpretable evidence of the model's reasoning. Thus, XAI enhances trust in AI diagnostics and aids medical professionals in understanding and validating the automated analysis, supporting more informed clinical decisions.

3   Experiments

The Saarbruecken Voice Database dataset is a significant source in the field of speech research [15]. The provided data includes voice recordings of the vowels /a/, /i/, and /u/, which were selected for research purposes in the sustained type of articulation. The database covers both healthy individuals and those with various speech system disorders. In our research, we considered classification between samples from healthy and diagnosed with a disease test subjects with a total number of files 2031 and 2289 correspondingly, equally distributed within three vowels. Diseases given in the dataset are: Dysphonie, Functional Dysphonie, Hyperfunctionelle Dysphonie, Laryngitis, and Rekurrensparese.

This work introduces multi-phases encompassing data preparation, enhancement, and model assessment. Detailed explanations of each step are provided in the subsequent subsections.

Data Preparation. In contrast to many previous studies, we investigated whether generating spectrograms through audio slicing with 400 ms long windows and 100 ms stride would improve model generalization compared to resizing the entire visual representation of the audio to 224×224 pixels. Our hypothesis suggested that some information is lost when resizing mel-spectrograms of varying lengths to the same size. Mel-spectrograms were created (Fig. 1) with the following parameters: number of samples between successive frames 512, number of samples in each window 2048, number of mel filter banks 128.

Our study investigated mel-spectrogram data augmentations technique, including time masking, frequency masking, combining both, and audio augmentation with noise [21]. The visual augmentations are shown in Fig. 2.

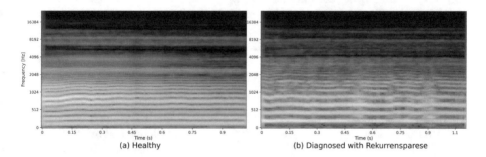

(a) Healthy   (b) Diagnosed with Rekurrensparese

Fig. 1. The figure compares the spectrogram of a healthy individual (a) with that of an individual diagnosed with Vocal cord paresis, called Rekurrensparese, in the dataset (b). The frequency patterns over time are distinct and aid in identifying and differentiating health conditions.

Model Evaluation. Each model instance was trained with an Adam optimizer using a different data augmentation technique, including no augmentation. We implemented one-fold stratified cross-validation with a 50% train-test split.

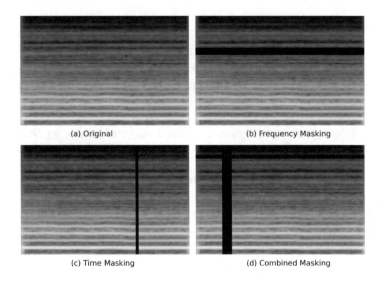

(a) Original

(b) Frequency Masking

(c) Time Masking

(d) Combined Masking

Fig. 2. The figure displays the original spectrogram (a) and its transformations through frequency masking (b), time masking (c), and a combination of both (d). These augmentations simulate variations in signal properties to enhance model robustness and performance.

During training, five different vowel configurations were used. Three configurations involved training separate models on different sounds, while the fourth configuration combined samples of /a/, /u/, and /i/ into one training set used for both training and validation (all vowels). In the last method, all samples (up to three) belonging to the patient were aggregated into one sample with three channels, described later as multichannel. Early stopping was implemented to overcome overfitting. Models fed with a signal divided to windows instead of a continuous signal, described later as Slicing, were trained with augmentation that added noise to the audio or without augmentation. We investigated two models: ResNet-18 [13] and VGG19 [12].

Results. The most effective classifiers are ResNet models trained on continuous data from the /a/ vowel with neither the use of slicing nor multiple channels. Both models achieved an f1 score of 0.76, one with unaugmented data and the other with noise added to the audio signal, as shown in Table 1. The best-performing VGG models lagged behind their ResNet counterparts, with the highest f1 score of 0.71 attained through the use of slicing, noise augmentation, and all vowels, but without combining data into multichannel. The second-best model reached f1 of 0.70 with neither the use of augmentations nor slicing and data from the /u/ vowel.

The ANOVA test revealed no significant differences in the effects of various augmentation methods ($p-value = 0.85$). Similarly, the Mann-Whitney U rank test showed no statistical differences between models trained using continuous signals and ones divided into windows. However, ResNet-18 models significantly outperformed their VGG counterparts, with Wilcoxon tests yielding ($p-value \ll 0.05$). The obtained results were as follows: ResNet-18 scored 71.51 ± 2.49, while VGG19 scored 68.51 ± 1.78. Both results are shown in Fig. 3.

The comparison of vowel significance was inconclusive. Table 2 shows which vowel or group of vowels (Superior vowel) outperformed (Inferior vowel) with statistical significance. VGG models performed worse on the /a/ vowel, and ResNet performed better.

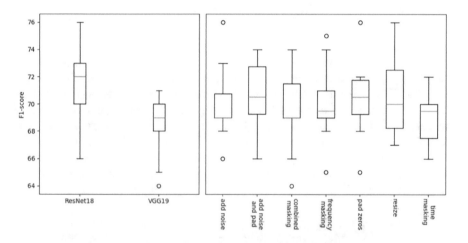

Fig. 3. F1 scores of specific models and augmentations

Table 1. Best Results for Each Model and Augmentation Type

Model	Multi Ch.	Slicing	Augmentation	Vowels	F1
ResNet-18	No	No	–	/a/	0.76
	No	No	Noise	/a/	0.76
	No	Yes	Combined Masking	/a/	0.74
	Yes	No	Frequency Masking	all	0.74
	Yes	No	Time Masking	all	0.72
VGG19	No	Yes	Noise	all	0.71
	No	No	–	/u/	0.70
	Yes	No	Frequency Masking	all	0.70
	No	No	Time Masking	/u/	0.69
	No	No	Combined Masking	/i/	0.70

Table 2. Selected significant differences between vowels divided to models

Model	Superior vowel	Inferior vowel
ResNet-18	a	multichannel
ResNet-18	multichannel	all vowels
ResNet-18	multichannel	i
ResNet-18	all vowels	u
ResNet-18	i	u
VGG19	u	i
VGG19	u	all vowels
VGG19	i	multichannel
VGG19	multichannel	a

The Grad-CAM method was used to demonstrate differences between normal and pathological conditions in mel spectrograms. However, instead of displaying the Grad-CAM on a few samples as shown in [22], we have chosen to average the results across the entire test set (see Fig. 4). The visual representation of the results indicates that model weights are more activated for the ill subjects. Pathology samples exhibit more varied and intense coloration for both multi-channel and single-channel data, indicating the areas of the image that the model found most relevant for identifying illness. In contrast, healthy subjects display a more uniform and less intense color pattern, suggesting fewer areas of concern.

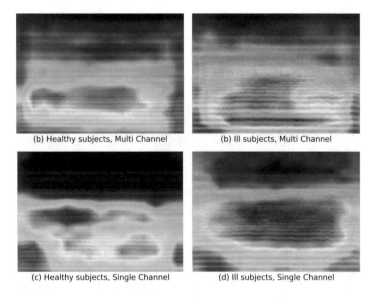

(b) Healthy subjects, Multi Channel   (b) Ill subjects, Multi Channel

(c) Healthy subjects, Single Channel   (d) Ill subjects, Single Channel

Fig. 4. Comparative Visualization of Model Attention using Grad-cam for Healthy and Pathological Subjects across Multi- and Single-Channel Data The focus areas of the neural network are highlighted.

4   Summary

The best-obtained results, measured in an F1-score, were 0.76 for ResNet18 and 0.71 for VGG19.

Analysis of the highest scores suggests that the most effective models, both ResNet and VGG, are those trained without augmentation or with the addition of noise. Surprisingly, the use of augmentations does not lead to significant score differences. This may be due to the best models benefiting from access to the entire spectrogram during training and masking, improving the results of other models. Similarly, there is no difference between a model trained on signals divided into windows and continuous ones.

Given the lower overall performance of VGG models, it can be inferred that the /u/ and /i/ vowels may not be reliable sources of medical information for machine learning models or, if collected, should be aggregated into multichannel data to allow the model to extract useful features. It seems to be essential to avoid using all available vowels as unaggregated samples in training, as this approach has been shown to hinder the results of ResNet18.

Grad-cam enabled the identification of differences between pathological and healthy speech by highlighting areas with more intense colors.

In future research, emphasis can be placed on employing XAI to analyze specific diseases and differentiate between them through classification.

References

1. Litjens, G., et al.: A survey on deep learning in medical image analysis. Med. Image Anal. **42**, 60–88 (2017)
2. Garg, A., Mago, V.: Role of machine learning in medical research: a survey. Comput. Sci. Rev. **40**, 100370 (2021)
3. Monir, R., Kostrzewa, D., Mrozek, D.: Singing voice detection: a survey. Entropy **24**(1), 114 (2022)
4. Sisman, B., et al.: An overview of voice conversion and its challenges: from statistical modeling to deep learning. IEEE/ACM Trans. Audio Speech Lang. Process. **29**, 132–157 (2020)
5. Keller, E.: The analysis of voice quality in speech processing. In: Chollet, G., Esposito, A., Faundez-Zanuy, M., Marinaro, M. (eds.) NN 2004. LNCS (LNAI), vol. 3445, pp. 54–73. Springer, Heidelberg (2005). https://doi.org/10.1007/11520153_4
6. Gamboa, J.C.B.: Deep learning for time-series analysis. arXiv preprint arXiv:1701.01887 (2017)
7. Li, J., et al.: LSTM time and frequency recurrence for automatic speech recognition. In: 2015 IEEE Workshop on Automatic Speech Recognition and Understanding (ASRU), pp. 187–191. IEEE (2015)
8. Sharan, R.V., Xiong, H., Berkovsky, S.: Benchmarking audio signal representation techniques for classification with convolutional neural networks. Sensors **21**(10), 3434 (2021)
9. Seo, S., Kim, C., Kim, J.-H.: Convolutional neural networks using log mel-spectrogram separation for audio event classification with unknown devices. J. Web Eng. **21**(2), 497–522 (2022)

10. Huzaifah, M.: Comparison of time-frequency representations for environmental sound classification using convolutional neural networks. arXiv preprint arXiv:1706.07156 (2017)

11. Li, Z., et al.: A survey of convolutional neural networks: analysis, applications, and prospects. IEEE Trans. Neural Netw. Learn. Syst. **33**(12), 6999–7019 (2021)

12. Simonyan, K., Zisserman, A.: Very deep convolutional networks for large-scale image recognition. arXiv preprint arXiv:1409.1556 (2014)

13. He, K., et al.: Deep residual learning for image recognition. In: Proceedings of the IEEE Conference on Computer Vision and Pattern Recognition, pp. 770–778 (2016)

14. Tan, M., Le, Q.: EfficientNet: rethinking model scaling for convolutional neural networks. In: International Conference on Machine Learning. PMLR, pp. 6105–6114 (2019)

15. Saarland University. Saarbruecken Voice Database. Database of voice recordings for speech and voice disorders research. https://stimmdb.coli.uni-saarland.de/help_en.php4

16. Vavrek, L., et al.: Deep convolutional neural network for detection of pathological speech. In: 2021 IEEE 19th World Symposium on Applied Machine Intelligence and Informatics (SAMI), pp. 000245–000250 (2021). https://doi.org/10.1109/SAMI50585.2021.9378656

17. Al-Dhief, F.T., et al.: Voice pathology detection and classification by adopting online sequential extreme learning machine. IEEE Access **9**, 77293–77306 (2021)

18. Ding, H., et al.: Deep connected attention (DCA) ResNet for robust voice pathology detection and classification. Biomed. Sig. Process. Control **70**, 102973 (2021)

19. Sheu, R.-K., Pardeshi, M.S.: A survey on medical explainable AI (XAI): recent progress, explainability approach, human interaction and scoring system. Sensors **22**(20), 8068 (2022)

20. Selvaraju, R.R., et al.: Grad-CAM: visual explanations from deep networks via gradient-based localization. In: Proceedings of the IEEE International Conference on Computer Vision, pp. 618–626 (2017)

21. Hwang, Y., et al.: Mel-spectrogram augmentation for sequence to sequence voice conversion. arXiv preprint arXiv:2001.01401 (2020). https://arxiv.org/abs/2001.01401

22. Jegan, R., Jayagowri, R.: Voice pathology detection using optimized convolutional neural networks and explainable artificial intelligence based analysis. Comput. Methods Biomech. Biomed. Eng. 1–17 (2023). https://doi.org/10.1080/10255842.2023.2270102

Preparing a Dataset of Ransomware BTC Addresses for Machine Learning Purpose

Przemysław Rodwald[(⊠)]

Department of Computer Science, Polish Naval Academy,
Śmidowicza 69, 81-127 Gdynia, Poland
p.rodwald@amw.gdynia.pl

Abstract. Ransomware is a class of malicious software that encrypts the files of an infected machine and demands payments, mainly in cryptocurrencies such as Bitcoin. One of the challenges faced by law enforcement is identifying suspected Bitcoin addresses. The aim of this article is to create a dataset of historical ransom addresses as training data for future research. This dataset consists of two elements: the first is a compilation of suspicious addresses from various data sources, and the second comprises custom-calculated parameters for these addresses. This dataset serves as a starting point for future research on the automatic identification of ransom addresses.

Keywords: bitcoin · address · ransomware · dataset

1 Introduction

A FinCEN report [9] states that, in the first six months of 2021, the number of reported ransomware payments made was 30% greater than in all of 2020 and the total of all payments made in the first half of 2021 was estimated to $590 million. Chainalysis [5], first, deemed 2020 the "Year of Ransomware" due to the huge growth in cryptocurrency extorted in ransomware attacks. Fortunately, after years of record-setting ransomware payouts, 2022 broke the mold. Chainalysis data shows ransomware attackers extorted at least $457 million from victims in 2022, down from $766 million the year before—a drop of 40.3%. Cipher-Trace analysis [6] proves that BTC is still the dominant cryptocurrency used among ransomware actors although demanding payment in privacy coins like Monero has been another trend. The European Union Agency for Cybersecurity (ENISA) [8] ranked ransomware as the "prime threat for 2020–2021". According to Statista [21], in the second quarter of 2023, 34% of ransomware attacks within global organizations resulted in a ransom payment, down from 45% in the previous quarter. Despite this, the average amount of ransom paid increased more than twice during the same period. The rising popularity of ransomware among cybercriminals has led to the emergence of a new business model known as Ransomware as a Service (RaaS). Financial institutions and law enforcement are grappling with the challenge of identifying suspected BTC addresses associated with ransomware and tracing funds from origin to destination.

In this paper, our contributions encompass the following aspects: 1. We collected, cleaned, and aggregated public datasets of ransomware addresses. 2. We analyzed the 'behavior' of suspected addresses and identified several characteristic patterns. 3. Based on these patterns, we identified and calculated characteristic metrics. 4. To facilitate future research in this area, we are making our newly created dataset of ransom Bitcoin addresses publicly available[1].

2   Related Works

After the release of the Nakamoto whitepaper [16], which introduced Bitcoin as a pseudoanonymous payment system, researchers across various domains [22] have been exploring ways to trace BTC addresses. All transactions are inherently public and stored in a shared ledger called the blockchain. Nakamoto decided to use UTXO (unspent transaction output) model in his proposal. Each transaction is represented by a list of inputs and outputs, reflecting the amount of Satoshis transferred to/from a specific address. A BTC address is an alphanumeric string derived from the public key using cryptographic hash functions, specifically SHA-256 and RIPEMD-160, and is Base-58 encoded at the end

2.1   Bitcoin Heuristics

While an address itself does not provide any information about its owner, the pseudoanonymity highlighted by Nakamoto[2] has prompted researchers to develop techniques for de-anonymizing Bitcoin users by associating them with real-world entities. The first widely adopted heuristic is called a co-spend or multiple-input [18]. It posits that two or more addresses used as inputs in the analysed transaction must be controlled by the same real entity. In non-CoinJoin transactions, it is assumed that the entity creating a transaction has control of all the private keys for input addresses [19] and, consequently, is the owner of those addresses.

Once at least one address owner has been identified, it becomes possible to create a cluster of addresses, thereby revealing other addresses belonging to the same actor. This cluster identification has been extended by change heuristics [2,14]. The concept of 'change' addresses, also referred to as 'shadow' addresses, in the Bitcoin network pertains to transactions with two output addresses. The change heuristic is more challenging and significantly less secure than the co-spend heuristic, and a wrong implementation could result in a high false-positive rate. The combination of these heuristics with tagging can lead to

---

[1] https://sydeus.rodwald.pl/datasets.

[2] "Some linking is still unavoidable with multi-input transactions, which necessarily reveal that their inputs were owned by the same owner. The risk is that if the owner of a key is revealed, linking could reveal other transactions that belonged to the same owner." [16].

the de-anonymization of a large fraction of BTC addresses. Tags may come from external sources such as forums, dedicated web services (e.g., walletexplorer, graphsense), scientific research, or commercial analytics tools (e.g., Chainalysis, Elliptic, CipherTrace, QLUE).

2.2 Ransomware Transactions

Some previous studies have investigated ransomware activity in the Bitcoin network. Kharaz et al. [12] analysed 1872 BTC addresses associated with the *CryptoLocker* ransomware family. They discovered that 68.93% of addresses had short (lower than 10 day) activity period, called lifetime and approximately 50% of Bitcoin addresses have zero to five days of active life. 72.9% of ransom transactions are associated with Bitcoin addresses having two transactions: the incoming transaction made by victims to pay the ransom, and the outgoing transaction conducted by the ransomware owner to transfer the Bitcoin to another address, making tracing infeasible. Another type of addresses had more transactions and remained active for a longer period (e.g., more than 10 days). These addresses were often utilized to aggregate the collected ransom fees. In 48.9% of the Bitcoin addresses they analyzed, a Bitcoin address received at most two Bitcoins. These transactions occurred when one Bitcoin was worth roughly 100 USD. The ransom fee demanded by cybercriminals for sending the decryption key was 200 USD. Liao et al. [13] also conducted an analysis of the *CryptoLocker* ransomware family. Beginning with two seed addresses collected from online forums, they generated a cluster of 968 addresses associated with the *CryptoLocker* entity. Building on findings from previous studies [20] and their own research, they concluded that the analyzed Bitcoin addresses exhibited similar transaction records. The amount of money requested by criminals varies among different ransomware families, but it is often rounded to a multiple of 100 USD [3] (e.g., 200, 300 USD) or to a decimal part of BTC[3] (e.g., 0.3 BTC, 0.5 BTC, 0.6 BTC, 1.0 BTC, 2.0 BTC). Through timestamp transaction analysis, researchers were able to make conjectures about regions where infections were prevalent. They assumed that a typical working day starts at 09:00 a.m. and ends at 05:00 p.m., and that ransom payments are processed during business hours. This allowed for a rough identification of the time zones of victims (e.g., UK, USA, Australia).

Huang et al. [11] tracked payments made by 19,750 potential victims of several ransomware families. They categorized these families into two payment categories. In the first category, malware generates a unique BTC address for each victim to automate the identification of paying victims (e.g., *Locky*, *Cerber*, *Spora*). Any Bitcoins received by each seed ransom address are associated with only a single victim who was provided with that unique ransom address. In the second category, the ransom address is reused among multiple victims (e.g., *WannaCry*, *CryptoDefence*). When addresses are reused, ransomware operators cannot identify paying victims. In such cases, they either require the victim to send them the payment transaction hash, or they simply do not decrypt the

---

[3] https://twitter.com/hasherezade/status/733586140364603392.

victim's files. Reports cited by the authors suggest that for *WannaCry* victims from multiple infections, the same ransom addresses are displayed, making the successful recovery of encrypted files questionable. The authors stated that ransom amounts are typically fixed, denominated in fiat money (e.g., 1000 USD for some *Cerber* strains) or Bitcoins (e.g., 0.5 BTC for some *Locky* strains). In their conclusions about payment timing, they state that 95.4% of *Cerber*'s ransom payments and 98.3% of *Locky*'s ransom payments were made in a single transaction. The extraction and categorization of timestamps from every single inflow transaction led them to some interesting findings in terms of days of the week and hours of the day. For instance, 22.2% of *Locky*'s ransom payments come from inflows on Thursdays. The distribution chart of ransom payments over a week reveals discrepancies between working days and the weekend.

Paquet et al. [17] distinguished six features for an address in their study: income, neighbors, weight, length, count, and loop. They proposed a data analytic framework to automatically detect new malicious addresses in a ransomware family. Income of an address is the total amount of coins output to it. Neighbors is the number of transactions which have this address as one of its output addresses. The next four address features are calculated in the defined 24 h time window. Weight is defined as the sum of fraction of coins that originate from a starter transaction and reach that address. Length is the number of non-starter transactions on its longest chain, where a chain is defined as an acyclic directed path originating from any starter transaction and ending at analysed address. Count is the number of starter transactions which are connected to investigated address through a chain. Loop of an address is the number of starter transactions which are connected to it with more than one directed path.

Michalski et al. [15], revealing the characteristics of nodes (mining pools, miners, coinjoins, gambling, exchange services), analyzed 149 features grouped into six families. These features include the number of addresses in transaction inputs and outputs, amounts of Bitcoins sent in transactions, numbers of coinbase transactions, numbers of transactions, lifetime, and address type[4].

3   Methodology

3.1   Identified Ransom BTC Addresses

One of the key aspects of the precise identification of ransom addresses is a reliable training source. However, the reliability of data from various sources will always be questionable due to the pseudo-anonymous nature of Bitcoin. This is one of the reasons why data from different sources often assigns different labels to the same address.

As the first source of BTC addresses, we decided to use the BitcoinHeist dataset [1]. The original dataset spans daily transactions from January 2009 to December 2018, with network edges featuring amounts less than 0.3 BTC

---

[4] Examples: Pay To Public Key Hash (P2PKH), Pay To Script Hash (P2SH), Pay To Public Key (P2PK), Pay To Witness Public Key Hash (P2WPKH).

removed, as ransomware payments were generally substantially higher during that time. Ransomware labels in this dataset are a union of datasets from three studies: Montreal [17], Padua [7], and Princeton [11]. After removing addresses from the shared dataset[5] that are not known to belong to any ransomware family, labeled as 'white' addresses, we acquired 41,413 ransom address appearances, among which were 20,849 unique addresses.

To demonstrate the earlier mentioned inconsistencies, we further checked selected addresses on services such as graphsense and walletexplorer. What was observed is a relatively high number of inconsistent labels among services. For example, the address 1LKeQAKkZy92jHy7eM98QdFh4QoKAXUFtS is identi-fied as montrealCryptoLocker in the BitcoinHeist dataset, as exchange BTC-e.com in walletexplorer, and is not labeled in graphsense.

For all labeled addresses from our initial source, using the JSON API designed by the author Aleš Janda, we cross-checked labels existing in walletexplorer. Only 2,890 addresses exist in both data sources. Walletexplorer labels them in three wallets: BTC-e.com-old (1,957), CoinJoinMess (27), and CryptoLocker (906).

As a second source of data, we utilized ransomwhe.re [4]. This is a crowdsourced dataset of ransomware payment addresses, containing payment addresses, transactions, and associated ransomware families. Anyone-whether a victim, a firm, or a security researcher-can contribute to expanding the data by submitting addresses of ransomware actors. From this source, we collected 10,448 records (address, ransom family).

As a third data source, we utilized records from the author's sydeus[6] system. This system includes labels for addresses identified as ransomware by wallet-explorer and those obtained from more sensitive sources, such as real criminal cases. From this source, we collected 38,149 records.

3.2   Ransom Addresses Behavior

To understand the flow of money coming from ransomware victims, in addition to examining the current literature, we conducted a deep analysis of sample addresses. In the first step, we employed a manual approach and observed visual graphs and some statistics in cryptocurrency investigation online tools: graph-sense[7] [10] and walletexplorer[8].

By examining the 'behavior' of addresses, we were able to generalize our observations and distinguish a few templates.

ORP Address: The first group is one-time ransom payment (ORP) addresses. In this scenario, malware generates a unique BTC address associated with a single victim. The ORP address appears in the blockchain in two transactions: once as an output address where the value of income is close to some equal value

---

[5] https://archive.ics.uci.edu/ml/datasets/BitcoinHeistRansomwareAddressDataset.

[6] https://sydeus.rodwald.pl/en.

[7] https://demo.graphsense.info/.

[8] https://www.walletexplorer.com/.

(e.g., 0.1 BTC; 0.7 BTC; 1.0 BTC; 1000 USD), and once as an input address among other similar-value addresses in a consolidation transaction. The activity period of an ORP address is typically a few days. Figure 1 illustrates the flow of money for a sample ORP address and the typical structure of an outgoing transaction.

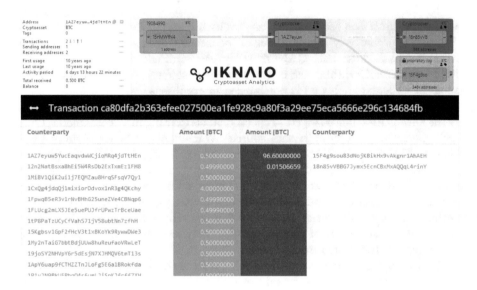

Fig. 1. Screenshots of the ORP address 1AZ7eyuw5YucEaqvdwWCjioMRq4jdTtHEn. (Sources: https://www.ikna.io and https://sydeus.rodwald.pl)

MRP Address: The second identified group is multiple ransom payment (MRP) addresses. In this scenario, the ransom address is reused among multiple victims. The MRP address appears in the blockchain in a few or a dozen transactions. The total number of transactions is usually even because outgoing transactions take place almost immediately (within a few next blocks) after each income. The value of incomes is close to some equal value (e.g., 0.1 BTC; 0.5 BTC; 2000 EUR). The outgoing transactions are very often used to consolidate money. The activity period of an MRP address is usually from a few to a dozen months. Figure 2 illustrates the flow of money for a sample MRP address and a typical table of in/out transactions.

AGG Address: An aggregation or consolidation address refers to an address used to consolidate or gather multiple smaller amounts of Bitcoin into a single address. After receiving payments from several or a dozen addresses, the total sum of cryptocurrency is immediately transferred further (Fig. 3).

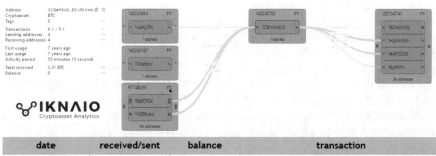

date	received/sent	balance	transaction
2016-04-16 10:03:39	-0.0099	0.	903d618d12c84cf87634de0605305e2e056a9507eb45ba077da197dbe08bb819
2016-04-16 10:00:07	-1.	0.0099	5bb33224ac6a5a44bb5ec123999a3659cb9256638f7451197bfdf6e2130a5fa4
2016-04-16 09:54:52	+0.0099	1.0099	0a291d650690eaf39171338f5565496e9be5f828754178f7de8af3cd906c995a
2016-04-16 09:54:52	-1.273556	1.	0221165c6d3732f8e9d5e7069a7b7be2f414ee13c1e3cd8d3304165561d18294
2016-04-16 09:52:08	+1.	2.273556	bb5451dd16bab5a03cd43a4e144551cda16606df69f5132af9ea156e2053ed52
2016-04-16 09:30:24	+1.273556	1.273556	cdc4c750e1013e0f29f97196aa82da0d394598363caab1ca8c835615ee7d19d8
2016-04-16 09:18:15	-0.22634376	0.	567f1121be9c7285faa7669614bddb224cdd067b3bf14872b123f31a5202b1ba
2016-04-16 09:08:24	+0.22634376	0.22634376	d796275c8bffb507ca855fbac399398f518c99cf4950ce074ee4ddea6db59d21

Fig. 2. Screenshots of the MRP address 1C8mM4UXRnvuTeDhVsKaaAVrt4DJv9t VVm. (Sources: https://www.ikna.io and https://walletexplorer.com)

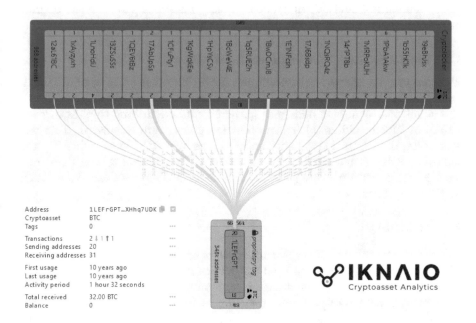

Fig. 3. Screenshot of the AGG address 1LEFrGPTNZC9Khmj6saiRJzWmnXHhq7 UDK. (Source: https://www.ikna.io)

Source of Purchased Bitcoins: While examining the source of Bitcoins, we observed that quite often, money for ransom is purchased on crypto-exchanges. A similar finding was also observed in [11], where authors, using Chainanalysis' API, determined the likely real-world identity of the source cluster. Huang et al. estimate that the source of payment in 37%–52% comes from crypto-exchanges (top ones: BTC-e, Bithumb, Coin.mx, Coinbase, Korbit), depending on the ransom family. They presented results for: WannaCry, Locky, CryptoDefense, CryptXXX, and Cerber. It shows that the source of money (if identifiable) could be an important factor in automatic ransom address identification.

Payment Timing: Huang et al. [11] also analyzed timestamps of ransom payments and reached the following conclusions: a) potential victims paid the ransom in a single transaction; b) once victims obtain bitcoins, they typically send them immediately to a ransom address, and c) roughly only 10% of all ransom payments are transferred on the weekends.

Destination of Ransom Payments: Huang et al. [11] examined the outflows from the ransomware clusters and observed that bitcoins are transferred to: a) crypto-exchanges (top ones: BTC-e, CoinOne, and LocalBitcoin - thus, the set is different than the source of purchased Bitcoins set), potentially to cash out money; b) bitcoin mixers (top-ones: Bitmixer.io and BitcoinFog - both no longer exist, which is typical in this area: some mixers are closed by legal authorities for money laundering, and others suddenly cease operations), to hide the original source of illegal funds; or c) not labeled wallets.

3.3   Identification of Characteristic Metrics

With the ransom address behavior in mind, the following, presented in Table 1, set of metrics was established.

For all 53,376 unique addresses gathered (after removing duplicates from a total of 71,916 addresses), metrics defined above were calculated. This was done using data from blockchain explorers, the WalletExplorer web API, and dedicated PHP scripts created for this purpose. In addition to raw data from the blockchain, certain values (bitcoin equivalents in USD) were calculated based on Yahoo Finance.

4   Summary

The ongoing popularity of cryptocurrencies, coupled with their exploitation for illicit activities, highlights the need for law enforcement to pinpoint suspected BTC addresses. The proposed dataset, made publicly available, containing identified ransom addresses and accompanying metrics, represents an initial step toward further advancements in this field. As part of his research and court expert activities, the author plans to delve deeper into this realm, with a primary focus on the automatic identification and categorization of ransom Bitcoin addresses.

Table 1. Labels and explanation of defined metrics.

Label[1]	Explanation	Source[2]
CTR_TRA_INC	total number of incoming transactions	[B]
CTR_TRA_OUT	total number of outcoming transactions	[B]
MMM_REC_BTC	MMM amount of received assets in BTC	[B, C]
MMM_REC_USD	MMM amount of received assets in USD	[B, C]
SUM_REC_BTC	total value of received assets in BTC	[B]
SUM_REC_USD	total value of received assets in USD	[B, C]
MMM_TRO_OUT	MMM amount of transaction outputs TO[3]	[W, C]
MMM_TRI_INP	MMM amount of transaction inputs TI[4]	[W, C]
FUL_TIM	lifespan of address[5]	[B]
MMM_TIM	period between deposit and withdrawal[6]	[B]
NAM_CLU_WAL	cluster identificator in walletexplorer	[W]
LAB_CLU_WAL	label of cluster in walletexplorer	[W]
CTR_CLU_WAL	total number of addresses in cluster	[W]

[1]MMM: minimum/mean/maximum (3 values)
[2][W] - walletexplorer.com, [B] - blockchain.info, [C] - value is calculated
[3]TO: when the address in question is an output of the transaction
[4]TI: when the address in question is an input of the transaction
[5]calculated as range of blocks between first and last occurrence in the blockchain
[6]calculated as range of blocks between deposit and withdrawal (FIFO aproach)

References

1. Akcora, C.G., Li, Y., Gel, Y.R., Kantarcioglu, M.: BitcoinHeist: topological data analysis for ransomware prediction on the bitcoin blockchain. In: Proceedings of the Twenty-Ninth International Joint Conference on Artificial Intelligence (2020)
2. Androulaki, E., Karame, G.O., Roeschlin, M., Scherer, T., Capkun, S.: Evaluating user privacy in bitcoin. In: Sadeghi, A.-R. (ed.) FC 2013. LNCS, vol. 7859, pp. 34–51. Springer, Heidelberg (2013). https://doi.org/10.1007/978-3-642-39884-1_4
3. Bursztein, E., McRoberts, K., Invernizzi, L.: Tracking desktop ransomware payments. Black Hat USA, Las Vegas, United States (2017)
4. Cable, J.: Ransomwhere: a crowdsourced ransomware payment dataset (2022). https://doi.org/10.5281/zenodo.6512123
5. Chainalysis: The 2023 crypto crime report (2023). https://blog.chainalysis.com/reports/. Accessed 10 Jan 2024
6. CipherTrace: Current trends in ransomware with special notes on monero usage (2022). https://ciphertrace.com/resources/. Accessed 10 Jan 2024
7. Conti, M., Gangwal, A., Ruj, S.: On the economic significance of ransomware campaigns: a bitcoin transactions perspective. Comput. Secur. **79**, 162–189 (2018)

8. ENISA: Threat landscape report 2021 (2022). https://www.enisa.europa.eu/publications/enisa-threat-landscape-for-ransomware-attacks. Accessed 10 Jan 2024

9. FinCEN: Ransomware trends in bank secrecy act data between January 2021 and June 2021. https://www.fincen.gov/sites/default/files/shared/FinancialTrendAnalysis_Ransomeware508FINAL.pdf. Accessed 10 Jan 2024

10. Haslhofer, B., Stütz, R., Romiti, M., King, R.: GraphSense: a general-purpose cryptoasset analytics platform. arXiv preprint arXiv:2102.13613 (2021)

11. Huang, D.Y., et al.: Tracking ransomware end-to-end. In: 2018 IEEE Symposium on Security and Privacy (SP), pp. 618–631. IEEE (2018)

12. Kharraz, A., Robertson, W., Balzarotti, D., Bilge, L., Kirda, E.: Cutting the gordian knot: a look under the hood of ransomware attacks. In: Almgren, M., Gulisano, V., Maggi, F. (eds.) DIMVA 2015. LNCS, vol. 9148, pp. 3–24. Springer, Cham (2015). https://doi.org/10.1007/978-3-319-20550-2_1

13. Liao, K., Zhao, Z., Doupé, A., Ahn, G.J.: Behind closed doors: measurement and analysis of cryptolocker ransoms in bitcoin. In: 2016 APWG Symposium on Electronic Crime Research (eCrime), pp. 1–13. IEEE (2016)

14. Meiklejohn, S., et al.: A fistful of bitcoins: characterizing payments among men with no names. In: Proceedings of the 2013 Conference on Internet Measurement Conference, pp. 127–140 (2013)

15. Michalski, R., Dziubałtowska, D., Macek, P.: Revealing the character of nodes in a blockchain with supervised learning. IEEE Access **8**, 109639–109647 (2020)

16. Nakamoto, S.: Bitcoin: a peer-to-peer electronic cash system. Decentralized Business Review, p. 21260 (2008)

17. Paquet-Clouston, M., Haslhofer, B., Dupont, B.: Ransomware payments in the bitcoin ecosystem. J. Cybersecurity **5**(1), tyz003 (2019)

18. Reid, F., Harrigan, M.: An analysis of anonymity in the bitcoin system. In: Altshuler, Y., Elovici, Y., Cremers, A., Aharony, N., Pentland, A. (eds.) Security and Privacy in Social Networks, pp. 197–223. Springer, New York (2013). https://doi.org/10.1007/978-1-4614-4139-7_10

19. Ruffing, T., Moreno-Sanchez, P., Kate, A.: CoinShuffle: practical decentralized coin mixing for bitcoin. In: Kutyłowski, M., Vaidya, J. (eds.) ESORICS 2014. LNCS, vol. 8713, pp. 345–364. Springer, Cham (2014). https://doi.org/10.1007/978-3-319-11212-1_20

20. Spagnuolo, M., Maggi, F., Zanero, S.: BitIodine: extracting intelligence from the bitcoin network. In: Christin, N., Safavi-Naini, R. (eds.) FC 2014. LNCS, vol. 8437, pp. 457–468. Springer, Heidelberg (2014). https://doi.org/10.1007/978-3-662-45472-5_29

21. Statista: Ransomware - statistics report about ransomware attacks worldwide (2023). https://www.statista.com/study/43873/ransomware-attacks-worldwide/. Accessed 10 Jan 2024

22. Yli-Huumo, J., Ko, D., Choi, S., Park, S., Smolander, K.: Where is current research on blockchain technology?-A systematic review. PLoS ONE **11**(10), e0163477 (2016)

Human Technology Frontier: A Retrospective and Challenges for the Future in the Era of Artificial Intelligence

Jerzy W. Rozenblit[(⊠)] [iD]

Department of Electrical and Computer Engineering, The University of Arizona, Tucson, AZ 85721, USA
jerzyr@arizona.edu

Abstract. As intelligent systems and technologies mature, a shift in the modality of use is occurring, namely computer-based systems are no longer an assistive extension of the human operator but are an inherent part of human endeavors. This, on the one hand, leads to augmented human performance, while on the other hand, presents challenges in managing the boundaries between human versus machine control, over- and under-reliance on automation, and artificial intelligence (AI). Social impacts such as dramatic changes in work paradigms, outsourcing, and human interactions are more and more pronounced. The presentation will explore the evolution of smart systems from fundamental models to advanced technologies today. It will discuss the human impact of the new advances on engineering, medicine, other endeavors, and the society at large. It will debate the question of "how much better off are we?" living in the technology driven universes and our interactions with the ubiquitous computer-based worlds.

Keywords: High technologies · Artificial Intelligence · Human computer interactions

1 Intelligent Systems

When I lecture on computer-based techniques in robot-assisted surgery (such as with the da Vinci surgical system [1]), I often pose the following question to the audience: "Would you let a robot operate on you by itself?" After some consternation, common responses are "it depends" and "when?" Clearly, the attendees seek a more detailed context to this dilemma and an understanding of the capabilities of the robot. The broader issues and design aspects of such a life-critical device – where the patient's safety and excellent operative outcomes are paramount to its use – are dependability, trustworthiness, and explainability. These aspects are equally important to both the patients and healthcare provides who use such robots and other, technologically very advanced, and complex, medical devices.

These concerns will become more acute as the systems that we develop embody higher and higher levels of autonomy, intelligence, and cognitive abilities. While in the

© The Author(s), under exclusive license to Springer Nature Switzerland AG 2024
W. Zamojski et al. (Eds.): DepCoS-RELCOMEX 2024, LNNS 1026, pp. 237–240, 2024.
https://doi.org/10.1007/978-3-031-61857-4_23

past decade – or even a shorter span of time – we have witnessed the revival of artificial intelligence (AI) and machine learning (ML), the emergence of generative AI/ML tools such as ChatGPT [2], and significant progress in the deployment of autonomous systems, for instance, self-driving cars, the foundational basis for intelligent systems design had been laid out well before the current century. In the presentation, we will provide a retrospective of the evolution of "smart" systems, beginning with the works of Wiener [3] and Ashby [4] who are recognized as fathers of cybernetics, a field of study of causal and feedback systems. We will examine the tenets of high autonomy systems and how they emerged in the mid-1980ties as part of NASA's Systems Autonomy Demonstration Project to support Space Station automation [5].

Autonomy is understood here as the ability of a system to function independently, without the intervention of a human operator [6]. This requires integration of complex perceptual (sensing) tasks, planning, real-time control as well as diagnosis and adaptation to ever changing environmental conditions. Early work in AI and Expert Systems leveraged from agent- and model-based approaches, where the agents would act/control the environment based of the precepts (information about the state of the system under control) and the overall operational goals (Fig. 1.).

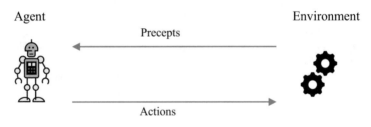

Fig. 1. Agent metaphor

Some examples of such systems are given in Table 1 below.

Table 1. Examples of agent-based systems (adopted from Russell and Norvig [7])

Agent Type	Percepts	Actions	Goals	Environment
Medical Diagnosis System	Symptoms, findings, patient's answers	Questions, tests, treatments	Healthy patient, minimize cost	Hospital, patient
Satellite image analysis system	Pixels of varying luminance, color	Classification of scenes	Correct classification	Image processing computers/satellites
Part picking robot	Pixels of varying intensity	Pick parts and sort into bins	Place parts in correct bins	Manufacturing system
Reactor controller	Temperature, pressure readings	Open, close valves, adjust pressure, water temperature	Maximize safety, power	Reactor

Agent-based paradigms are the foundational basis for building cognitive systems defined as autonomous systems that *"can perceive [their] environment, learn from experience, anticipate the outcome of events, act to pursue goals, and adapt to changing circumstances."* [8]. This leads to a debate if such systems "know what they are doing" and to what extent they will incorporate attributes of intelligence that we typically associate with humans. More specifically, humans perceive and classify events, phenomena, and objects; they also abstract common properties – current neural networks-based models do that relatively well. Humans have mental states. We often allude to a person thinking about something, believing that something is true or false – intelligent systems have states of computation and can automatically assert logical values of assertions given axioms and sound rules of reasoning. We learn, acquire new knowledge, can solve new problems. We use language and, while communication is not unique to humans, we are the only ones who disseminate knowledge and outcomes of creative endeavors in a variety of forms – machines do not do that, yet; they are a medium for dissemination. Humans make and use models, internal representations of the world to make decisions and predict consequences of their actions – currently, computational systems rely on models encoded in them. We will debate to what degree current cognitive systems integrate all the above abilities at the level of a creative, sentient human.

However, as stated at the outset of this presentation, of more immediate concern is the dependability of the emerging smart systems in their interactions with end users and human beneficiaries. We address this issue in the next section.

2   Life Critical Systems

"If it fails, people die," is a quote often used to describe computing systems whose malfunction may lead to catastrophic consequences such as loss of life, severe damage to the environment, and extensive harm to other systems. Life-critical systems are ubiquitous: they are commonly deployed in the aerospace, defense, medical, security, communications, and consumer sectors. They are complex hardware and software computing systems that assist in flying planes, controlling power plants, that work in dangerous and inaccessible environments, run heart-lung machines, defibrillators, and surgical robots. The key question always remains: "how do we know they are safe?"

Despite the rapid growth of innovative and powerful technologies for hardware and software design, networked computation, sensing, and control, many life-critical systems design and verification issues remain open:

- The complexity of the emerging systems is very high and thus the time needed to develop and verify them has increased.
- The effort needed to validate and certify them is significant.
- The high complexity calls for methods to validate and assure dependability, explainability, and trustworthiness.
- A large variety of formalisms and tools used in the computing and physical elements of existing systems may force conservative design decisions and constraints that may potentially limit or degrade performance and robustness.
- In deployed systems, fears of unpredictable side-effects may forestall modifications.

These issues present major research and development challenges that we need address and work on, with the ultimate goal of making life-critical systems safe.

3 Human Impact

We will close the presentation by raising the questions of our place in the technology driven universe. The exponential advances in the enabling hardware, software, and networking as well as the new AI computational paradigms offer promise, anxiety, and potential for significant disruption. A shift is occurring in how smart systems are used and operated. Namely, the types of interactions between the humans and systems are changing from the asymmetric communication and control where we still serve in a supervisory role, to that of a federation like setting, where we imagine the systems to work hand in hand with us and provide augmentation of performance with a level of decision making capabilities that exceed those of a human, to fully autonomous behaviors (e.g., self-driving cars).

We will also ponder the disruptive and malicious potential of smart systems, the impact on privacy, social interactions (social media effects), financial and political aspects, creativity, and the need for government regulation.

The future of work in the context of how intelligent systems will be deployed in virtually every aspect of our lives, will be addressed in regard to our values, modes of professional and social engagements, and adaptation to technological change.

References

1. Intuitive   Homepage.   https://www.intuitive.com/en-us/products-and-services/da-vinci. Accessed 23 Mar 2024
2. Open AI Homepage. https://openai.com/blog/chatgpt. Accessed 24 Mar 2024
3. Wiener, N.: Cybernetics or Control and Communication in the Animal and the Machine, 1st edn. The MIT Press, Cambridge (1965)
4. Ashby, R.W.: An Introduction to Cybernetics. Chapman and Hall Ltd., London (1956)
5. Erickson, W., Cheesman, P.: Issues in the design of an executive controller shell for space station automation. Opt. Eng. **25**, 1195 (1986)
6. Rozenblit, J.W.: Design for autonomy: an overview. Appl. Artif. Intell. **6**(1), 1–18 (1992). https://doi.org/10.1080/08839519208949939
7. Russell, S., Norvig, P.: Artificial Intelligence: A Modern Approach. Prentice Hall, Upper Saddle River (1995)
8. Vernon, D.: Cognitive system. In: Ikeuchi, K. (ed.) Computer Vision: A Reference Guide, pp. 157–164. Springer, Cham (2021). https://doi.org/10.1007/978-3-030-63416-2_82

Styles for Describing Reliable Finite State Machines in Verilog HDL

Valery Salauyou[(⊠)] [iD]

Bialystok University of Technology, Wiejska 45A, 15-351 Bialystok, Poland
v.salauyou@pb.edu.pl

Abstract. Improving the reliability and designing of fault-tolerant finite state machines (FSMs) is an important task. There are many known methods for designing reliability FSMs. In these methods, a change of one or two bits in the FSM state register is most often considered as a fault. Structural models of robust Moore FSM have been presented in [9], which allow us to detect invalid input and output vectors both in each state and for the whole FSM, an invalid code of the present and next state, and invalid transitions between states. This paper discusses the Verilog HDL description styles of these structural models of robust FSMs, as well as the combined structural models. Experimental studies have shown that the considered structural models insignificantly increase the area of the original FSM, and for some structures the area even decreases. At the same time, the performance of the original FSM changes insignificantly, and for some structures it increases by 35–37%.

Keywords: finite state machine (FSM) · reliability · fault tolerance · hardware description language (HDL) · Verilog · field programmable gate array (FPGA) · unmanned aerial vehicle (UAV)

1 Introduction

This paper presents styles for describing structural models of reliable finite state machines (FSMs) in Verilog hardware description language (HDL). The reliability of FSMs consists in detecting faults and preventing their negative impact on the controlled object. The considered structural models of FSMs are most effective when they are used as control devices for unmanned aerial vehicles (UAVs).

In [1], using embedded memory blocks of field programmable gate array (FPGA) is proposed to improve the FSM reliability that is susceptible to ion radiation. In [2], two methods considered to improve the FSM reliability: the duplication with self-checking and the triple modular redundancy. In [2] it is also pointed out the need to use methods in HDL to ensure the FSM reliability. In [3], a method for detecting and correcting single event upsets (SEU) in the FSM state register based on parity codes is presented. In [4], a tool for identifying FSM illegal states at the register transfer level is discussed. In [5], FSM architectures for cryptographic devices that prevent strong fault injection attacks are analyzed. In [6], state encoding algorithms for asynchronous

© The Author(s), under exclusive license to Springer Nature Switzerland AG 2024
W. Zamojski et al. (Eds.): DepCoS-RELCOMEX 2024, LNNS 1026, pp. 241–251, 2024.
https://doi.org/10.1007/978-3-031-61857-4_24

FSMs for robust systems are provided. In [7], ways to design fault-tolerant FSMs using Trahtman's theorem are discussed. In [8], design problems of reliable FSMs such as protection against illegal states and invalid transitions, as well as the impact of optimizers of synthesis tools on the reliability of FSMs are discussed.

In [9], the problem of designing reliable FSMs is solved at the structural level, when the additional combinational circuits and the output register R_O are introduced into the structure of the Moore FSM to detect faults and prevent their negative impact on the controlled object.

However, a description of structural models of robust FSM in HDL is not considered in [9]. This paper addresses to styles for description in Verilog HDL of structural models, which have been presented in [9].

The paper is organized as follows. Section 1 discusses the transitions of an FSM from illegal states. The way of describing the transitions of an FSM from each state is presented in Sect. 2. Section 3 analyzes the styles of describing output vectors for reliable FSMs. Section 4 presents the description styles for reliable FSMs. Experimental studies and their analysis are given in Sect. 5. The main results and the direction of further research are summarized in Conclusion.

2 Styles for Describing Robust FSMs

The structural models of robust Moore FSMs and the demonstration example are given in [9]. For fault detection, the following combinational circuits are added into the traditional structural model of a Moore FSM: TVI (total valid input) to detect invalid input vectors for the whole FSM, VI (valid input) to detect invalid input vectors in each state, TVO (total valid output) to detect invalid output vectors for the whole FSM, VO (valid output) to detect invalid output vectors in each state, VS (valid state) to detect an invalid code of a present state, VNS (valid next state) to detect an invalid code of a next state, and VT (valid transition) to detect invalid FSM transitions.

The considered structural models function as follows. Each additional combinational circuit generates a diagnostic signal of the same name (tvi, vi, tvo, vo vs, vns, and vt), the value 1 of which indicates the absence of a fault.

To prevent a negative impact of the failure on the operation of the FSM, the input CE (clock enable) of the state register R is controlled by the diagnostic signals tvi, vi, tvo, vo vs, vns, and vt. As a result, if the failure occurs, the FSM does not switch to the next state, and the FSM remains in the current state until the fault is resolved. In addition, to prevent the negative impact of the failure on the controlled object, the output register R_O is added to the structures TVO, VO, and VS. The input CE of the register R_O is also controlled by the diagnostic signals vto, vo, and vs. So in case of a failure the register R_O will not be switched, and the outputs of the FSM will remain the last correct value of the output vector.

The functioning of the combinational circuits TVO, VO, VS, VNS, and VT is determined from the FSM specification, while the functioning of the combinational circuits TVI and VI is determined by the developer.

The traditional description in Verilog of the demonstration example of the Moore FSM from [9] using three processes is as follows.

```verilog
module based_FSM(
    input clk, reset,        // control signals
    input [2:0] x,           // inputs
    output reg [2:0] y);     // outputs
    reg [2:0] state, next;   // state variables
    localparam [2:0] s0=0, s1=1, s2=2, s3=3, s4=4;    // states
    always @(posedge clk, negedge reset)     // describing state register
        if (~reset)  state <= s0;
        else     state <= next;
    always @(*)                              // describing transitions
        case (state)
        s0:casex (x)
            3'b0??:  next = s0;
            3'b100:  next = s1;
            default: next = s0;
          endcase
        s1:casex (x)
            3'b?01:  next = s2;
            3'b?11:  next = s4;
            default: next = s1;
          endcase
        s2:          next = s3;
        s3:   casex (x)
            3'b0?0:  next = s4;
            3'b0?1:  next = s0;
            default: next = s3;
          endcase
        s4:casex (x)
            3'b0?0:  next = s2;
            3'b0?1:  next = s0;
            default: next = s4;
          endcase
          default:   next = s0;
        endcase
    always @(*)                              // describing outputs
        case (state)
          s0: y = 3'b100;
          s1: y = 3'b001;
          s2: y = 3'b010;
          s3: y = 3'b011;
          s4: y = 3'b001;
          default: y = 3'b000;
        endcase
endmodule
```

The *safe0* description style from [10] was used here, where the initial transition state is defined in the construction **default** of the statement **case** and the null vector is formed in illegal states.

The description of the TVI structure is as follows.

```
module TVI_FSM(…);
    … // state variables and states
    reg tvi;
    always @(posedge clk, negedge reset) // describing state register
        if (~reset)  state <= s0;
        else if(tvi) state <= next;
        else      state <= state;
    … // describing transitions
    … // describing outputs
    always @(*)
        case (x)
            3'b110: tvi = 1'b0;
            3'b111: tvi = 1'b0;
            default: tvi = 1'b1;
        endcase
endmodule
```

Here, the statement **case** in the fourth process defines the value of the variable tvi for invalid input vectors and the construct **default** defines the value of the variable tvi for valid input vectors.

The description of the VI structure is as follows.

```
module VI_FSM(...);
    ... // state variables and states
    reg vi;
    always @(posedge clk, negedge reset)
        if (~reset)  state <= s0;
        else if (vi)  state <= next;
        else      state <= state;
    ... // describing transitions
    ... // describing outputs
    always @(*)
        case (state)
        s0:            vi = 1'b1;
        s1:casex(x)
              3'b??1:  vi = 1'b1;
              default: vi = 1'b0;
            endcase
        s2:            vi = 1'b1;
        s3:casex(x)
              3'b0??:  vi = 1'b1;
              default: vi = 1'b0;
            endcase
        s4:casex(x)
              3'b0??:  vi = 1'b1;
              default: vi = 1'b0;
            endcase
        default:       vi = 1'b0;
        endcase
endmodule
```

Here, the statement **case** in the fourth process determine the values of the variable vi at each transition when the input vector is valid, and the constructs **default** determine the values of the variable vi at each transition when the input vector is invalid.

The description of the TVO structure is as follows:

```
module TVO_FSM (input clk, reset, input [2:0] x, output reg [2:0] y_out);
    ... // state variables and states
    reg [2:0] y;
    reg tvo;
    ... // describing state register as in based_FSM
    ... // describing transitions
    ... // describing outputs
    always @(*)
        case (y)
            3'b100: tvo = 1'b1;
            3'b001: tvo = 1'b1;
            3'b010: tvo = 1'b1;
            3'b011: tvo = 1'b1;
            default: tvo = 1'b0;
        endcase
    always @(posedge clk, negedge reset)
        if (~reset)  y_out <= 0;
        else if(tvo)    y_out <= y;
        else       y_out <= y_out;
endmodule
```

Here, an additional variable y is declared to define the values of the outputs, and the outputs of the FSM are the port y_out. The fourth process defines the value of the variable tvo, and the fifth process describes the output register R_O.

The description of the VO structure is as follows.

```
module VO_FSM(... /* the ports as in TVO */);
   ... // state variables and states
   reg [2:0] y;
   reg vo;
   always @(posedge clk, negedge reset)
      if (~reset)  state <= s0;
      else if (vo)    state <= next;
      else      state <= state;
   ... // describing transitions
   ... // describing outputs
   always @(*)
      case (state)
      s0:case(y)
            3'b100:  vo = 1'b1;
            default: vo = 1'b0;
         endcase

         ...

      s4:case(y)
            3'b001:  vo = 1'b1;
            default: vo = 1'b0;
         endcase
      default:      vo = 1'b0;
      endcase
   always @(posedge clk, negedge reset)
      if (~reset)  y_out <= 0;
      else if(vo)  y_out <= y;
      else      y_out <= y_out;
endmodule
```

Here, the fourth process generates the value of the variable vo in each state, and the fifth process repeats the similar process to describe the TVO structure.

The description of the VS structure is as follows.

```
module VS_FSM(... /* the ports as in TVO */);
    ... // state variables and states
    reg [2:0] y;
    reg vs;
    always @(posedge clk, negedge reset)
        if (~reset)  state <= s0;
        else if(vs)  state <= next;
        else      state <= state;
    ... // describing transitions
    ... // describing outputs
    always @(*)
        case (state)
            s0:      vs = 1'b1;
            ...
            s4:      vs = 1'b1;
            default: vs = 1'b0;
        endcase
    always @(posedge clk, negedge reset)
        if (~reset)  y_out <= 0;
        else if(vs)  y_out <= y;
        else      y_out <= y_out;
endmodule
```

Here, the fourth process generates the value of the variable vs, and the fifth process repeats the similar process of describing the TVO structure.

The description of the VS structure is as follows.

```
module VNS_FSM(...);
    ... // state variables and states
    reg vns;
    always @(posedge clk, negedge reset)
        if (~reset)  state <= s0;
        else if(vns)state <= next;
        else      state <= state;
    ... // describing transitions
    ... // describing outputs
    always @(*)
        case (next)
            s0:      vns = 1'b1;
            ...
            s4:      vns = 1'b1;
            default: vns = 1'b0;
        endcase
endmodule
```

Here, the fourth process generates the value of the variable vns in each state.

The description of the VT structure is as follows.

```
module VT_FSM(...);
    ... // state variables and states
    reg vt;
    always @(posedge clk, negedge reset)
        if (~reset)  state <= s0;
        else if (vt)  state <= next;
        else       state <= state;
    ... // describing transitions
    ... // describing outputs
    always @(*)
        case (state)
        s0:case(next)
                s0:      vt = 1'b1;
                s1:      vt = 1'b1;
                default: vt = 1'b0;
            endcase
        ...
        s4:case(next)
                s0:      vt = 1'b1;
                s2:      vt = 1'b1;
                default: vt = 1'b0;
            endcase
            default:     vt = 1'b0;
        endcase
endmodule
```

Here, the fourth process generates the value of the variable vt at each FSM transition.

3   Description of Combined Structures of Robust FSMs

For more effective detection of the faulty element of the FSM, the considered structure can be combined into one structure. Note that the structure TVI is covered by the structure VI, i.e. if a fault is detected using the structure TVI, the fault will certainly be detected by the structure VI. Similarly, the structure TVO is covered by the structure VO, which in turn is covered by the structure VT. In addition, the structures VS and VNS are covered by the structure VT. For example, the VI_VT structure, which combines the VI and VT structures, allows to detect errors of the input vector in each state and for the whole FSM, invalid transitions, invalid codes of the present and next states.

It is also possible to combine all the considered structures into one structure V_ALL.

4 Experimental Results

In order to estimate the area and the performance, the FSM designs has been synthesized using Quartus Prime tool (version 23.1) from Intel in the Cyclone 10 LP FPGA for User-Encoded state encoding mode. The experimental results are shown in Table 1, where L is the number of used LUTs (look-up table) or area; F is the maximum operating frequency (in megahertz) or performance; L_b and F_b are similar parameters for the basic (traditional) structural model; L/L_b and F/F_b are relations of corresponding parameters.

Table 1. The results of experimental studies of structural models for our example of the Moore FSM.

FSM	L	L/Lb	F	F/F_b
Based_FSM	10	1	349	1
TVI-FSM	11	1.1	372	1.066
VI_FSM	16	1.6	292	0.837
TVO_FSM	9	0.9	472	1.352
VO_FSM	9	0.9	478	1.370
VS_FSM	9	0.9	472	1.352
VNS_FSM	10	1	349	1
VT_FSM	11	1.1	370	1.061
VI_VT_FSM	15	1.5	268	0.768
VI_VT_VO_FSM	15	1.5	342	0.980
V_ALL_FSM	17	1.7	349	1

Table 1 shows that the structure VNS does not increase the area of the original FSM, and the structures TVO, VO, and VS even decrease the area by 10%. The structures TVI and VT increase the area by 10%, and the structure VI increases the area by 60%. The combined structures VI_VT and VI_VT_VO increase the area by 50%, and the structure V_ALL increases the area by 70%.

Table 1 also shows that the structures TVI, VT, and V_ALL do not change the performance of the original FSM, the structures VI, VI_VT, and VI_VT_VO decrease the performance, and the structures TVO, VO, and VS even increase the performance of the original FSM by 35–37%.

5 Conclusions

This paper presented the styles of describing in Verilog HDL the structural models of robust Moore FSMs as well as the combined structures of robust FSMs. The experimental studies have shown that the considered structural models slightly increase the area of the original FSM, and some structures even decrease the area of the original FSM. A noticeable increase in the area (by 50–70%) is observed only for the combined structures and structure VI. The experimental studies also showed that the performance of the original FSM changes insignificantly, and for the structures TVO, VO, and VS even increases by 35–37%.

A promising direction for future research seems to be the development of new styles and ways of FSM describing, aimed at improving the FSM parameters (an area, a performance and a power consumption), as well as aimed at improving the reliability and the fault tolerance of FSMs.

Acknowledgments. The present study was supported by the a grant WZ/WI-III/5/2023 from Bialystok University of Technology and founded from the resources for research by Ministry of Science and Higher Education.

References

1. Tiwari, A., Tomko, K.A.: Enhanced reliability of finite-state machines in FPGA through efficient fault detection and correction. IEEE Trans. Reliab. **54**(3), 459–467 (2005)
2. Cassel, M., Lima, F.: Evaluating one-hot encoding finite state machines for SEU reliability in SRAM-based FPGAs. In: 12th IEEE International On-Line Testing Symposium, pp. 6–pp. IEEE (2006)
3. Singh, J., Mathew, J., Hosseinabady, M., Pradhan, D.K.: Single event upset detection and correction. In: 10th International Conference on Information Technology, pp. 13–18. IEEE (2007)
4. Hobeika, C., Thibeault, C., Boland, J.F.: Illegal state extraction from register transfer level. In: Proceedings of the 8th IEEE International NEWCAS Conference, pp. 245–248. IEEE (2010)
5. Wang, Z., Karpovsky, M.: Robust FSMs for cryptographic devices resilient to strong fault injection attacks. In: 16th International On-Line Testing Symposium. IEEE (2010)
6. Bychko, V., Yershov, R., Gulyi, Y., Zhydko, M.: Automation of anti-race state encoding of asynchronous FSM for robust systems. In: International Conference on Problems of Infocommunications. Science and Technology, pp. 501–506. IEEE (2020)
7. Kharchenko, V., Tyurin, S., Fesenko, H., Goncharovskij, O.: The fault tolerant Černý finite state machine: a concept and VHDL models. In: 11th IEEE International Conference on Intelligent Data Acquisition and Advanced Computing Systems: Technology and Applications, vol. 2, pp. 1163–1169. IEEE (2021)
8. Liszewski, K., McDonley, T.: Understanding tool synthesis behavior and safe finite state machine design. arXiv preprint arXiv:2108.04042 (2021)
9. Salauyou, V.: Structural models for fault detection of Moore finite state machines. In: Zamojski, W., Mazurkiewicz, J., Sugier, J., Walkowiak, T., Kacprzyk, J. (eds.) Dependable Computer Systems and Networks: Proceedings of the Eighteenth International Conference on Dependability of Computer Systems DepCoS-RELCOMEX, July 3–7, 2023, Brunów, Poland, pp. 231–241. Springer, Cham (2023). https://doi.org/10.1007/978-3-031-37720-4_21
10. Salauyou, V.: Description styles of fault-tolerant finite state machines for unmanned aerial vehicles. Radioelectronic Comput. Syst. **2024**(1), 196–206 (2024)

Smartphone-Based Biometric System Involving Multiple Data Acquisition Sessions

A. Sawicki[1]([⊠]) [iD] and K. Saeed[1,2] [iD]

[1] Faculty of Computer Science, Bialystok University of Technology, Bialystok, Poland
{a.sawicki,k.saeed}@pb.edu.pl

[2] Department of Computer Science and Electronics, Universidad de la Costa, Barranquilla, Colombia

Abstract. This paper presents an analysis of a mobile phone-based gait biometrics system that involves several data acquisition sessions. The conducted work verified the influence of the number of training sessions on the performance of the biometric system. Experiments were conducted using a publicly available 14-person database in which individuals attended three data acquisition sessions. A CNN with an attentional mechanism architecture was used as a baseline classifier. According to the experimental results, the system based on one motion tracking session achieved a performance of approximately 0.50 and 0.62 F1-score for the raw and processed data, respectively. In contrast, for two motion tracking sessions, the performance of the system was 0.60 and 0.85 F1-score for raw and processed data, respectively. The study showed that accuracy greater than 0.8 F1 score could be achieved when data preprocessing was applied and measured data from two motion sessions were incorporated into the training process.

Keywords: biometrics · gait · accelerometer · CNN

1 Introduction

Recently, there has been a growing interest in using accelerometer and gyroscope signals to recognize users based on their physical activity, such as walking [1]. This approach has several significant advantages over techniques based on a fingerprint or retinal scan. Gait can be measured remotely and in a non-invasive manner that does not require active interaction by the individual. By using behavioral characteristics, biometric authentication is more difficult to capture or imitate [2].

On the other hand, the approach of using smartphone devices in the acquisition process has several significant drawbacks, primarily in the development of a consistent and reliable biometric signal in real-life scenarios. One important aspect to mention is the impact of the orientation problem (i.e. turning the device e.g. in a pocket) on the readings of the sensors. Another is the issue of carrying the device off the body, e.g. by women in a handbag [1].

In general, a very big problem in the gait analysis is the so called distribution shift, which is especially visible in the so called cross day analysis. In this configuration, training samples are collected on one day and validation samples are collected on another day. In this case, the main factors influencing the gait pattern may be the subject's fatigue, differences in footwear, or the type of surface on which the subject is walking. Gait patterns may also be affected by changes in body weight or muscle growth due to exercise if gait samples are collected over time [3]. Some published approaches recommend updating gait samples over a period of nine months [4].

The focus of this paper is on gait analysis using three sessions of data collection.

The paper is organized as follows: Sect. 2 reviews the current state of the art in the use of multiple motion tracking sessions in the gait biometrics process, Sect. 3 describes the methodology with illustrations of the signals used as well as a description of the classifiers applied. The results of the performed experiments are shown in Sect. 4. Conclusions and future research directions are presented in Sect. 5.

2   Related Works

In the field of behavioral biometrics, the nature of the acquired signals is time-varying. Electrode-based solutions, such as EEG, use data from up to 15 acquisition sessions [5]. In the case of signal analysis from so-called wearable sensors, multi-session solutions are much less common. Very often, published methods use data sets from a single day (SD) (which has been criticized for lack of applicability in real-world scenarios [1]). Relatively less common are published solutions where decision models are trained on one day and tested on another day, the so-called CR (cross-day). Solutions that use data from multiple days are very rare.

Apart from the number of data collection events, their distribution over time has to take into account. In [4], the authors pointed out that in the case of human gait analysis using video media, a significant decrease in the effectiveness of the biometric system is observed after a period of 9 months. The authors suggested updating the reference sample after such a period.

Two publications that address the topic of validation using multiple tracking sessions are particularly noteworthy. In [1], accelerometer and gyroscope data from a smartwatch device were used. Each of the 60 participants participated in six tracking sessions over a period of three weeks. The experiment involved the participant comfortably walking on a hard surface and a fixed route, with the user having to stop to open a door and make a few turns to get a more realistic scenario.

In each session, a window of 2 min was sampled using the sliding window technique. The selected signal features were used as the feature vector of a classical MLP network. For the single-day analysis, the training set represented 60% of the available data and the test set represented the remaining 40%. In contrast, for the cross-day analysis, the first day's data were selected as training data and the second day's data were selected as test data. In this case, despite the availability of four additional sessions, the idea of a system based on multi-day data was completely ignored. Nevertheless, it is important to emphasize that an EER of 0.15 was obtained for the accelerometer in the case of single-day analysis and as high as 0.93 in the case of cross-day analysis. This indicates a very strong influence of the distribution shift between acquisition days.

In contrast, work [6] used a dedicated proprietary prototype with a 9-axis inertial measurement unit (BNO055, BOSCH, Germany). The unit consisted of two such sensors placed around the left wrist and right thigh. Twenty participants took part in the experiment and the acquisition process was performed under quasi-realistic conditions. The gait was recorded over a fixed distance of 50 m over a period of seven days. Three such walking trials were performed in the experiment.

The pipeline developed was much more complex and included an algorithm for segmenting gait cycles and a method for minimizing the effect of sensor orientation on the measurements [7]. In this solution, feature extraction and an SVM model were used in the classification process. However, the evaluation of the developed classifiers themselves is not clear, the description says that a 10-fold cross-validation was used. At the same time, reference is made to CD validation with 70% from the first day and 30% from the remaining day. In addition, there were no experiments to investigate the effect of combining the data from the first and second tracking sessions on the classification results. Due to the small time gap in the acquisition process, the results between cross-day 1 and cross-day 2 are also very similar. An efficiency of 0.976 was obtained for the SD experiment, 0.896 for CD1, and 0.869 for CD 2.

Both [1] and [6] indicate a decrease in effectiveness for cross-day validation compared to single-day validation. However, there is no precise summary of what percentage increase in effectiveness a biometric system can expect when using two days of data as a combined training set.

3   Methodology

Data Corpus and Data Processing

The study used a publicly available data corpus from the SIGNET group (University of Padova). The data corpus is extensive and allows analysis in several scenarios. The same 14 participants participated in three sessions. There were changes in both footwear and clothing between data collection sessions. The subjects were asked to move/walk freely for a period of about 5 min. The corpus itself was built up over a period of up to 6 months. Unfortunately, there is no clear description of the length of time for the selected participants.

The data was collected using a smartphone placed in the right pocket of the trousers. The database contains signals from the phone's built-in motion sensor and includes readings from the triaxial accelerometer and gyroscope, as well as the orientation signals determined with them.

A manual inspection based on accelerometer signals was used to divide the raw signal blocks into gait cycles. A gait cycle was defined as the time from when the right foot touches the ground to when it touches the ground again.

As a result of manual segmentation, the following numbers of gait cycles were extracted:

- during day I 302.21 ± 138.25 per participant (total 4231);
- during day II 272.43 ± 119.78 per participant (total 3814);
- during day III 317.86 ± 127.42 per participant (total 4450).

The scenario of using an IMU sensor embedded in a mobile phone has the inconvenience of the sensor orientation affecting the readings. Figure 1 shows the raw measurement data according to the data acquisition sessions. For the accelerometer data and day I, significant positive values are observed for the X axis. For the other days, positive values are observed for the Y axis. This is a consequence of the fact that in the case of the first day, the phone in the pocket was rotated with a different side than in the case of the other days.

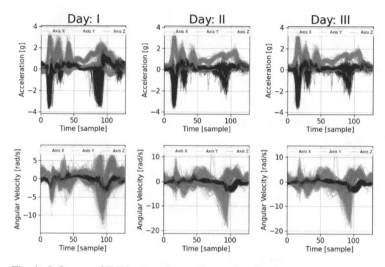

Fig. 1. Influence of IMU orientation on Sensor Reading between sessions/days

This problem was solved by using one of the additional processing methods, which will lead to the conversion of the coordinate system and ultimately minimize the impact of sensor orientation on the measurement values [7]. Figure 2 shows an analogous statement, for the same participant after applying the algorithm.

Classification

Classification was carried out in a 10-times repeated simple validation configuration. In this study, two types of CNN classifiers were used: One with an attentional mechanism [8] and a multi-input CNN [9], which had shown promising results in previous work. The accelerometer and gyroscope readings, which were interpolated to a fixed length, were used as the input data. For both models, cross entropy was used as a function of classifying cost. The Adam algorithm with a learning rate of 0.01 was chosen to adjust the network weights. The network learning process lasted for a period of 100 epochs. The general architecture of the neural networks used in the experiments is shown in Fig. 3.

The experiments were conducted in a configuration where a training set was created with one or two tracking sessions. The test set remained constant. A detailed evaluation of the developed solution was carried out as follows: Day I samples formed the training

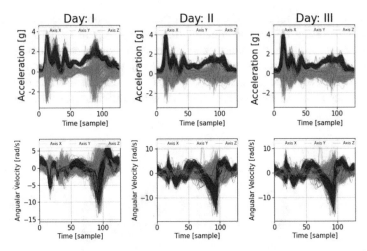

Fig. 2. Impact of Gadaleta et al.'s processing algorithm on sensor measurement values (with reference to Fig. 1)

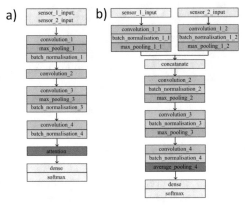

Fig. 3. Block diagram of the architecture of a) CNN with attention mechanism b) Multi-Input CNN

set. Day III samples formed the reference test set;The Day I samples combined with the Day II data formed the training set. Day III samples formed the reference test set.

4 Results

Figure 4 presents the effect of the number of sessions on the performance of the biometric system in the form of a box plot. The vertical axis presents the accuracy of classification in the form of F1-score metric, the horizontal shows the number of motion capture sessions occurring in the training set. With the help of two colours, the variants of raw data (blue colour) and processed data (orange colour) are marked. Two separate graphs show results for different types of classifiers.

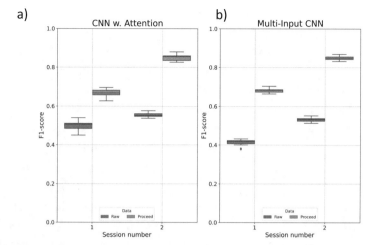

Fig. 4. Performance of biometrics systems for different classifier architectures: CNN with attentional mechanism (a) CNN with multiple inputs (b)

Figure 4 a) shows the following dependencies:

- There is an improvement in recognition metrics when data processing is applied, regardless of the number of training sessions. Therefore, pre-processing should be a mandatory part of the pipelining process.
- Higher classification rates are achieved with one session and preprocessing than with two sessions and raw data. In this particular case, preprocessing the data provides a greater efficiency gain than increasing the number of data acquisition sessions.
- Finally, it should be noted that an efficiency of more than 0.85 F1 can only be achieved with the presence of two tracking sessions and the inclusion of pre-processing.

The following observation can be made from Fig. 4 b):

- In the case of a classifier with a multi-input CNN architecture, the same relationships apply as in the case of a CNN classifier with an attention mechanism.

A single F1 score metric aggregates the metrics for individual participants in the experiment. It is worth attempting to answer the question of whether the increase in performance of the biometric systems is related to the overall average improvement in metrics for all participants, or to the individual improvement in identification for selected participants.

Table 1 contains information on the F1-score of the classification, and the individual rows contain metrics on each participant. The left-hand side of the table contains information on the number of gait cycles obtained in each motion capture session. The right side shows the F1-score metrics for the two types of classifiers.

Several observations can be made from the table:

- The identification of individual participants is very uneven, with some participants (e.g. participants "0", "1") having very high identification rates, while in some cases no model learning occurs at all (e.g. participant "6").

Table 1. F1-score metrics

part. ID	Number of gait samples			F1-score CNN w. Attention				F1-score Multi-Input CNN			
				1 Session		2 Sessions		1 Session		2 Sessions	
	Day I	Day II	Day III	Raw	Proc.	Raw	Proc.	Raw	Proc.	Raw	Proc.
0	230	200	214	0,99	0,99	0,98	0,99	0,99	0,94	0,99	0,98
1	402	516	393	1,00	0,88	1,00	0,98	1,00	0,96	0,92	0,87
2	231	231	240	0,99	0,96	0,98	0,97	0,96	0,83	0,99	0,97
3	731	502	440	**0,00**	**0,78**	**0,00**	**0,96**	**0,00**	**0,84**	**0,01**	**1,00**
4	116	256	296	0,91	0,84	0,99	0,96	0,74	0,89	0,98	0,93
5	322	335	409	0,92	0,98	0,98	1,00	0,53	0,99	0,93	1,00
6	284	197	196	**0,00**	**0,00**	**0,00**	**0,00**	**0,00**	**0,00**	**0,00**	**0,00**
7	196	243	208	0,52	0,61	0,67	0,67	0,63	0,47	0,67	0,67
8	314	339	356	**0,00**	**0,59**	**0,00**	**0,98**	**0,00**	**0,91**	0,00	0,94
9	352	116	301	**0,00**	**0,03**	0,91	0,71	**0,00**	**0,02**	0,91	0,93
10	214	336	252	0,98	0,93	0,95	0,97	0,93	0,86	0,72	0,94
11	332	154	198	**0,00**	**0,04**	0,41	0,90	**0,00**	**0,24**	0,37	0,74
12	273	117	680	0,43	0,68	**0,06**	**0,68**	0,18	0,62	**0,08**	**0,75**
13	234	272	267	0,35	0,52	0,59	0,94	0,30	0,39	0,60	0,84

- For participants "3", "8", "12", data pre-processing helps to significantly improve the identification rates.
- Participant '6' seems to be a special case, where both increasing the number of sessions and using preprocessing do not improve the measures obtained;
- The best achieved overall rate of 0.85 (Figure) is mainly due to the lack of identification of participant 6. The very large impact of individual identifications on the overall measure of system performance is a consequence of using a corpus with relatively few participants.
- The number of detected gait cycles for participant "6" in motion capture sessions is not significantly different from, for example, participant "1". The low performance of participant 6 is not due to sample size limitations.

Regardless of which classifier was used, no recognition at all was observed for participant "6". From Table 1 it is not possible to conclude whether the decision model recognised this participant as a specific person or as several other participants. It was therefore decided to present the results in the form of a confusion matrix.

The misidentification of participant "6" is entirely related to the assignment of its labels as participant "7", as shown in the confusion matrix in Fig. 5. Manual inspection of the data revealed that the same samples were assigned as 'test' to participant 7 and as 'training' (day II) to participant 6. This case illustrates the need to evaluate existing databases and to approach them with a degree of caution. The problem of identifying

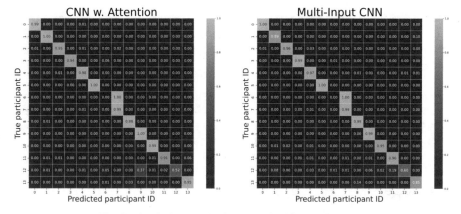

Fig. 5. Confusion Matrix for Two Classifier Architecture

participant 12 is more complex. Further analysis would require a deeper look into the CNN's decision making process, e.g. analysis of trained kernels, which is beyond the scope of this study.

5   Conclusions

This paper presents a case study of a mobile phone-based gait biometrics system. The influence of the number of data acquisition sessions on the accuracy of participant recognition was the focus of the research. The experiments were conducted on a publicly available data corpus of 14 people who participated in three motion tracking sessions. Two classifiers, a CNN with an attentional mechanism and a multi-input CNN, were used in the decision process.

The paper focuses on the aspect of using multiple data collection sessions in the development and validation of a decision model. All experiments were conducted in cross-day configurations, i.e. training and test data were collected on separate days. The experiments were carried out using one and two data collection sessions to train the classifiers. The main purpose of the work was to verify whether the use of additional training data would significantly increase the quality of the developed decision model. The experiments were conducted in the baseline case, where the input samples were raw sensor readings, as well as in the processed case, where an algorithm [7] was used to compensate for the effect of sensor mounting on the signal.

The study (Fig. 4) shows that: for one tracking session, the effectiveness of the system is about 0.50 and 0.62 F1 scores for raw and processed data, respectively. For two tracking sessions, system effectiveness is approximately 0.60 and 0.85 F1 scores for raw and processed data. The use of pre-processing should be mandatory, as the results for a single tracking session are comparable to the results for two tracking sessions in the case of raw data. Finally, it should be noted that the 0.85 F1 score metric was only achieved by implementing processing and using data from two tracking sessions.

Possible further experiments in ongoing work could include the implantation of additional deep learning model analysis methods of the LIME or SHAP type. These would allow for a deeper understanding of the decision process. Furthermore, it would be desirable to repeat the experiments on another error-free data set. Unfortunately, other data corpora containing multi-session sets, such as the one in [1], have not been made publicly available.

Acknowledgment. This work was supported by grant 2021/41/N/ST6/02505 from Białystok University of Technology and funded with resources for research by National Science Centre, Poland. For the purpose of Open Access, the author has applied a CC-BY public copyright license to any Author Accepted Manuscript (AAM) version arising from this submission.

References

1. Al-Naffakh, N., Clarke, N., Li, F.: Continuous user authentication using smartwatch motion sensor data. In: Gal-Oz, N., Lewis, P.R. (eds.) IFIPTM 2018. IAICT, vol. 528, pp. 15–28. Springer, Cham (2018). https://doi.org/10.1007/978-3-319-95276-5_2
2. Vajdi, A., Zaghian, M.R., et al.: Human gait database for normal walk collected by smartphone accelerometer. arXiv:1905.03109v3
3. Boyd, J.E., Little, J.J.: Biometric gait recognition. In: Tistarelli, M., Bigun, J., Grosso, E. (eds.) Advanced Studies in Biometrics. LNCS, vol. 3161, pp. 19–42. Springer, Heidelberg (2005). https://doi.org/10.1007/11493648_2
4. Matovski, D.S., Nixon, M.S., Mahmoodi, S., et al.: The effect of time on gait recognition performance. Trans. Inf. Forensics Secur. 7(2), 543–552 (2011). https://doi.org/10.1109/TIFS.2011.2176118
5. Plucińska, R., Jędrzejewski, K., Malinowska, U., Rogala, J.: Influence of feature scaling and number of training sessions on EEG spectral-based person verification with artificial neural networks. In: Jędrzejewski, K. (ed.) Proceedings of 2023 Signal Processing Symposium (SPSympo 2023), pp. 139–143 (2023). https://doi.org/10.23919/SPSympo57300.2023.10302695
6. Lee, S., Lee, S., Park, E., Lee, J., Kim, I.Y.: Gait-based continuous authentication using a novel sensor compensation algorithm and geometric features extracted from wearable sensors. IEEE Access **10**, 120122–120135 (2022). https://doi.org/10.1109/ACCESS.2022.3221813
7. Gadaleta, M., Merelli, L., Rossi, M.: Human authentication from ankle motion data using convolutional neural networks. In: 2016 IEEE Statistical Signal Processing Workshop (SSP), Palma de Mallorca, Spain, pp. 1–5 (2016). https://doi.org/10.1109/SSP.2016.7551815
8. Sawicki, A., Saeed, K.: Gait-based biometrics system. In: De Francisci Morales, G., Perlich, C., Ruchansky, N., Kourtellis, N., Baralis, E., Bonchi, F. (eds.) ECML PKDD 2023. LNCS, vol. 14175, pp. 350–355. Springer, Cham (2023). https://doi.org/10.1007/978-3-031-43430-3_29
9. Delgado-Escano, R., Castro, F.M., Cozar, J.R., et al.: An end-to-end multi-task and fusion CNN for inertial-based gait recognition. IEEE Access **7**, 1897–1908 (2019). https://doi.org/10.1109/ACCESS.2018.2886899

An Ontology for the Fashion Domain Based on Knowledge Retrieval

Karolina Selwon🆔 and Julian Szymański(✉)🆔

Gdansk University of Technology, Narutowicza 11/12, 80-233 Gdansk, Poland
{karselwo,julian.szymanski}@pg.edu.pl

Abstract. Building a large-scale knowledge graph in the fashion domain can be advantageous for further applications such as recommendations and searches. The paper proposes an approach to building an ontology for the field of fashion. First, a literature review on the construction and applications of ontologies in the fashion domain has been performed. Then we analyze fashion-related entities and attributes and make knowledge acquisition for domain ontology construction. The ontology has been formally defined according to the RDF standard. Finally, the ontology was evaluated with datasets consisting of fashion products.

Keywords: fashion ontology · knowledge graph · knowledge retrieval

1 Introduction

The increasing amount of valuable data and Internet users simultaneously resulted in straightforward access to knowledge and information overload in many areas. With the growing number of online clothing purchase transactions, researchers are interested in using fashion industry data to recognize fashion elements, search for products, and recommendation-based systems. Machine learning algorithms and knowledge mining techniques became popular to interpret massive fashion data's rich semantics automatically. Therefore, it opened the way to an area for research on building a knowledge graph for the fashion domain. Ontologies enable to capture of classes of objects with their properties, define relationships between items in the graph, standardize terminology, and provide classification guidelines. However, the ontology development process requires constant feedback from domain experts and incremental refinement as domain information grows.

Motivation. The development of fashion ontologies can significantly impact the Semantic Web community as it addresses the challenge of providing context for web searches, recommendation systems, fashion recognition, and conversational agents tasks. In addition, interest in fashion ontologies is growing. Therefore, based on the fashion ontologies analysis, this paper proposes an approach to creating an ontology that could be applied to the tasks mentioned above, for instance, e-commerce products search.

W. Zamojski et al. (Eds.): DepCoS-RELCOMEX 2024, LNNS 1026, pp. 261–271, 2024.
https://doi.org/10.1007/978-3-031-61857-4_26

Contribution. This paper focuses specifically on designing an ontology dedicated to represent the clothing domain. This paper uses fashion-related information from selected sources to create an ontology. This approach adopts vocabulary from existing ontologies and information from an e-commerce website. Ontology was created by extending new information obtained based on the analysis of descriptions of texts from online shops. In this method, we got a solution that corresponds to the real business problem and is useful for representing information in the system resources.

The rest of the paper is organized as follows. Section 2 describes the related works on ontology development and applications in the fashion domain. Section 3 introduces the methodology to build a domain ontology, provides insights into the resources, and considers qualitative and quantitative aspects of the proposed ontology. Finally, Sect. 4 proposes directions for further work.

2   Related Works

Researchers combine ontology with deep learning to improve fashion recognition, recommendation, and search systems. Fashion ontologies can enhance e-commerce user experience by providing a structured view of products.

State of the Art Fashion Ontologies. The Europeana Fashion project [14] contains a multilingual thesaurus in 10 languages for fashion and fashion-related concepts. The main concepts of the project were content aggregation from different fashion collection sources, complementing the standard metadata descriptions, and providing access to this digital content through the dedicated fashion portal. The final version of the Europeana Fashion portal was launched in 2014. Ontology is publicly available. VETIVOC [15] ontology combines the knowledge of several experts on domains and the terminologist's knowledge on, and also from users of e-commerce websites via manufacturers. The authors provide a resource in French and apply ontology as an industrial project. The ontology has modules on four levels: top ontology, core ontology, specific modules, and consolidation modules. Due to modular terminal-ontological resources, the ontology can be integrated into semantic information retrieval and recommendation. Unfortunately, ontology is not publicly available.

Fashion Recognition Applications. In the fashion recognition task, the knowledge represented in the ontology can be integrated with a deep learning approach to analyze the visual features of fashion items, such as images, to extract information about their attributes. Kuang et al. [2] used ontology to train a deep learning model, allowing it to learn a hierarchical clothing structure and recognize clothing items at different levels of ontology with high accuracy. The training dataset has images of clothing items and corresponding labels generated with the ontology. Ly et al. [9] present an approach that uses a hierarchical ontology structure to object retrieval model, which predicts the object's class and location in the image, allowing the system to perform object classification and localization. Jia et al. [5] developed the ontology, named Fashionpedia, by manually annotating images with concepts. The proposed dataset includes images of clothing items from various categories and the labels information. The authors following paper [6] use ontology to the model for the segmentation and localization of fashion attributes on a dataset annotated with segmentation masks.

Recommendations. The recommendation system can use the knowledge represented in the ontology to make more accurate and personalized user recommendations. Jung et al. [11] developed a recommendation system that uses ontology to discover user preferences by analyzing their past purchase history. Their ontology includes classes, such as clothing items, properties, such as material and brand, and relationships between classes. Ajmani et al. [8] presents an ontology and a system that integrates content-based and collaborative filtering techniques to recommend clothing items to users based on their preferences and styles. Usip et al. [10] use ontology to create a rule-based reasoning system to make occasion-specific apparel recommendations based on user input. Yethindra et al. [1] presents a method for recommending clothing items to users based on their preferences and style using a logistic regression model and an ontology-based approach. The model learns to predict the likelihood of a user liking a clothing item based on the user's preferences and attributes, represented by the ontology's concepts.

Products Search. In a product search system, the ontology can improve the accuracy of the search results by matching the user's query to the most relevant concepts and attributes. Nazir et al. [12] presents a chatbot system designed to understand natural language queries from users and provide them with relevant information about fashion brands using ontology. The authors evaluated the chatbot system by conducting a user study in which participants were asked to interact with the chatbot and provide feedback. The results show that the chatbot can understand the user's intent and provide relevant information. Their ontology represents the knowledge about fashion brands, products, and attributes based on data from an e-commerce website. Sultani et al. [13] used manual annotation and machine learning techniques to extract data to populate the ontology with instances of the different classes and properties. Then, the authors created an ontology-based search engine that allows users to search for brands and fashion items based on various criteria, such as demographics or fashion item details. Finally, the authors evaluate ontology by comparing the results of a user study with the results of a traditional keyword-based search.

3   Ontology Development

3.1   Scope Definition

Most previous works organize the concept of ontology in hierarchical classes. Kuang et al. [2] proposed a fashion recognition approach based on an ontology of fine-grained fashion classes categorized into coarse-grained groups. In work [6], authors with fashion experts built an ontology that contains apparel categories, fine-grained attributes, and relationships. The Europeana [14] ontology contains

1071 concepts. Lai et al. [4] use theme-aware features for fashion recommendation, which define subjective attributes related to an outfit's style, occasion, and culture. These features are commonly used in online stores due to user preferences. Most of the mentioned concepts were applied to the approach proposed in this article. The selected concepts are hierarchical ontology structure, visual features, and preference features such as style and occasion. Features and elements not directly related to clothing have not been considered.

Challenges. The first problem is to extract the most representative classes and features for many classes. Utilizing too many features can result in multiple correlations [1]. Therefore extracting the minimum number of features that suits our purpose is essential. The second challenge is the computational cost of flat hierarchy or too many attributes in the knowledge base that increases linearly with the number of classes [2]. Therefore, it is crucial to narrow down the representation of features and obtain optimal computational complexity. Hierarchically organization of large numbers of fine-grained classes can lower the computational complexity and provide concise feature representation. Furthermore, the concepts should be defined in a way understandable to domain experts and users of the ontology.

Concepts and Hierarchy Definition. After research consideration, this study defines the main concepts in the domain by analyzing the categorization of clothing features proposed in previous works [2,6]. As an extension to the related work, the approach presented in this paper organizes categories in hierarchical representation. First, categories were classified into the coarse-grained and the fine-grained classes. Next, the most general classes from the coarse-grained ones were determined. Then, the separated fine-grained classes were matched to the best representative coarse-grained classes as sub-classes. Finally, the most specific concepts were defined as labels. In addition to the hierarchy and attributes, definitions and aliases representing similar phrases were collected. For an example of an attribute: the label collar is a subclass of the class called detail and the class attribute. For an example of clothing, the label tracksuit bottoms are a subclass of the trousers class and the class called Lowerbody. As a result, this research collected a list of concepts that define garment categories, attributes, and possible relationships.

3.2   Information Extraction Process

In this stage, merged different sources of fashion-related information. The information retrieved in this process was analyzed with attention to aspects such as categorizing fashion types and their attributes, definitions, the ambiguity of features, and relations between clothing items and their attributes. This article proposes to combine domain knowledge from several sources to create an ontology: from Wikidata concepts, the Europeana [14], and entities extracted from text descriptions from the shopping website. The information search in Wikidata was performed by creating SPARQL language for querying RDF data. The queries were structured with the identification numbers of the attributes and categories of clothes and obtained information about ontology concepts and new potentially related concepts by extracting each element's identification number, descriptions, and aliases. Collected concepts were assigned into classes, with their descriptions and aliases. However, it turned out that it is impossible to isolate defined relations. Therefore, this step was skipped at this stage. Some concepts, such as color and material from Europeana Fashion Thesaurus, were merged with extracted concepts from Wikidata. The fashion objects were divided according to the specified two-level hierarchy to be consistent with data from other sources. For e-commerce data, the keyword analysis was carried out in text descriptions from a popular online clothing store. The corpus consists of 50000 unique fashion product descriptions. In the texts, analysis searched for concepts that define clothing characteristics, which were not included in the previous Wikidata and Thesaurus sources. This way we obtained 98 additional concepts. In addition, definitions and possible aliases have bxeen prepared for each concept. In this step, a proprietary tool was used for automatic data extraction and combining with previously acquired concepts.

Data Refinement. At this stage, keeping only concepts relevant to fashion ontology was critical. After the automatic data extraction, information that was not domain related or would not be used, such as mixed color types or materials not found in textiles, has been found. Therefore, all concepts irrelevant or may not belong to the fashion domain were removed from the ontology. That helped to solve the domain complexity problems and to avoid entity ambiguity. In addition, all extracted aliases were verified for semantic correctness. Words that did not clearly define the item of clothing have been removed. Moreover, extracted definitions were automatically corrected for user readability. Also, brand names were not considered since they may change, i.e., outdated or new ones may be created.

Integration. Data integration to create an ontology involves combining data from different sources into a coherent representation of knowledge. First, we collected data from various sources were cleaned by removing duplicates and correcting errors. Then, the data was mapped to a common ontology and added additional information to the data, such as descriptions and aliases, to make a complete and consistent representation of knowledge. Finally, cleaned data were integrated into a single ontology by creating relationships between the concepts, classes, and properties. Furthermore, the definition of relationships between clothing items and attributes has been assigned at this stage. This process involved using the Protégé tool and Python programming language to automate the integration and data manipulation.

3.3 Ontology Evaluation

The qualitative evaluation focuses on the issues regarding the content of the ontology in the context of compliance with the domain, the structure of the arrangement of concepts, and limitations. A quantitative evaluation was carried out to check the coverage and completeness of the ontology. The experiment used NLP techniques to develop a data annotation and evaluation program.

Domain Applicability. Domain compliance is determined with attention to different data sources. Firstly, existing research on ontologies and their applications. Secondly, information from different sources was used. In this approach, various views of the fashion domain are adopted. This process increases the ability of the ontology to represent the concepts and relationships in the domain accurately and completely. In addition, the ontologies and nomenclature appearing in the datasets of product descriptions from online stores have been supplemented to ensure that the ontology is applicable.

Structure and Concepts Identification. The hierarchical structure of the ontology has been taken into account due to the categorization rules in fashion datasets used in previous works and the categorization of products in online stores was also considered. In addition, the goal of the work was the ability to be interpreted by both humans and computability for the algorithm. Therefore, there is a need to have definitions of concepts written in natural language and relationships defining each relationship between concepts. Each concept has a short description, and most concepts also have aliases, i.e., phrases with similar meanings. As a result, the ontology the logical structure enables to understand and navigate for domain experts and users.

Consistency. Each concept is unique regarding high-level and detailed concepts. The ontology should not contain any contradictions or logical inconsistencies. However, there may be ambiguity for some concepts. For example, the garment's name may also mean the fabric from which it is made, e.g., jeans or denim trousers. Another problem arising from the naming is the inseparable connection between the name of the garment and its feature, for example, a mini-skirt. The concept of mini means length or hemline, while skirt means category. These are cases that occur notoriously in data catalogs and are generally accepted concepts used by many retailers. Therefore, these concepts were not considered as wrong concepts for building the ontology but were included in the ontology as separate concepts. In this way, they will remain identifiable both together and separately. However, this approach may need to change over time towards a unified concept due to dynamic changes in domain terminology.

Limitations. The proposed ontology has some limitations identified in the development phase. First, the field of fashion is vast and includes various stakeholders such as consumers and specialists in producing textiles and clothing. Second, formal definitions require expertise in various subfields. Therefore, the proposed ontology was the result of reconciling these different points of view. Third, the target application of the ontology is e-commerce. Therefore, the concepts in the ontology are limited to those used by sellers and understood by consumers. Therefore, this resource can be considered objective but still a subjective representation of this domain.

Scalability and Reusability. The ontology should be designed for reuse in other applications or ontologies, handle large amounts of data, and perform well regarding response time and memory usage. For this study, data was collected through extraction and analysis using natural language processing tools. This ontology is realistic in the sense that concepts are clearly defined through the use of these techniques. However, with the evolution of domain knowledge and its applications, vocabulary can change over time. The proposed structure can be modified by adding new labels into categories and attribute concepts. This operation will not affect the loss of compatibility as long as the relationships as categories classification are preserved.

Method. In the first step, publicly available datasets with clothing descriptions for each defined clothing category were collected. A corpus consisted of text descriptions from data sets on clothing products. The collected clothing descriptions and concepts from the ontology were subjected to processes and normalized to facilitate further interpretation of the texts. The concepts and their aliases from the ontology were brought to their root forms with the steaming process, and then the two-part concepts were combined, for example, *t-shirt* as *tshirt* or *long-sleeve* as *longsleev*. The next step was automatically annotating the data corpus using the concept from the designed ontology. The program aims to identify all terms representing concepts from ontologies present in the corpus. The exact process was applied to texts by finding concepts from ontology in exact or root form in English. During the first run of the experiment, phrases meaningful in the context of fashion were selected, and the ontology and phrases were extended. The extension consisted of the following steps. First, if the phrase was similar to an existing concept in the ontology, that phrase was matched as an alias. Conversely, a new concept was created if the phrase was not similar to any of the concepts. Finally, the experiment was repeated after supplementing the ontology and additional concepts.

Results. The data corpus for evaluation was prepared from two datasets that contained actual text descriptions of fashion products. The Polyvore dataset [3] contains descriptions and titles of fashion products with equivalent categories to each high class in ontology. Only titles with detailed and concise information were used for the evaluation and assigned to accessories, lingerie, lowerbody, upperbody, and wholebody categories. The Fashion200K [7] collection contains only titles for the mid-class categories: skirt, pants, dress, jacket, and top. Therefore, they were assigned to the high classes: lowerbody, upperbody, and wholebody. For the Polyvore dataset covered 40% of tokens from 50000 product descriptions by concepts from ontology. For the Fashion200K dataset covered 68% of tokens from 61729 product descriptions by concepts from ontology. The analyzing words not included in the ontology demonstrate that the titles of products most often concern brand names or a rare clothing features. Brand names have not been included in the ontology due to the vast of possible clothing brands and the fact that new ones are being created all the time, or many may not exist in a given time. Furthermore, regarding the rare features of clothing, its analysis requires expanding expert knowledge. Therefore, the evaluation results of this ontology are valid for a certain period and selected datasets.

3.4   Final Version

The proposed fashion ontology was developed with the Protégé tool using the Ontology Web Language (OWL in version 2.0.) and the Resource Description Framework (RDF) format. The general classes Clothes and Attribute are built under the OWL *Thing* base class. The nested classes are built under the *subClassOf* property. Relationships are built as the *ObjectProperty*. Each class contains a label, description, main class, subclass, and alias properties. The ontology file is available at https://github.com/karolinaselwon/fashion-ontology.

Statistics. The final version of the generated ontology contains concepts including 5 coarse garment categories, 20 fine-grained garment categories, 181 fashion items with 964 aliases, 1508 attributes categorized in 18 classes with 1454 aliases. Details about the hierarchy of categories and attributes with a numbers of concepts are as follows:

- **Fashion Categories:** Accessories: bag, eyewear, hairstyle, jewelry, waist, watch (138 concepts); Lingerie: bra, hosiery, suit, undergarment, overshoes (139 concepts); Lowerbody: shoe, shorts, skirts, trousers (242 concepts); Upperbody: handwear, headgear, neckwear, top (341 concepts); Wholebody: dress, outerwears (221 concepts).

Annotations: Q380339	Annotations: Q10715829
Annotations ⊕	Annotations ⊕
alias [type: xsd:string]	alias [type: xsd:string]
thong sandal	cold weather clothing
alias [type: xsd:string]	alias [type: xsd:string]
thongs	winter clothes
description [type: xsd:string]	description [type: xsd:string]
Type of scandal.	Costume for cold weather.
high_class [type: xsd:string]	high_class [type: xsd:string]
lowerbody	attribute
label [type: xsd:string]	label [type: xsd:string]
flip-flops	winter clothing
mid_class [type: xsd:string]	mid_class [type: xsd:string]
shoe	season
(a) Example of clothing concept.	(b) Example of attribute concept.

Fig. 1. Examples of concepts from the ontology. The images contains one of the concepts from the clothing category and one of the concept of an attribute.

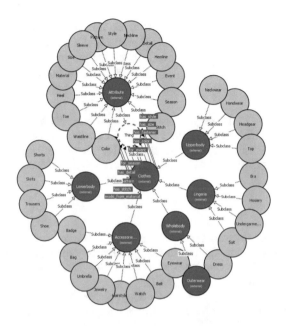

Fig. 2. Structure of concepts and relations in ontology.

- **Attributes:** Color: 497 concepts; Detail: 61 concepts; Event: 38 concepts; Fastener: 57 concepts; Gender: 5 concepts; Heel: 8 concepts; Hemline: 29 concepts; Material: 338 concepts; Neckline: 27 concepts; Pattern: 81 concepts; Season: 7 concepts; Size: 2 concepts; Sleeve: 24 concepts; Stitch: 22 concepts; Style: 74 concepts; Technique: 220 concepts; Toe: 5 concepts; Waistline: 13 concepts.

Demonstration. Figure 1 presents two examples of concepts from ontology. Each concept in the ontology contains a description, aliases if any, and is a high-class and mid-class sub-class. Moreover, Fig. 2 shows the structure of concepts and relations defined as properties of concepts. The nodes represent the concepts of clothes and attributes, while the edges represent their relationships. The ontology visualization was created with the VOWL Plugin for Protégé.

4 Conclusions and Future Works

The proposed ontology was developed by analyzing concepts in existing solutions and adding concepts based on data from online stores. In the mentioned applications, the combination of fashion ontology and deep learning approaches can improve the performance of fashion recommendations and recognition systems by analyzing the features of fashion items and providing a structured representation of knowledge about the fashion domain. In addition, it can be used to understand the user's intent and preferences to make more accurate and efficient recommendations and searches. Therefore, the ontology will be used and validated to search for shopping products in future research. In addition, the scalability of the ontology will be performed by introducing new concepts and translating them to other languages.

Acknowledgments. The work has been supported partially by funds of the Department of Computer Architecture of the Faculty of Electronics, Telecommunication and Informatics and Gdańsk University of Technology under the Excellence Initiative - Research University, DEC - 06/2023/IDUB/II.1/AMERICIUM.

References

1. Yethindra, D.N., Deepak, G.: A semantic approach for fashion recommendation using logistic regression and ontologies. In: International Conference on Innovative Computing, Intelligent Communication and Smart Electrical Systems (ICSES), 24 September 2021, pp. 1–6. IEEE (2021)
2. Kuang, Z., Yu, J., Yu, Z., Fan, J.: Ontology-driven hierarchical deep learning for fashion recognition. In: 2018 IEEE Conference on Multimedia Information Processing and Retrieval (MIPR), 10 April 2018, pp. 19–24. IEEE (2018)
3. Han, X., Wu, Z., Jiang, Y.G., Davis, L.S.: Learning fashion compatibility with bidirectional LSTMs. In: Proceedings of the 25th ACM International Conference on Multimedia, 19 October 2017, pp. 1078–1086 (2017)
4. Lai, J.H., Wu, B., Wang, X., Zeng, D., Mei, T., Liu, J.: Theme-matters: fashion compatibility learning via theme attention. arXiv preprint arXiv:1912.06227, 12 December 2019
5. Jia, M., et al.: The fashionpedia ontology and fashion segmentation dataset. Cornell University (2019)
6. Jia, M., et al.: Fashionpedia: ontology, segmentation, and an attribute localization dataset. In: Vedaldi, A., Bischof, H., Brox, T., Frahm, J.-M. (eds.) ECCV 2020. LNCS, vol. 12346, pp. 316–332. Springer, Cham (2020). https://doi.org/10.1007/978-3-030-58452-8_19
7. Han, X., et al.: Automatic spatially-aware fashion concept discovery. In: IEEE International Conference on Computer Vision, pp. 1463–1471 (2017)
8. Ajmani, S., Ghosh, H., Mallik, A., Chaudhury, S.: An ontology based personalized garment recommendation system. In: 2013 IEEE/WIC/ACM International Joint Conferences on Web Intelligence (WI) and Intelligent Agent Technologies (IAT), 17 November 2013, vol. 3, pp. 17–20. IEEE (2013)
9. Ly, N.Q., Do, T.K., Nguyen, B.X.: Large-scale coarse-to-fine object retrieval ontology and deep local multitask learning. Comput. Intell. Neurosci. **18**, 2019 (2019)
10. Usip, P.U., Osang, F.B., Konyeha, S.: An ontology-driven fashion recommender system for occasion-specific apparels. J. Adv. Math. Comput. Sci. **8**(1), 67–76 (2020)
11. Jung, K.-Y., Na, Y.-J., Park, D.-H., Lee, J.-H.: Discovery knowledge of user preferences: ontologies in fashion design recommender agent system. In: Laganá, A., Gavrilova, M.L., Kumar, V., Mun, Y., Tan, C.J.K., Gervasi, O. (eds.) ICCSA 2004. LNCS, vol. 3044, pp. 863–872. Springer, Heidelberg (2004). https://doi.org/10.1007/978-3-540-24709-8_91

12. Nazir, A., Khan, M.Y., Ahmed, T., Jami, S.I., Wasi, S.: A novel approach for ontology-driven information retrieving chatbot for fashion brands. Int. J. Adv. Comput. Sci. Appl. IJACSA **10**(9), 546–552 (2019)
13. Sultani Sadrabadi, S., Kahani, M.: InfOnto: an ontology for fashion influencer marketing based on Instagram. In: International Conference on Computer and Knowledge Engineering (2022)
14. Van Steen, N.: Europeana Fashion Thesaurus v1
15. Aimé, X., George, S., Hornung, J.: VetiVoc: a modular ontology for the fashion, textile and clothing domain. Appl. Ontol. **11**(1), 1–28 (2016)

Comparative Efficiency Study of Protective Relays Schemes in Wind Energy Conversion Systems

Robert Adam Sobolewski[✉] [ORCID]

Faculty of Electrical Engineering, Bialystok University of Technology, Wiejska 45D Street, 15-351 Białystok, Poland
r.sobolewski@pb.edu.pl

Abstract. Since the performance of wind energy conversion systems is unsatisfactory, one of the main concerns about systems planning and operation should be better reliability of the components and their operation. Among others, the attention should affect protection systems, and relays incorporated to. Relays have two alternative ways in which they can be unreliable, i.e., may fail to operate when they are expected to, or may operate when it is undesired. This leads to a two-pronged definition of protective relay, i.e., it must be dependable and secure. One of a few options of relays reliability enhancement can be use of logical redundancy, i.e. series, parallel, 2-out-of-3. The advisability of using a given type of redundancy (instead of single relay), can be evaluated relying on efficiency of relays operation. Both the structural reliability and the operation reliability must be taken into consideration. The efficiency of different types of redundancy can be measured based on economical criterion and incorporate the costs of incorrect operation, i.e., failure to trip caused by undetected fault, undesired trip, and repair of faulted relay(s). The efficiency of different protective relays scheme can be investigated relying on semi-Markov models. In the paper, four models are investigated in detail. As an example, the comparative efficiency study of these schemes are presented.

Keywords: Protective relay · Efficiency · Wind energy · Reliability · Semi-Markov model

1 Introduction

In present power system scenarios, renewable energy sources becoming most paramount energy sources due to increased awareness on environmental issues. The most common renewable energy sources are wind energy conversion systems (WECS). Their performance is still unsatisfactory and is much lower as compared to conventional generation units. Thus, one of the main concern about WECS planning and operation should be more reliable both the components and their operation. Among others, the attention should affect protection systems (PS). In general, they consist of few main components, i.e., instrument transformers, relays, and circuit breakers. Their role is to detect and clear faults that occurs in protected WECS components (e.g. generators, cables,

© The Author(s), under exclusive license to Springer Nature Switzerland AG 2024
W. Zamojski et al. (Eds.): DepCoS-RELCOMEX 2024, LNNS 1026, pp. 272–282, 2024.
https://doi.org/10.1007/978-3-031-61857-4_27

busbars, transformers, inverters). Current-based, voltage-based, frequency-based, and impedance-based methods are among the well-known PS. They must meet a few requirements, i.e. reliability, selectivity, speed, economy, and simplicity [1]. Reliability of PS is one of the most crucial and should be investigated intensively since they influence the WECS performance essentially. Relays have two alternative ways in which they can be unreliable, i.e., may fail to operate when they are expected to, or may operate when it is undesired [2]. This leads to a two-pronged definition of a relay, i.e., it must be dependable and secure. Dependability is defined as the measure of certainty that relay will operate correctly for all the faults in protected component for which they are designed to operate. Security is defined as the measure of the certainty that the relay will not operate incorrectly for any faults in other components.

Depending on a WECS size (number of wind turbines, their rated capacity), internal collection grid arrangement, site of operation (onshore and offshore), it can be equipped with more than one PS, i.e. PS of individual wind turbine generators, PS of section feeders, PS of inter-tie link that interconnects collector busbar of WECS and point of common coupling. They act as primary protection function and some of them – back-up protection while a primary PS are unavailable. One of a few options of PS' reliability enhancement can be application of logical redundancy of relays incorporated into primary PS. One can investigate three types of logical redundancy, i.e. series arrangement, parallel arrangement, 2-out-of-3 arrangement. The advisability of using a given type of redundancy (instead of single relay), can be evaluated relying on efficiency of PS operation. Both the structural reliability and the operation reliability must be taken into consideration. The efficiency can be measured based on economical criterion and incorporate the costs of and incorrect operation, i.e., failure to trip caused by undetected fault, undesired trip, and repair of faulted relay(s). The costs of failure to trip and undesired trip concern the loses of energy served.

The relevant literature offers some approaches that can be used for WECS protection systems reliability analysis [3–6]. In [3] the semi-Markov reliability model of internal electrical collection grid is presented. The model enables calculating of expected energy not served by WECS caused by faults in protected components and failures and incorrect operation of PS. In [4] a probabilistic modelling of reliability and maintenance of PS incorporated into internal collection grid of WECS is shown. The semi-Markov model is formulated and applied for calculating the availability of PS taking into account the faults of protected components, failures and incorrect operation of PS. In [5] a foundational Markovian model for a relay-based protected component that can be incrementally updated to represent more advanced behaviours, i.e. self-checking, routine test and continuous monitoring. The paper [6] develops a Markov reliability model to categorize the possible functional states of protected component and assess both protected component reliability and reliability analysis of PS. Although, these models concern the reliability and maintainability study of protective relays, they are not designed for investigation of logical redundancy of PS.

In the paper, an approach to evaluate the efficiency of different types of relays logical redundancy is proposed. Since the time to failure of a relay, time to fault in the wind turbine/farm components are of stochastic nature, the approach relies on semi-Markov model which enables the deriving the number of entries into each reliability state within fixed period relay(s) operation time. Combining the numbers of entries into the states and unit costs of failure to trip, undesired trip, and repair of relay(s) gives the total costs of relays operation. The resultant lower cost of relay(s) incorrect operation the higher efficiency of either single relay or logical redundancy incorporated into more than one relay, and finally the higher performance of WECS. As an example, the efficiency of the logical redundancy incorporated into relays in comparison to the single relay is investigated along different both failure rates and probabilities of detecting the failures, is calculated relying on the model and presented in the case study.

2   Types of Protective Relays Redundancy

The logical redundancy incorporated into protective relays (instead of single relay) can be introduced to improve the dependability and security of PS. The redundancy can be established among contact members of the relays in questions. Contact member is a conductive part designed to co-act with another to close or open the output circuit (control circuit of circuit breaker). The most common types of logical redundancy are the following: (a) series, (b) parallel, and (c) 2-out-of-3.

In series arrangement (see Fig. 1b) two relays must operate correctly for the system success. Such arrangement assures high security, i.e., undesired trip is possible only while both relays operate simultaneously. Otherwise, sending the signal for opening the circuit breakers is blocked. However, the arrangement is less dependable as compared to other types of redundancy because failure to trip is more likely while the fault occurs in protected component. In parallel arrangement (see Fig. 1c) all the relays operate simultaneously and at least one such unit must work correctly for the PS success. Such arrangement assures high dependability, i.e., a fault is always cleared by some relay. Its tendency to become less secure increases because unnecessary operation of at least one relay causes the undesired trip of protected component.

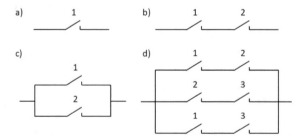

Fig. 1. Arrangements of contact members of single relay (a), series (b), parallel (c), and 2-out-of-3 (d). 1, 2, and 3 – relay number

2-out-of-3 (see Fig. 1d) is another form of redundancy in which at least two relays out a total of three units must work correctly for the PS success. Furthermore, all the units in the PS are active. This arrangement simultaneously assures both the dependability and the security of PS better than series and parallel ones.

3 Reliability Models

The total cost (in monetary unit, m.u.) of relay(s) incorrect operation within a period of time t can be expressed by the following expression

$$TC(t) = C_D \cdot \sum_{i \in I} \beta_i(t) + C_S \cdot \sum_{j \in J} \beta_j(t) + C_R \cdot \sum_{k \in K} \beta_k(t), \tag{1}$$

where: C_D, C_S, C_R – unit cost of failure to trip of relay(s) per occur., unit cost of undesired trip per occur., unit cost of repair per occur., respectively, $\beta_i(t), \beta_j(t), \beta_k(t)$ – number of entries into the state i, j and k within the operation duration t, respectively, I, J, K – the set of states that refer failure to trip, undesired trip and repair, respectively.

The costs C_D and C_S concern the expected cost of energy not served by WECS because of incorrect trip of protective relay(s), whereas the cost C_R refers to the expected cost of energy not served during repair and cost of relay(s) corrective maintenance. $\beta_i(t), \beta_j(t), \beta_k(t)$ can be calculated relying on the general formula [7]

$$\beta_x(t) = \frac{\pi_x}{\sum_{y=1}^{L} \pi_y \cdot ET_y} \cdot t, \tag{2}$$

where: π_x – stationary probability of state x, $x \in I, J, K$, ET_y – mean waiting time in the state y, L – total number of the states in reliability model.

To calculate the probability π_x for single relay and each type of logical redundancy in the relays the semi-Markov models are formulated. π_x is an element of the vector π, and can be derived based on formula $\pi \cdot P = \pi$ [8], where P is a matrix of transition probabilities p_{a_b}, $(a, b) \in L$, $a \neq b$. . Let assume, the random variables: time to fault in protected component, time to relay failure, and time to fault in other components (protected by other relays) are distributed exponentially. Thus, λ is a fault rate of protected component rate, θ is a failure rate of relay, and γ is a fault rate of other components. The fault rate is the following $\gamma = \sum_n \gamma_n$, where n is the component number (excluding protected one) in question. The component n is selected while its fault can cause the undesired trip of the primary relay dedicated to protected component. The probability of failure to trip is q_D (and $p_D = 1 - q_D$), whereas the probability of undesired trip is q_S (and $p_S = 1 - q_S$). Let assume, the cost of relay(s) repair is independent on the number relays to be failed. And finally, the rest of PS components (instrument transformers and circuit breakers) are fully reliable.

3.1 Single Relay

Transition diagram is depicted in Fig. 2. The reliability states are as follows:

  S0 – the component in service, the relay reliable,
  S1 – the component faulted, the relay reliable,
  S2 – the component in service, other components faulted, the relay reliable,
  S3 – the component in service, the relay failed,
  S4 – switch off the faulted component, the relay reliable,
  S5 – failure to trip of the faulted component, the relay reliable (failure to trip),
  S6 – switch of the component in service, the relay reliable (undesired trip),

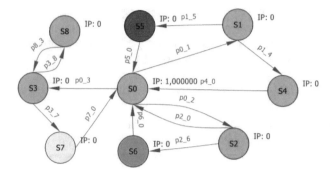

Fig. 2. Transition diagram for reliability model of single relay

S7 – failure to trip of the faulted component, the relay failed (relay repair),
S8 – the component in service, other components faulted, the relay reliable.
Transition probabilities are the following:

$$p_{0\_1} = \tfrac{\lambda}{\lambda+\theta+\gamma}, \quad p_{0\_2} = \tfrac{\gamma}{\lambda+\theta+\gamma}, \quad p_{0\_3} = \tfrac{\theta}{\lambda+\theta+\gamma}, \quad p_{1\_4} = p_D, \quad p_{1\_5} = q_D, \quad p_{2\_0} = p_S,$$
$$p_{2\_6} = q_S, \quad p_{3\_7} = \tfrac{\lambda}{\lambda+\gamma}, \quad p_{3\_8} = \tfrac{\gamma}{\lambda+\gamma}, \quad p_{4\_0} = p_{5\_0} = p_{6\_0} = p_{7\_0} = p_{8\_3} = 1. \tag{3}$$

Mean waiting times in the state S0 and S3, are as follows, respectively:

$$ET_0 = \frac{1}{\lambda + \theta + \gamma}, \quad ET_3 = \frac{1}{\lambda + \gamma}. \tag{4}$$

Following the general formula (1) and (2) the total cost of inaccurate operation of the relay within a period of time t is the following

$$TC(t) = \frac{\lambda \cdot t \cdot (q_D \cdot \lambda \cdot C_D + q_S \cdot \gamma \cdot C_S + \theta \cdot C_R)}{\lambda + \theta}. \tag{5}$$

3.2 Series Arrangement

Transition diagram is depicted in Fig. 3. The reliability states are as follows [7]:

S0 – the component in service, two relays reliable,
S1 – the component faulted, two relays reliable,
S2 – the component in service, other components faulted, two relays reliable,
S3 – the component in service, one relay reliable and second relay failed,
S4 – switch off the faulted component by PS since two relays are reliable,
S5 – failure to trip of at least one relay while the component faulted, two relays reliable (failure to trip),
S6 – switch off the component in service by PS since two relays are reliable, other components faulted (undesired trip),
S7 – failure to trip of the faulted component, one relay reliable and one relay failed (relay repair),
S8 – the component in service, other components faulted, one relay reliable and one relay failed,

S9 – the component in service, two relays failed,
S10 – the component faulted, two relays failed (relays repair),
S11 – the component in service, other components faulted, two relays failed.

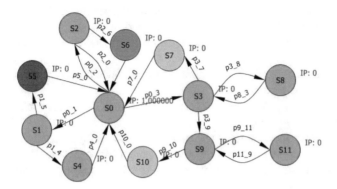

Fig. 3. Transition diagram for reliability model of series arrangement

Transition probabilities are the following:

$$p_{0\_1} = \frac{\lambda}{\lambda+2\theta+\gamma}, \quad p_{0\_2} = \frac{\gamma}{\lambda+2\theta+\gamma}, \quad p_{0\_3} = \frac{2\theta}{\lambda+2\theta+\gamma}, \quad p_{1\_4} = p_D^2, \quad p_{1\_5} = 1 - p_D^2,$$
$$p_{2\_0} = 1 - q_S^2, \quad p_{2\_6} = q_S^2, \quad p_{3\_7} = \frac{\lambda}{\lambda+\theta+\gamma}, \quad p_{3\_8} = \frac{\gamma}{\lambda+\theta+\gamma}, p_{3\_9} = \frac{\theta}{\lambda+\theta+\gamma},$$
$$p_{4\_0} = p_{5\_0} = p_{6\_0} = p_{7\_0} = p_{10\_0} = p_{8\_3} = p_{11\_9} = 1,$$
$$p_{9\_10} = \frac{\gamma}{\lambda+\gamma}, \quad p_{9\_11} = \frac{\lambda}{\lambda+\gamma}.$$

$$(6)$$

Mean waiting times in the state S0, S3, S9, are as follows, respectively:

$$ET_0 = \frac{1}{\lambda + 2\theta + \gamma}, \quad ET_3 = \frac{1}{\lambda + \theta + \gamma}, \quad ET_9 = \frac{1}{\lambda + \gamma}. \tag{7}$$

Following the general formula (1) and (2) the total cost of inaccurate operation of two relays within a period of time t is the following

$$TC(t) = \frac{t \cdot \{(1 - p_D^2) \cdot \lambda \cdot \gamma \cdot (\lambda + \theta) \cdot C_D + q_S^2 \cdot \gamma^2 \cdot (\lambda + \theta) \cdot C_S + 2\theta \cdot \gamma \cdot (\lambda + \theta) \cdot C_R\}}{2\theta \cdot (\theta + \gamma) + \gamma \cdot (\lambda + \theta)}. \tag{8}$$

3.3 Parallel Arrangement

Transition diagram is depicted in Fig. 4. The reliability states are as follows [7]:
    S0 – the component in service, two relays reliable,
    S1 – the component faulted, two relays reliable,
    S2 – the component in service, other components faulted, two relays reliable,
    S3 – the component in service, one relay reliable, one relay failed,
    S4 – switch off the faulted component by PS since two relays are reliable,

S5 – failure to trip of two relays while the component faulted, two relays reliable (failure to trip),

S6 – switch off the component in service, other components faulted, two relays reliable, a trip of at least one relay (undesired trip),

S7 – the component faulted, one relay reliable, one relay failed,

S8 – the component in service, two relays failed,

S9 – the component in service, other components faulted, one relay reliable, one relay failed,

S10 – switch off the faulted component by PS since one relay is reliable, second relay is failed (relay repair),

S11 – failure to trip of one relay while the component faulted, one relay reliable, one relay failed (failure to trip),

S12 – the component faulted, two relays failed (two relays repair),

S13 – the component in service, other components faulted, two relays failed,

S14 – switch off the component in service, other components faulted, one relay reliable, one relay failed (undesired trip).

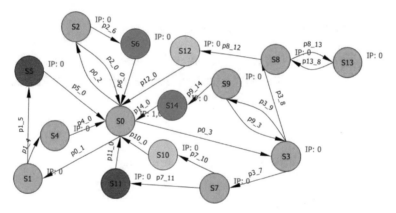

Fig. 4. Transition diagram for reliability model of parallel arrangement

Transition probabilities are the following:

$$
\begin{aligned}
&p_{0\_1} = \tfrac{\lambda}{\lambda+2\theta+\gamma}, \quad p_{0\_2} = \tfrac{\gamma}{\lambda+2\theta+\gamma}, \quad p_{0\_3} = \tfrac{2\theta}{\lambda+2\theta+\gamma}, \quad p_{1\_4} = 1 - q_D^2, \quad p_{1\_5} = q_D^2, \\
&p_{2\_0} = p_S^2, \quad p_{2\_6} = 1 - p_S^2, \quad p_{3\_7} = \tfrac{\lambda}{\lambda+\theta+\gamma}, \quad p_{3\_8} = \tfrac{\gamma}{\lambda+\theta+\gamma}, \quad p_{3\_9} = \tfrac{\theta}{\lambda+\theta+\gamma}, \\
&p_{4\_0} = p_{5\_0} = p_{6\_0} = p_{10\_0} = p_{11\_0} = p_{12\_0} = p_{14\_0} = p_{13\_8} = 1, \\
&p_{7\_10} = p_D, \quad p_{7\_11} = q_D, \quad p_{8\_12} = \tfrac{\lambda}{\lambda+\gamma}, \quad p_{8\_13} = \tfrac{\gamma}{\lambda+\gamma}, \quad p_{9\_3} = p_S, \quad p_{9\_14} = q_S.
\end{aligned}
\tag{9}
$$

Mean waiting times in the state S0, S3, S8 are as follows, respectively:

$$
ET_0 = \frac{1}{\lambda + 2\theta + \gamma}, \quad ET_3 = \frac{1}{\lambda + \theta + \gamma}, \quad ET_8 = \frac{1}{\lambda + \gamma}.
\tag{10}
$$

Following the general formula (1) and (2) the total cost of inaccurate operation of two relays within period of time t is the following

$$TC(t) = \frac{\lambda \cdot t}{\lambda \cdot (\lambda + \theta + q_L \cdot \gamma) + 2\theta \cdot (\lambda + \theta)} \cdot \{ q_D \cdot \lambda \cdot [q_D \cdot (\lambda + \theta + q_S \cdot \gamma) + 2\theta] \cdot C_D$$
$$+ \gamma \cdot \left[\left(1 - p_S^2\right) \cdot (\lambda + \theta + q_S \cdot \gamma) + 2q_S \cdot \theta \right] \cdot C_S + 2\theta \cdot (p_D \cdot \lambda + \theta) \cdot C_R \} \tag{11}$$

3.4 2-out-of-3 Arrangement

Transition diagram is depicted in Fig. 5. The reliability states are as follows [7]:

S0 – the component in service, three relays reliable,

S1 – the component faulted, three relays reliable,

S2 – the component in service, other components faulted, three relays reliable,

S3 – the component in service, two relays reliable, one relay failed,

S4 – switch of the faulted component, three relays reliable,

S5 – failure to trip of three relays while the component faulted, three relays reliable (failure to trip),

S6 – switch off the component in service, other components faulted, three relays reliable, at least two relays trip (undesired trip),

S7 – the component faulted, two relays reliable, one relay failed,

S8 – the component in service, other components faulted, two relays reliable, one relay failed,

S9 – the component in service, one relay reliable, two relays failed,

S10 – switch off the faulted component, two relays reliable, one relay failed (relay repair),

S11 – failure to trip of two relays while the component faulted, two relays reliable, one relay failed (failure to trip),

S12 – switch off the component in service, other components faulted, two relays reliable, one relay failed, trip of two relays (undesired trip),

S13 – the component faulted, one relay reliable, two relays failed (relays repair),

S14 – the component in service, other components faulted, one relay reliable, two relays failed,

S15 – the component in service, three relays failed,

S16 – failure to trip of three failed relays while the component faulted (relays repair),

S17 – the component in service, other components faulted, three relays failed.

Transition probabilities are the following:

$$p_{0\_1} = \tfrac{\lambda}{\lambda + 3\theta + \gamma}, \quad p_{0\_2} = \tfrac{\gamma}{\lambda + 3\theta + \gamma}, \quad p_{0\_3} = \tfrac{3\theta}{\lambda + 3\theta + \gamma}, \quad p_{1\_4} = p_D^2 \cdot (3 - 2p_D),$$
$$p_{1\_5} = 1 - p_D^2 \cdot (3 - 2p_D), \quad p_{2\_0} = p_S^2 \cdot (3 - 2p_S), \quad p_{2\_6} = 1 - p_S^2 \cdot (3 - 2p_S),$$
$$p_{3\_7} = \tfrac{\lambda}{\lambda + 2\theta + \gamma}, \quad p_{3\_8} = \tfrac{\gamma}{\lambda + 2\theta + \gamma}, \quad p_{3\_9} = \tfrac{2\theta}{\lambda + 2\theta + \gamma}, \tag{12}$$
$$p_{4\_0} = p_{5\_0} = p_{6\_0} = p_{10\_0} = p_{11\_0} = p_{12\_0} = p_{13\_0} = p_{16\_0} = p_{14\_9} = p_{17\_15} = 1,$$
$$p_{7\_10} = p_D^2, \quad p_{7\_11} = 1 - p_D^2, \quad p_{8\_3} = 1 - q_S^2, \quad p_{8\_12} = q_S^2, \quad p_{9\_13} = \tfrac{\lambda}{\lambda + \theta + \gamma},$$
$$p_{9\_14} = \tfrac{\gamma}{\lambda + \theta + \gamma}, \quad p_{9\_15} = \tfrac{\theta}{\lambda + \theta + \gamma}, \quad p_{15\_16} = \tfrac{\lambda}{\lambda + \gamma}, \quad p_{15\_17} = \tfrac{\gamma}{\lambda + \gamma}.$$

Mean waiting times in the state S0, S3, S9, and S15, are as follows, respectively:

$$ET_0 = \frac{1}{\lambda + 3\theta + \gamma}, \quad ET_3 = \frac{1}{\lambda + 2\theta + \gamma}, \quad ET_9 = \frac{1}{\lambda + \theta + \gamma}, \quad ET_{15} = \frac{1}{\lambda + \gamma}. \tag{13}$$

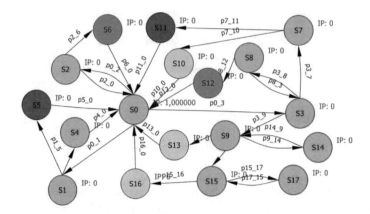

Fig. 5. Transition diagram for reliability model of 2-out-of-3 arrangement

Following the general formula (1) and (2) the total cost of inaccurate operation of three relays within period of time t is the following

$$TC(t) = \frac{\lambda \cdot t}{\lambda \cdot \left(\lambda + 2\theta + q_S^2 \cdot \gamma\right) + 3\theta \cdot (\lambda + 2\theta)} \cdot \left\{\lambda \cdot \left\{\left[1 - p_D^2 \cdot (3 - 2p_D)\right] \cdot \left(\lambda + 2\theta + q_S^2 \cdot \gamma\right) + 3\theta \cdot \left(1 - p_D^2\right)\right\} \cdot C_D\right.$$
$$+ \gamma \cdot \left\{\left[1 - p_S^2 \cdot (3 - 2p_S)\right] \cdot \left(\lambda + 2\theta + q_S^2 \cdot \gamma\right) + 3\theta \cdot q_S^2\right\} \cdot C_S + 3\theta \cdot \left(\lambda \cdot p_D^2 + 2\theta\right) \cdot C_R\right\}. \tag{14}$$

4   Case Study

The efficiency of single relay and three arrangements of logical redundancy incorporated into relays (see Sect. 2) can be investigated based on expressions (5), (8), (11), and (14). Let assume, the parameters of the expressions are as follows: $\lambda = 0.01$ 1/a, $\theta = 0.001$ 1/a, $\gamma = \gamma_1 + \gamma_2 = 0.06 + 0.04 = 0.1$ 1/a, $C_D = 20$ m.u., $C_S = 10$ m.u., $C_R = 0.2$ m.u., and $t = 20$ years. The investigation concerns the calculation of the total cost of relay(s) incorrect operation for different values of q_D (= 0.0, 0.1, 0.2, 0.3, and 0.4) and q_S (from 0.0 up to 0.4).

The of $TC(t)$ against q_S for different q_D is presented in Fig. 6. According to the figure, the efficiency of both the single relay and every type of redundancy is like each other within $q_D = q_S = 0$ (see Fig. 6a). The single relay's $TC(t)$ is the lowest one. Thus, one does not need introducing of any logical redundancy.

As the q_D increases ($q_S = 0$), the parallel arrangement becomes efficient in most (see Fig. 6b–6e). The reason is the sufficient correct operation of at least one of the two relays when the fault occurs in protected component. The higher q_D, the better the efficiency of this arrangement compared to others. The highest $TC(t)$ concerns the series arrangement. This is due to the need for two relays to simultaneously detect the fault. As the q_S increases ($q_D > 0$) the best efficiency concerns 2-out-of-3 arrangement. This arrangement is in a way like parallel one but requires correct operation of two relays when the fault occurs in other components (instead of one in parallel arrangement). A further increase in the q_S makes the series arrangement the most efficient one. Since the

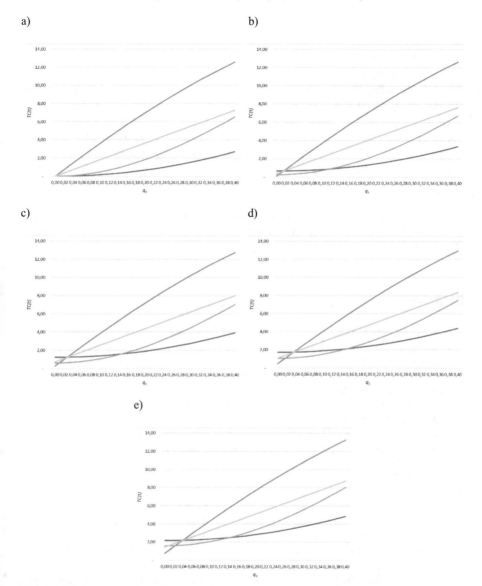

Fig. 6. $TC(t)$ against q_S for different q_D : a) $q_D = 0$, b) $q_D = 0, 1$, c) $q_D = 0, 2$, d) $q_D = 0, 3$, and e) $q_D = 0, 4$. Plots: series arrangement – in blue, 2-out-of-3 arrangement – in grey, single relay – in yellow, and parallel arrangement – in orange

undesired trip is increasingly likely the series arrangement is most efficient, because it requires correct operation of two relays when the fault occurs in other components.

5 Conclusions

Four reliability models of relays incorporated into WECS are presented. They rely on semi-Markov approach. The models consider all key features that determine correct and incorrect operation of relays, i.e. hardware failures, failure to trip, and undesired trip. The case study shows, the comparative efficiency study of four protective relay schemes can be investigated. This study can answer the question which scheme assures the best efficiency given: the failure rates of the relays, fault rates of both protected and other components, the probability of failure to trip and probability of undesired trip, the unit cost of failure to trip, the unit cost of undesired trip, the unit cost of repair, and period of relay(s) operation duration.

Acknowledgement. The work is financially supported by research project WZ/WE-IA/7/2023 being carried out at Department of Electrotechnics, Power Electronics and Electric Power Engineering in Bialystok University of Technology, Poland.

References

1. Glover, J.D., Sarma, M.S., Overbye, T.J.: Power System Analysis and Design, 5th edn. Cengage Learning (2011)
2. Horowitz, S.H., Phadke, A.G.V.: Power System Relaying, 3rd edn. Wiley, Hoboken (2008)
3. Sobolewski, R.A.: Semi-Markov reliability model of internal electrical collection grid of on-shore wind farm. In: Zamojski, W., Mazurkiewicz, J., Sugier, J., Walkowiak, T., Kacprzyk, J. (eds.) DepCoS-RELCOMEX 2019. AISC, vol. 987, pp. 466–477. Springer, Cham (2020). https://doi.org/10.1007/978-3-030-19501-4_46
4. Sobolewski, R.A.: Probabilistic modelling of reliability and maintenance of protection systems incorporated into internal collection grid of a wind farm. In: Zamojski, W., Mazurkiewicz, J., Sugier, J., Walkowiak, T., Kacprzyk, J. (eds.) DepCoS-RELCOMEX 2020. AISC, vol. 1173, pp. 585–595. Springer, Cham (2020). https://doi.org/10.1007/978-3-030-48256-5_57
5. Khurram, A., Ali, H., Tariq, A., Hasan, O.: Formal reliability analysis of protective relays in power distribution systems. In: Pecheur, C., Dierkes, M. (eds.) FMICS 2013. LNCS, vol. 8187, pp. 169–183. Springer, Heidelberg (2013). https://doi.org/10.1007/978-3-642-41010-9_12
6. Jedrzejczak, J., Anders, G.J., Fotuhi-Firuzabad, M., Farzin, H., Aminifar, F.: Reliability assessment of protective relays in harmonic-polluted power systems. IEEE Trans. Power Deliv. **32**(1), 556–564 (2017)
7. Bazilenko, O.K., Szor, E.J.: Werojatnostnyje metody ocenki effektivnosti zaszczitnych ustrojstv. Izdatielstvo 'Sztinica', Kiszyniew (1975)
8. Howard, R.A.: Dynamic Probabilistic Systems. Volume II: Semi-Markov and Decision Processes. Dover Publications, Inc., Mineola, New York (2007)

Dedicated FPGA Resources in Improving Power Efficiency of Implementations of BLAKE3 Hash Function

Jarosław Sugier[(⊠)] [iD]

Department of Computer Engineering, Wrocław University of Science and Technology, 11/17 Janiszewskiego Str., 50-372 Wrocław, Poland
`jaroslaw.sugier@pwr.edu.pl`

Abstract. The BLAKE cryptographic hash functions can be efficiently coded in software but their hardware realizations are not as fast and power-effective as those of alternative algorithms. This paper evaluates possible reductions in power consumption of BLAKE3 FPGA implementations thanks to realization of the binary summations with dedicated DSP resources rather than with logic cells inside the programmable array. The analysis took into account different possible organizations of the method in hardware: starting from the typical iterative architecture (with one instance of the cipher round), derived high-throughput versions with 2, 4 and 6-stage pipelining were also analyzed. The results came from simulation of the designs fully implemented in a Spartan-7 device; they indicate that replacing the "in-the-array" adders with 7-series's DSP48E1 elements can significantly reduce high dynamic power consumption which plagued the standard non-pipelined BLAKE3 architecture, but also reveal limitations of this approach in other cases. In addition to discussing one possible way of power reduction in this specific cipher, the paper can also be seen as another case study of an improvement in FPGA implementation made possible by taking advantage of (otherwise unused) specialized resources available beside the programmable array.

Keywords: Cryptographic Hash Function · BLAKE3 · FPGA · DSP slice

1 Introduction

Low power consumption is an important advantage of a hardware implementation of any cryptographic algorithm. The BLAKE3 hash function – a method which was initially proposed for the SHA-3 contest ([1, 2, 6]) but still remains under active development ([4, 7, 8]) – can be realized very efficiently in software but its hardware implementations are not as fast and power-effective as those of other alternative hashes. In this work we investigate a method for reducing power consumption of an FPGA implementation of this algorithm which consists in realization of the binary additions (used intensively in the cipher transformations) with dedicated DSP resources rather than with logic cells inside the programmable array. The paper presents a practical case study where the BLAKE3 compression function is translated to hardware in different organizations: in

W. Zamojski et al. (Eds.): DepCoS-RELCOMEX 2024, LNNS 1026, pp. 283–295, 2024.
https://doi.org/10.1007/978-3-031-61857-4_28

the typical iterative architecture (with one cipher round instantiated in hardware) and with 2, 4 and 6-stage pipelines. This set of architectures is implemented with DSP resources replacing simple binary adders and obtained speed and power parameters are compared with an analogous studies of standard implementations discussed in [10] and [11]. The results are computed through simulation: once the designs are fully implemented in a Spartan-7 device, the power estimation is calculated by the implementation tools from signal activity traces generated during timing simulation. As they indicate, replacing the "in-the-array" adders with 7-series's DSP48E1 resources can significantly reduce high dynamic power consumption which plagued the standard BLAKE3 architecture, but brings less or even no improvement in the highly pipelined ones.

The same set of BLAKE3 architectures was discussed initially in [10] where the iterative and pipelined organizations were evaluated with regard to their size and speed efficiency, indicating remarkable improvements brought by the third version of the cipher as compared to the previous one. In [11] the evaluation was extended with analysis of power requirements and severe problems caused by extremely high signal glitches were identified; it was shown that pipelining can significantly eliminate the transients thus decreasing power losses. Additional analysis of signal behavior and origin of the glitches was added in [9]. In this work we investigate another way of tackling this problem: replacing the typical, carry-propagation adders in the cells of the programmable array (suspected to be one of the main sources and amplifiers of the glitches) with dedicated DSP blocks located outside the array.

The text is organized as follows. The internals of BLAKE3 processing are briefly introduced in Sect. 2, along with explanation how the DSP resources (which are available in the selected Spartan-7 device but are of little use in cryptographic computations) can replace the standard, in-the-array adders. Section 3 presents implementation results of the new designs and differences in their basic speed and size characteristics. Then, in Sect. 4, the power analysis is performed and efficiency (or lack thereof) of the new solution is evaluated, while the conclusions summarize the contributions in Sect. 5.

2   The Algorithm and Possible Adaptation of DSP Logic

2.1   The BLAKE Compression Function

Like the most other round-based ciphers, data processing in BLAKE3 is organized around repetitive application of specific round transformations to a set of state words. In this case, the state consists of 16 words $v_0 \div v_{15}$, each 32-bit long (to the total of 512 bits), which are transformed in a series of $n_R = 7$ identical rounds ([7]). Realization of the round in hardware is the essence of cipher implementation – the surrounding logic performs only elementary multiplexing of the state or input words and has negligible impact on the final performance and power characteristics.

Inside the round (see the right part of Fig. 1), the 16 state words are transformed twice by so called G function. Each function reads and outputs 4 words (using additionally two words m_i of the message being hashed) so its four instances need to operate in parallel to comprise all the state. A cascade of two such 4 parallel G instances ($G_{0 \div 3} \rightarrow G_{4 \div 7}$) make up the round transformation. All BLAKE3 architectures which are considered in this paper require such a cascade of 8 G modules to be instantiated in hardware. Each G

instance is internally identical and the difference is only in specific permutations of the v_i/m_i words on its inputs.

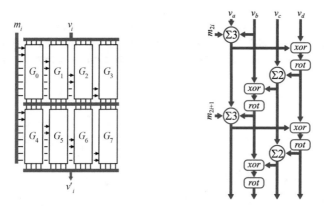

Fig. 1. Internal structure of the BLAKE round (left) and processing of the four state words inside one G function (right).

2.2 Introducing the DSP Slices as 32-Bit Adders

It was identified in [9] that increased power consumption of the BLAKE3 cores in FPGA implementations can be attributed to extremely high numbers of transient signal switching which, in the long propagation lines, consequently leads to excessive losses in the routing. With an average of over 1000 glitches per clock period in the most unstable signals of the standard iterative architecture, this parameter remains practically the same as it was in BLAKE2 (despite significant reductions in size and faster speed) and by far exceeds the values observed e.g. in equivalent architecture of Keccak. The study in [11] has shown that one remedy for this problem is pipelining: with the round processing split into stages (the evaluation included cases with 2, 4 or 6 stages) the glitches do not propagate through the pipeline registers and the avalanche effect of their multiplication is blocked at the stage boundaries. While this effect in BLAKE implementation is an excellent example of power reduction being a side-effect of pipelining (like it is discussed in the literature e.g. in [3, 5] and [12]), this work proposes a method for direct reduction of glitches at their source: in resources which implement summations appearing among cipher transformations.

The internals of word processing inside a G module are presented in the right part of Fig. 1. The set of elementary transformations include bitwise exclusive-or, rotations by constant offsets (which in hardware are accomplished entirely in routing and do not add to resource occupation) and 32-bit arithmetic additions of the state words. Summations in FPGAs are normally realized with dedicated but very simple primitives located in each logic cell of the array: next to a Look-Up Table (LUT) used for generation of combinatorial functions, each cell holds a 2:1 multiplexer and a 2-input XOR gate which create a carry propagation path; together with the LUT they can be configured as a 1-bit full adder. In summations of the BLAKE's state vectors 32 such adders would create the complete adder circuitry and, albeit this hardware can be sufficiently fast thanks to dedicated routing in the carry path, the serial nature of carry propagation through 32 levels of an adder generates numerous transient states on the output before the result settles to the final value. With other elementary transformations being only bitwise XOR (which do not generate any transients by themselves) this was the main source of glitches in previous – standard – implementations of the cipher which have been analyzed so far in [10] and [11].

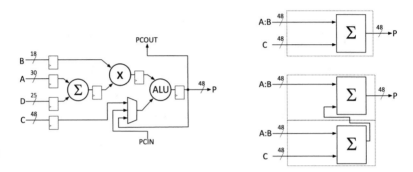

Fig. 2. The general structure of a DSP48E1 slice (left) and its configurations in 2 or 3-input 48b adders (right).

The central idea of this work is to replace the glitch-prone, in-the-array implementations of the adders with auxiliary resources which are available beside the programmable matrix. Since Virtex-II generation of FPGAs from Xilinx these devices offer dedicated hard-wired logic – so called DSP slices ([13]) – suited for "multiply and accumulate" calculations typical for signal processing algorithms. Each slice (see the left part of Fig. 2) includes a pre-adder, full 25×18 multiplier and a 48-bit ALU capable of add/subtract/accumulate or logical operations. The slices are arranged in vertical columns with fast dedicated routing between the neighbors (PCOUT to PCIN paths). While these resources can be configured for very efficient calculation of e.g. complex FIR filters, in this application only a small part of its potential is used: with all the entry preprocessing bypassed, the needed summation is done in the ALU. As the Fig. 1 reveals each implementation of the G function requires two 2-input adders (each would occupy a separate DSP slice) and two 3-input adders (each taking two adjacent DSP slices, see the right part of Fig. 2). As a result, the total of 6 DSP slices is needed for implementation of one G function and 48 slices for the whole round.

3 Implementations

3.1 The Architectures

To keep the test set identical to the previous study, the following 6 architectures of the BLAKE3 compression function were implemented, now with DSP resources:

- the standard iterative organization (denoted as X1) with an instance of one cipher round without pipelining, i.e. producing the output after $n_R = 7$ clock cycles;
- the derived variant with two pipeline stages (denoted as P2), with stage boundary between the $G_{0\div3}$ and $G_{4\div7}$ modules, generating two hashes after 14 clock cycles;
- the variant with 4 pipeline stages (P4) with additional stage boundaries inside the G instances just before the second 3-input adder in the v_a path, which produced 4 hashes every 28 clock cycles when working continuously;
- three organizations with 6 pipeline stages (P6a, P6b and P6c) capable of generating a set of six hashes every 42 clock cycles.

The three versions of the P6 architecture were repeated after their initial proposition in [10]. Briefly, the simplest variant P6a was created with even division of the 6 adders between the three pipeline stages $(2 + 2 + 2)$ but this led to splitting of the 3-input summation in the v_a path; in order to avoid this splitting, the variants P6b and P6c kept this adder in the middle stage and proposed different separation of the first two stages $(2 + 3 + 1$ or $3 + 2 + 1)$. The details of these projects were discussed in [10].

All 6 organizations required one instance of the round logic and needed a total of 48 DSP slices for the 32-bit summations of the state words. Of the Spartan-7 family which have been the platform of power analysis so far, the smallest device with sufficient number of DSP resources was the XC7S25 part and this (in csga225 package and -2 speed grade) was selected as the implementation target ([14]). Also, to comply with the previous methodology, all 6 cipher units were equipped with basic serial in – parallel out circuitry in order to keep the I/O pin count within reasonable levels (because otherwise the ports of the hashing unit would require 1152 pins) and such completely functional designs were implemented with Vivado 2023.2 tools.

Two kinds of studies were performed. First, for each architecture its maximum operational frequency was determined in the same way as in [11]. Then, each design was implemented with a target frequency lowered by some margin in order to ensure stable operation, and such an implementation was the subject of further power analysis. Parameters of these implementations are given in Table 1. They describe implementation size as a number of occupied logic slices, LUTs and registers, confirm utilization of 48 DSP elements across all the cases, and illustrate speed efficiency by listing the maximum frequency of operation, requested and actual (achieved) frequency of the implementation prepared for power analysis, and parameters of its longest path: percentage of delay generated by routing (the rest is attributed to logic) and the number of logic levels, including the DSP elements found among them.

Table 1. Size and speed parameters of the implementations examined in power analysis.

	X1	P2	P4	P6a	P6b	P6c
Slices	827	831	1107	1227	1234	1263
LUTs	1693	1397	1717	2306	2280	2264
Registers	2185	3206	4743	5854	5768	5768
DSP48E1	48	48	48	48	48	48
Max F_{clk} [MHz]	29.2	57.8	110.8	145.0	124.2	125.2
Target F_{clk} [MHz]	28	55	100	133	115	115
Actual F_{clk} [MHz]	28.1	51.4	104.7	135.2	116.3	118.9
Route delay	39%	42%	42%	45%	35%	34%
Logic levels	21	11	5	3	4	4
(incl. DSP)	(12)	(6)	(3)	(2)	(3)	(3)

3.2 Implementations Results

Before moving to power analysis, some important points should be made about new implementation efficiency observed after moving part of the logic to the DSP resources. The changes in basic size and speed parameters as compared to the values obtained with the standard implementation of the adders in [11], are visualized in Fig. 3.

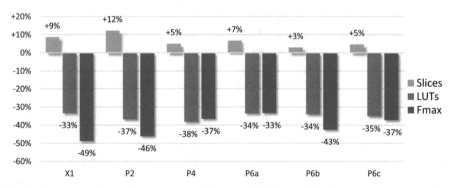

Fig. 3. Changes in the numbers of occupied slices and LUTs as well as in the maximum operating frequency, resulting from moving the summation logic to the dedicated DSP elements.

Comparing size characteristics (the first two categories in the chart), it is clearly seen that the modification affected not as much array (slice) occupancy as density of its utilization: despite the fact that the LUT numbers are reduced by $33 \div 38\%$ across all the 6 architectures (which is fully expected because the summation logic is now transferred from the tables to the DSP primitives), the numbers of occupied slices are not lower at all, and they even moderately increase by $3 \div 12\%$. This leads to a noticeable drop from 2.95 to 1.80 in the average LUT per slice ratio. The reduced logic density in the programmable array translates to longer propagation paths and does not help in power and speed efficiency. Indeed, the maximal operating frequencies of all the architectures are lower by 33 ÷ 49%, which should be noted as a significant deterioration.

To explain this effect, one must look at the spatial distribution of occupied resources in the FPGA array which is presented in Fig. 4. The new implementation (the left part of the figure) must use 48 DSP elements which are available in the device in two columns, each with 40 items. It is seen that the implementation tool tried to optimally place the 48 elements by locating half of them in each column but this nonetheless occupied 60% of the total matrix height, subsequently causing the design – which otherwise would be much more concentrated, as the right part of the figure shows – to be spread over larger area.

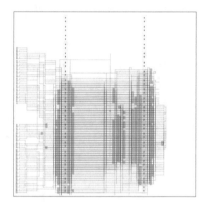 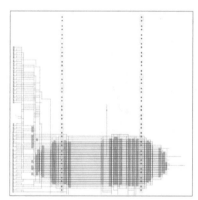

Fig. 4. A XC7S225 device with the iterative (X1) core implemented using the DSP resources (left) compared to the previous standard implementation (right); the two shaded columns mark locations of the DSP elements available in this chip.

4 Power Calculations

4.1 The Methodology

Like in [11], in order to get the most accurate power results which the tools were able to calculate, first the post-route simulation of the fully implemented designs was run in order to trace activity of all internal signals of the array, the traces were written to the SAIF files and only then the power estimation procedures were executed with precise data describing signals behavior ([15]). Because the tools could only measure an average power losses over the whole simulation period, special scenarios were prepared for testbenches which kept the cipher cores fully busy in all clock cycles (loading a new batch of input instantly as the previous one is passed to the outputs, maintaining disparity between data in adjacent pipeline stages to eliminate signal repetitions, avoiding idle cycles during generation of a new input, etc.).

The results of this evaluation for all the 6 architectures operating at their nominal frequencies are listed in Table 2. Like previously, the analysis will be limited exclusively to the dynamic dissipated power because this is the only component determined by the clock frequency and the internal structure of the implemented circuit. The static power – the other component in the total power losses – depends on characteristics of the idle (unprogrammed) matrix and other external parameters, like ambient temperature, cooling efficiency, etc. Ignoring the static power is justified in this analysis because it is focused on the power characteristics of various BLAKE3 implementations, but it should be noted that it would only show a part of the picture if the total power consumption of a real FPGA device would have to be calculated. The table, apart from the total dynamic power, gives also its components divided between the main types of array resources: clock distribution, configurable blocks, DSP elements, routing, and I/O.

Table 2. Results of dynamic power calculations estimated for the modules operating with their maximal frequencies.

	X1 (28 MHz)	P2 (55 MHz)	P4 (100 MHz)	P6a (133 MHz)	P6b (115 MHz)	P6c (115 MHz)
Total power [mW]	579	334	286	331	287	296
Clocks	2	7	18	27	23	24
Slice Logic	92	42	34	39	33	30
LUT	92	41	31	33	29	24
Register	0	1	3	6	4	4
DSP	167	109	96	112	104	103
Routing	317	173	133	149	121	134
I/O	1	2	4	4	4	4
Efficiency metrics:						
E_h (nJ/hash)	144.7	42.6	20.0	17.4	17.4	18.0
P_{MHz} (mW/MHz)	20.7	6.1	2.86	2.49	2.49	2.57

After implementation of each architecture for its nominal speed, the analysis was repeated for different clock frequencies. The results are visualized in Fig. 5 where the dynamic power is presented as a function of F_{clk}. The chart confirms the general rule that the power losses are linearly proportional to this parameter which additionally validates the estimation method and its results. Because of linearity, from the slope of the graphs additional synthetic characteristics can be calculated which are independent from the clock speed:

- E_h = an average energy consumed for computation of one hash value (in nJ per hash), and
- P_{MHz} = total dynamic power per clock frequency (in mW per MHz).

The values of these parameters are given in the last two rows of the table. The E_h values can alternatively be interpreted as the dynamic power per hashing speed, with the nJ/hash unit translated to mW/Mhps (mega hashes per second).

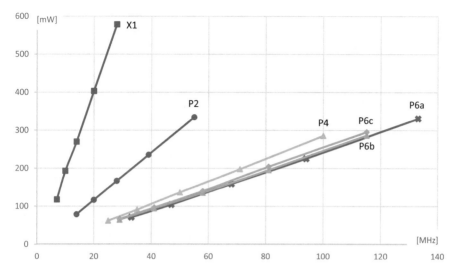

Fig. 5. Dynamic power of the designs as functions of clock frequency.

4.2 Analysis of Power Results

The distribution of power losses in different components of the FPGA array is presented in Fig. 6. Compared to the results from [11], the obvious difference is the carry logic (now not appearing at all) being replaced by the DSP elements but the scale of this replacement should be underlined: while the carry generated less than 7% of the losses with in-the-array adders, now the DSP logic takes from 29 to 38% of the power. These numbers indicate how large part of the standard adders was previously implemented in LUTs and how summation-intensive is BLAKE processing. More importantly, moving the adders out of the array significantly reduced stress on the LUT elements which previously also took part in in-the-array summations: their share is now from 9 to 16% which is from 2 to 3 times less. Still, the routing remains the most power-consuming resource, although for the first time it loses the first place to combined logic (LUT+DSP) in all architectures apart from the X1. Overall, the observed changes indicate importance of the DSP adders in the power analysis.

The E_h metrics of the architectures are visualized in the left part of Fig. 7. It should be noted that although the X1 case is still significantly the most power-hungry of all the organizations, now the difference between it and the P2 case is not as big as it was in the standard implementation: the observed ratio is now 7:2 instead of approx. 20:1. This smaller difference shows that the regular X1 architecture is now more predictable in terms of consumed power.

On the other hand, obviously the pipelining again does help in power reduction, albeit the decrease is not as crucial as it was before. In particular, it can be seen that the difference between the P4 and P6x cases is becoming less important: the variation within the three P6x organizations is almost comparable to the advantage they offer over the P4 case, despite the significantly more pipeline stages and, consequently, more complicated internal structure and longer latencies of data observed on the module ports.

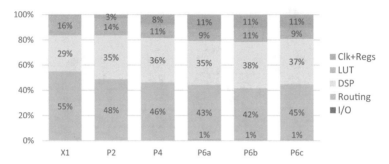

Fig. 6. Structure of power losses: components generated by different FPGA resources.

The saturation of the pipelining effect starts sooner, i.e. with less stages, and now it can be questioned if the power advantages justify increasing number of pipeline stages from 4 to 6.

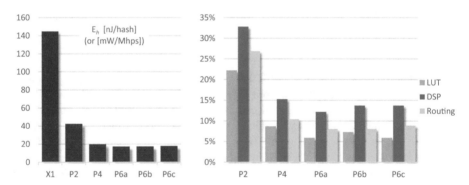

Fig. 7. Energy per one hash calculation (E_h, left) and the main components of the dynamic power in the pipelined architectures as fractions of the X1 case (right).

Another aspect of the changes brought by the pipelining can be analyzed in the right part of Fig. 7 where the absolute values of the dynamic power in the main FPGA components – LUT, DSP and routing – are computed as percentages of the X1 values. The graph shows that the reductions with increasing pipeline length are a little more visible in routing than in the elements performing actual logic transformations, i.e. in LUTs and DSP adders. Looking at the proportions within each individual architecture, in all cases the biggest power drop is observed in the LUT generators, while the smallest difference is in the DSP adders. This indicates that the summations, although do not generate the biggest part of the losses, are the elements with the most intense signal switching caused by their native internal processing and the pipelining does not help much in their operation. The rest of BLAKE logic which remained in the LUT generators, as well as the routing, are the areas where blocking the glitches at the pipeline registers causes larger reductions.

4.3   Comparison to the Standard Implementations

Finally, Fig. 8 evaluates the effects of application of the DSP elements by showing the changes in the synthetic E_h metrics, with the left graph referring to its overall value. It indicates excellent improvement in the X1 organization where the unit energy is reduced more than tenfold, to 9%. This confirms the initial assumption that application of a dedicated circuitry for 32-bit summations in place of the standard, in-the-array adders can improve signal stability and reduce power consumption. Nevertheless, the graph also shows that this positive result is counter-weighted by other effects which increase with longer pipeline: while in the P2 organization the reduction is still by a remarkable 41%, the P6x cases see a rise instead of a drop, with the P4 architecture showing (almost) the perfect balance and repeating the previous result.

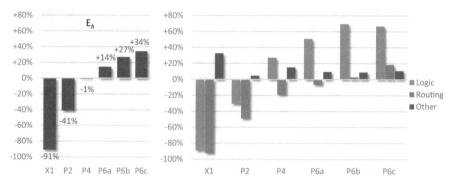

Fig. 8. Changes after introducing the DSP adders: in the total energy per hash (left) and in its components (right).

To analyze this counter-effect, the right part of Fig. 8 presents summarized changes in individual E_h components divided into the three basic categories: combinational logic (LUT generators plus DSP or carry elements), routing, and all the remaining resources (clock networks, registers and I/O blocks).The graph reveals that – despite glitch reduction – power losses of the logic components are not reduced with pipelining as much as it was in the previous implementations. This shows that the DSP sites, with their complex, 48-bit internal structure and strong computational potential, are not efficient when power preservation aspect is concerned. The summations implemented there do help in reductions of glitches and – subsequently – dissipated power but they lose in power efficiency to the standard in-the-array adders as soon as signal stability is sufficiently improved with pipelining – i.e. already in the P4 case, where the increase is by 27%. In the P6x cases the losses generated by logic are higher by 51% to even 69% than they were previously.

Although a similar tendency is seen in the routing component, the effect is much less intense: an increase by 3 and 18% is seen only in the last two P6x cases. This is despite the fact that the routing in the new designs is significantly longer than previously, as the discussion about Fig. 4 explained, which only confirms that the main effect – reductions of transient glitches in signal transmission – was indeed achieved by introducing the

DSP adders. The widespread design placement as seen in Fig. 4, on the other hand, amounted to increased power losses in long clock distribution lines: the "Other" component (which includes first of all clocking losses) is higher in all 6 architectures, regardless of pipelining.

5 Conclusions

The paper proposed specific approach to the problem of power reduction in FPGA implementations of the BLAKE3 cryptographic core. The method consisted in elimination of the standard, in-the-array 32-bit adders and replacing them with dedicated summation circuits which are available in DSP slices located beside the programmable array. The results confirm that this solution can greatly improve power requirements in the standard iterative implementation which suffers from very high level of transient glitches in internal signals, caused mainly by sequential carry propagation in the traditional adders. On the other hand, the DSP elements are optimized for computational and not power efficiency, so the advantages they bring can be counter-weighted in those pipelined architectures where the glitches are already sufficiently reduced by registers at pipeline stages. As a result, the unit hash energy (which is a fair metric across different architectures and operating frequencies) remains the same in the design with 4 pipeline stages, and deteriorates for 6 stages.

The analysis indicated that in this particular case the DSP elements were internally too complex and do not correspond to the much higher density of the cipher logic: with the core requiring 48 adders, they alone are located across 60% of the entire matrix height, while the full cipher circuitry can be much more condensed. With design layout unnecessarily stretched, the long propagation lines impair attainable speed as seen in reduction of the maximum clock frequency by approx. 1/3. On the other hand, this approach has proved to offer very high and unquestionable power optimization in the cases of iterative or 2-stage pipelined architectures.

While the presented study offers a practical method of power optimization in the particular cases of BLAKE3 hardware realizations, from a more general perspective it can also serve as another example of an improvement in FPGA implementations made possible by taking advantage of (otherwise idle) specialized resources available beside the programmable array.

References

1. Aumasson, J.P., Henzen, L., Meier, W., Phan, R.C.-W.: SHA-3 proposal BLAKE, version 1.3 (2010). https://www.aumasson.jp/blake/blake.pdf. Accessed Mar 2024
2. Aumasson, J.-P., Neves, S., Wilcox-O'Hearn, Z., Winnerlein, C.: BLAKE2: simpler, smaller, fast as MD5. In: Jacobson, M., Locasto, M., Mohassel, P., Safavi-Naini, R. (eds.) ACNS 2013. LNCS, vol. 7954, pp. 119–135. Springer, Heidelberg (2013). https://doi.org/10.1007/978-3-642-38980-1_8
3. Boemo, E., Oliver, J., Caffarena, G.:. Tracking the pipelining-power rule along the FPGA technical literature. In: 10th FPGAworld Conference - Academic Proceedings 2013, FPGAworld (2013). https://doi.org/10.1145/2513683.2513692

4. Ciocan, I.T., Kelesidis, E.A., Maimuț, D., Morogan, L.: A modified Argon2i using a tweaked variant of Blake3. In: 2021 26th IEEE Asia-Pacific Conference on Communications (APCC), 11–13 October 2021, Kuala Lumpur, Malaysia, pp. 271–274. IEEE Xplore (2021). https://doi.org/10.1109/APCC49754.2021.9609933

5. Grover, N., Soni, M.K.: Reduction of power consumption in FPGAs - an overview. Int. J. Inf. Eng. Electron. Bus. **4**(5), 50–69 (2012). https://doi.org/10.5815/ijieeb.2012.05.07

6. National Institute of Standards and Technology: SHA-3 Standard: Permutation-Based Hash and Extendable-Output Functions (FIPS 202). https://csrc.nist.gov/publications/detail/fips/202/final. Accessed Mar 2024. https://doi.org/10.6028/NIST.FIPS.202

7. O'Connor, J., Aumasson, J.P., Neves, S., Wilcox-O'Hearn, Z.: BLAKE3: one function, fast everywhere. Real World Crypto 2020 (lightning talk) (2020). https://github.com/BLAKE3-team/BLAKE3-specs/blob/master/blake3.pdf. Accessed Mar 2024

8. Sinha, S., Anand, S., Krishna Prakasha, K.: Improving smart contract transaction performance in hyperledger fabric. In: 2021 Emerging Trends in Industry 4.0 (ETI 4.0), 19–21 May 2021, pp. 1–6. IEEE Xplore (2021). https://doi.org/10.1109/ETI4.051663.2021.9619202

9. Sugier, J.: Comparison of power consumption in pipelined implementations of the Blake3 cipher in FPGA devices. Int. J. Electron. Telecommun. **70**(1), 23–30 (2024)

10. Sugier, J.: FPGA implementations of BLAKE3 compression function with intra-round pipelining. In: Zamojski, W., Mazurkiewicz, J., Sugier, J., Walkowiak, T., Kacprzyk, J. (eds.) DepCoS-RELCOMEX 2022. LNNS, vol. 484, pp. 319–330. Springer, Cham (2022). https://doi.org/10.1007/978-3-031-06746-4_31

11. Sugier, J.: Power analysis of BLAKE3 pipelined implementations in FPGA devices. In: Zamojski, W., Mazurkiewicz, J., Sugier, J., Walkowiak, T., Kacprzyk, J. (eds.) DepCoS-RELCOMEX 2023. LNNS, vol. 737, pp. 295–308. Springer, Cham (2023). https://doi.org/10.1007/978-3-031-37720-4_27

12. Wilton, S.J.E., Ang, S.-S., Luk, W.: The impact of pipelining on energy per operation in field-programmable gate arrays. In: Becker, J., Platzner, M., Vernalde, S. (eds.) FPL 2004. LNCS, vol. 3203, pp. 719–728. Springer, Heidelberg (2004). https://doi.org/10.1007/978-3-540-30117-2_73

13. Xilinx, Inc.: 7 Series DSP48E1 Slice: User Guide. UG479.PDF. www.amd.com. Accessed Mar 2024

14. Xilinx, Inc.: 7 Series FPGAs Data Sheet: Overview. DS180.PDF. www.xilinx.com. Accessed Mar 2024

15. Xilinx, Inc.: Vivado Design Suite User Guide: Power Analysis and Optimization. UG907.PDF. www.amd.com. Accessed Mar 2024

Assessing Inference Time in Large Language Models

Bartosz Walkowiak[ID] and Tomasz Walkowiak[(✉)][ID]

Wroclaw University of Science and Technology, Wrocław, Poland
{bartosz.walkowiak,tomasz.walkowiak}@pwr.edu.pl

Abstract. Large Language Models have transformed the field of artificial intelligence, yet they are often associated with elitism and inaccessibility. This is primarily due to the large number of their parameters, ranging from 1 billion to 70 billion, making inference on these models costly and resource intensive. To tackle this challenge, various solutions have emerged with the goal of enabling efficient, fast, and resource-constrained inference. This study aims to review and compare these available solutions. The authors conducted a series of experiments that compared the inference speed of the basic HuggingFace transformers library, the HuggingFace Text Generation Inference server, and the open source vLLM library. The findings reveal that vLLM outperforms the other approaches examined. Additionally, the results highlight how relatively straightforward techniques, such as continuous batching, can significantly accelerate inference for large batch sizes.

Keywords: Large language models · model deployment · continous batching

1 Introduction

Despite the quality of the model, response time is a critical factor for users. Large language models (LLM) are based on autoregressive generation of responses, where the network uses input queries and previous responses to generate the next token based on estimated probabilities. This process continues to generate subsequent output tokens. Due to the operational mechanism and the large number of parameters, significantly exceeding the number of tokens of older transformer-based models such as T5 [13] and BERT [4], response generation is time-consuming and affected by the length of the output text. Therefore, it is important to implement these models effectively in production systems, including mechanisms that allow to speed up the process.

What is more, due to the trend started by OpenAI, language models have also started to be provided as a service, via an API behind which is a network of machines handling the requests. Despite its many benefits, including zero computational cost on the client side, this type of use of models also has its disadvantages. The biggest drawbacks of using models via an API are the costs

W. Zamojski et al. (Eds.): DepCoS-RELCOMEX 2024, LNNS 1026, pp. 296–305, 2024.
https://doi.org/10.1007/978-3-031-61857-4_29

imposed on the internet connection, the financial outlay needed to gain access and, in the case of interfaces such as chatGPT, the limited control over model parameters.

The aim of this paper is to compare the performance of inference on generative models due to the run-time technique used and to examine methods ranging from the standard CasualLM module provided by HuggingFace in the transformers [17] library, through Text Generation Inference [7] also by HuggingFace, to the rapidly improving vLLM [8, 14] library developed within the vllm-project. In addition, experiments will be carried out on models of different sizes in order to test the scalability of the tested methods in terms of the different memory requirements resulting from the number of parameters.

The paper is structured as follows: we begin with an overview of related works, followed by a description of the inference speed-up techniques utilized. This is then followed by a detailed account of the experiments conducted and a discussion of the results obtained.

2   Related Works

The problem of too slow or expensive inference arises for many users who wish to use the rapidly growing group of LLMs on their own equipment, a phenomenon caused by the change in both the capabilities and the size of the available models. Several branches of solutions come to the rescue to optimize and speed up the inference process.

Among the solutions, there are some that focus on reducing the requirements for the runtime environment; one such project is llama-cpp [6], using the GGML [5] format, which has developed software that allows even large models to run efficiently on the CPU. Another model storage format that significantly affects inference performance is ONNX [2], whose use speeds up the process of generating successive tokens when GPU or CPU are used.

Mention should also be made of techniques such as quantization, by which the precision at which calculations operate can be reduced and therefore the cost of these operations. However, as research [15] has shown, despite reducing the memory requirements of the graphics card, quantization significantly slows the generation speed and, in extreme cases, can lead to a huge degradation in the quality of the model output. Additionally, not all solutions on the market support quantization; e.g. vLLM has little support for the technique so far.

Another group of technologies that support rapid inferencing is those that derive from methods for accelerating and parallelizing the training of large language models. Among these methods, we should mention the DeepSpeed [1] library prepared by Microsoft, whose great advantage is the reduction of GPU memory consumption thanks to the revolutionary ZeRO (Zero Redundancy Optimizer) technology. This technique is based on combining the weights of the model on the CPU or NVMe, depending on the available resources, and then passing the weights to the GPU for inference. This approach to the static part of the model makes more efficient use of GPU memory and allows larger models to be run. Another very useful resource is the toolkit provided by Colossal-AI [9], which includes support for a number of strategies for enhancing computation speed through parallelization, including Sequence-Parallelism [10], Auto-Parallelism [12], and the previously discussed ZeRO approach.

The most relevant to the end user are methods that accelerate inference by optimizing how the GPU's memory is used, i.e., methods such as FlashAttention [3] and PagedAttention [8], an analysis of which can be found in Sect. 3.

3   Inference Methods

3.1   Inference Time Assesment

Before investigating inference methods, it is important to note the available approaches to measuring generation speed, among which we can find the following: Time To First Token, which is an important indicator of responsiveness; Time Per Output Token, i.e. how long it takes on average to generate one token; Throughput, which takes into account the acceleration that is precipitated in computing systems by using different techniques to combine multiple queries. However, the just mentioned techniques do not provide complete information on performance, since, according to the study mentioned in [15], the efficiency of the system should be measured in the generation speed, expressed in the average number of tokens per second, for a small, medium, and large number of tokens per output. This way of reporting efficiency makes it possible to estimate the shape of the speed versus output length relationship curve, which is important because this relationship is usually not linear and can take shapes such as logarithmic or exponential. This approach to reporting system efficiency makes the results independent of the biases resulting from averaging the generation speed regardless of the length of the model response, and we also adopt this approach as our efficiency metric.

During the process of testing the inference efficiency of various methods, we take into account the total query processing time, which consists of the time the task was queued for execution and the time of the actual processing and calculation of the result. Such an approach is dictated by the desire to check the real, perceived efficiency of the tested methods by the end user; measuring processing time alone would be a certain misrepresentation that does not provide full information about the tested technology.

3.2   FlashAttention

The effectiveness of the FlashAttention algorithm [3] is based on reducing the frequency of I/O operations to High Bandwidth Memory, which is achieved through the use of tiling. The operation of splitting the computation into blocks avoids the requirement to load the entire data matrix to compute the attentions. An additional improvement is the placement of the softmax normalization factor in SRAM, this memory has a small capacity but its great advantage is that it is located in the compute chip which provides very fast access and further reduces the input-output operations when calculating the attentions during the backward transition.

3.3 PagedAttention

The space used during computation can be divided into two main parts; a static one, which is used to store the loaded model weights, and a dynamic one, intended for the tensors that serve as context for the token just being computed. The PagedAttention mechanism [8], approaches dynamically changing memory containing key and value tensors, also known as the KV cache, differently from other methods. Existing and frequently used solutions are based on pre-allocation of part of the memory for the KV cache, usually the size of this memory is equal to the maximum length of the possible tensors, for example 2048 tokens. Another drawback of standard dynamic memory area approaches is the lack of memory sharing, as the KV cache for each sequence is stored separately in a continuous space.

The PagedAttention algorithm solves the aforementioned shortcomings of the state-of-the-art methods by using KV cache partitioning into blocks of fixed size that are not stored in a continuous memory area. This solution enables a much more efficient use of memory, as KV blocks can be shared and do not require pre-allocation.

3.4 Batching Techniques

Batching is a widely used technique in visual and language models running on GPUs. It involves grouping multiple inputs together and processing them simultaneously during inference or training. GPUs are designed with a SIMD (Single Instruction, Multiple Data) architecture, making them well suited for handling large amounts of data efficiently. The traditional approach, known as static batching (see Fig. 1), involves starting a batch all at once and waiting for all inputs to complete their calculations. However, this method can lead to inefficiencies, particularly in generative models, where the number of tokens generated per request is unpredictable. This results in idle time slots on the GPU and under-utilization of resources. A more effective solution, as suggested in [18], is to continuously schedule new requests as the previous ones are completed and slots become available. This approach helps maximize GPU utilization and improves overall efficiency in processing generative models.

3.5 Analysed Methods

For the study we have chosen advanced methods that use FalshAttention and PagedAttention, namely HuggingFace Text Generation Inference and vLLM, Moreover, we check the performance of and HuggingFace CasualLM as a popular method to serve as a reference. A comparison of the features of the selected methods can be found in Table 1.

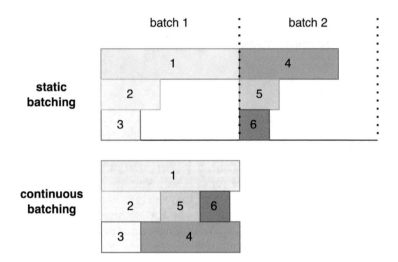

Fig. 1. Static vs. continous batching.

Table 1. Summary of inference method features, a + sign indicates the presence of a feature, while a - sign indicates its absence.

Feature	Method		
	HF CasualLM	HF Text Generation Inference	vLLM
Support for all popular models	+	+	−
Use of PagedAttention	−	+	+
Continous Batching	−	+	+
Open licence	+	−	+

HF CasualLM. The standard approach to running a deep textual language model is to use the transformers library; among its vast set of supported architectures are those for generative text generation. The library responsible for handling generative models has a section dedicated to causal language models. Such models predict the next tokens in the sequence and use left-handed attention.

HuggingFace Text Generation Inference. Text Generation Inference (TGI) was created by HuggingFace as a toolkit for efficient text generation using generative large language models. The technology provides tools that enable, among other things, the simple launch of an LLM service, the use of FlashAttention and PagedAttention, and quantization with bitsandbites.

vLLM. What sets vLLM apart from the current state of the art is its efficient approach to managing attention key and value memory, thanks to the PagedAttenion algorithm. This technology optimizes the memory usage of the graphics card, among other things, by enabling shared computation on KV blocks. Another important technique used is continuous batching, which ensures that the model does not experience significant slowdowns during heavy query loads, as request blocks are allocated efficiently in available memory.

4   Experiments and Results

4.1   Prompt Dataset

The dataset consists of a selected set of TruthfulQA [11] questions extended with customized short answer questions. The prepared set contains 50 questions that provide short to long answers from language models. Each question was run multiple times to decouple the results from randomness due to the generative nature of the models tested.

4.2   Models Testeted

In order to get an overview of the inference rate for the different methods, models of various sizes were evaluated, from relatively small models of one billion parameters, i.e., TinyLlama [19], throughout the widely-used, reasonably sized OpenChat [16] model with 7B parameters, to the Vicuna-13B [20] model.

4.3   Speed for a Single Prompt

In the initial set of experiments, we analyzed the inference time for a single prompt. Following the methodology proposed in [15], we measured the average time taken to generate the output for three distinct ranges of LLM output (1–30 tokens, 31–300 tokens, and more than 300 tokens). The experiments were carried out using the A100 80 GB card and the results are presented in Table 2. It is evident that the vLLM library provides the fastest responses. Additionally, the response speed of vLLM is consistent regardless of size, whereas with TGI, the speed increases significantly as the output size grows.

4.4   Throughput for Varying Batch Sizes

In the second set of experiments, we tested the throughput of the models according to batch size. We generated a set of 2,500 prompts (50 sets of 50 prompts) and arranged them randomly. We measured the processing time of all prompts using different model implementations while keeping the batch sizes consistent across one experiment. The results, which show the average speed obtained as the size of all outputs over the measured time, are presented in Table 3.

The results highlight two key findings. Firstly, the basic HF CausalLM implementation demonstrates poor scalability. While the speed of HF is 3–6 times slower for a batch size of one, this difference increases significantly for larger batches, reaching up to 30 times slower. This effect can be attributed to the

Table 2. Average speed [tokens/s] in function of output size for three models of different size and three analyzed methods of inference.

Model	Size	Method	Output size in tokens		
			1–30	31–300	301-
TinyLlama	1B	HF	51	51	52
		TGI	15	60	95
		vLLM	260	290	280
OpenChat	7B	HF	38	33	26
		TGI	4	48	70
		vLLM	83	85	83
Vicuna	13B	HF	34	30	15
		TGI	6	34	44
		VLLM	50	50	49

Table 3. Average speed [tokens/s] in function of batch size for three models of different size and three analyzed methods of inference. In case of HF we got out-of-memory errors for large batches.

Model	Size	Method	Batch size							
			1	2	16	32	64	128	256	512
TinyLlama	1B	HF	52	98	189	149	127	106	OoM	OoM
		TGI	117	216	1127	1732	2434	2837	2837	2819
		vLLM	283	527	2683	3940	5145	5358	5565	5550
OpenChat	7B	HF	30	36	29	29	26	24	OoM	OoM
		TGI	82	155	762	1348	2110	2344	2364	2005
		vLLM	85	162	1024	1681	2488	3093	3411	3507
Vicuna	13B	HF	17	16	11	11	10	9	OoM	OoM
		TGI	49	95	542	842	1171	1167	1164	1167
		VLLM	50	97	608	984	1468	1795	1874	1924

Table 4. VLLM relative to TGI speed for varing batch size across three models of differing sizes. It is apparent that for no or small batch sizes, both vLLM and TGI exhibit similar performance with 7B and 13B models. However, as the batch size increases, vLLM outperforms TGI by a factor of 1.65 to 1.97.

Model	Size	Batch size							
		1	2	16	32	64	128	256	512
TinyLlama	1B	2.41	2.44	2.38	2.28	2.11	1.89	1.96	1.97
OpenChat	7B	1.04	1.05	1.34	1.25	1.18	1.32	1.44	1.75
Vicuna	13B	1.02	1.02	1.12	1.17	1.25	1.54	1.61	1.65

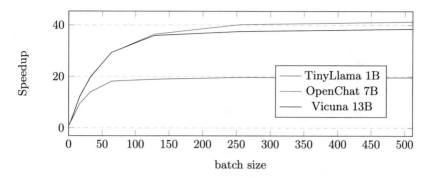

Fig. 2. Speedup for the vLLM inference method for models of different sizes.

use of different batching methods. The static batching approach, as discussed in Sect. 4, results in all prompts waiting for the longest (slowest) responses, leading to significant performance degradation with larger batch sizes. Second, vLLM, although it employs mechanisms similar to TGI (such as FlashAttention, PagedAttention, and continuous batching), consistently outperforms TGI. For larger batches, the performance relative difference (as indicated in Table 4) can be as high as 1.6–1.9 times, depending on the size of the model.

Figure 2 illustrates the speedup for the vLLM inference method, which represents the speed of a given batch compared to the speed of batch one. It is evident that the speedup for the smallest model (TinyLlama) is approximately half that of the larger models.

5 Conclusions

The paper presents a performance analysis of three inference methods for Large Language Models: the basic HuggingFace transformers library, the HuggingFace Text Generation Inference server, and the open-source vLLM library. Tests carried out on different batch sizes and models of varying sizes demonstrate that vLLM outperforms the other approaches analyzed. Furthermore, the results illustrate how relatively simple approaches, such as continuous batching, can accelerate inference for large batch sizes. Furthermore, minor implementation details play a crucial role. Although vLLM and TGI use similar techniques such as FlashAttention, PagedAttention, and continuous batching, vLLM consistently outperforms TGI, achieving almost 1.6 to 1.9 times faster responses when using long batches.

Acknowledgements. The work was financed as part of the investment: "CLARIN ERIC - European Research Infrastructure Consortium: Common Language Resources and Technology Infrastructure" (period: 2024-2026) funded by the Polish Ministry of Science and Higher Education (Programme: "Support for the participation of Polish scientific teams in international research infrastructure projects"), agreement number 2024/WK/01.

References

1. Aminabadi, R.Y., et al.: DeepSpeed-inference: enabling efficient inference of transformer models at unprecedented scale. In: Proceedings of the International Conference on High Performance Computing, Networking, Storage and Analysis, SC 2022. IEEE Press (2022)
2. Bai, J., et al.: ONNX: Open neural network exchange (2019). https://github.com/onnx/onnx
3. Dao, T., Fu, D.Y., Ermon, S., Rudra, A., Re, C.: Flashattention: fast and memory-efficient exact attention with IO-awareness. In: Oh, A.H., Agarwal, A., Belgrave, D., Cho, K. (eds.) Advances in Neural Information Processing Systems (2022). https://openreview.net/forum?id=H4DqfPSibmx
4. Devlin, J., Chang, M., Lee, K., Toutanova, K.: BERT: pre-training of deep bidirectional transformers for language understanding. In: Burstein, J., Doran, C., Solorio, T. (eds.) Proceedings of the 2019 Conference of the North American Chapter of the Association for Computational Linguistics: Human Language Technologies, NAACL-HLT 2019, Minneapolis, MN, USA, 2–7 June 2019, Volume 1 (Long and Short Papers), pp. 4171–4186. Association for Computational Linguistics (2019). https://doi.org/10.18653/v1/n19-1423
5. Gerganov, G.: GGML - tensor library for machine learning (2023). https://github.com/ggerganov/ggml
6. Gerganov, G.: Inference of LLaMA model in pure C/C++ (2023). https://github.com/ggerganov/llama.cpp
7. HuggingFace: HuggingFace text generation inference (2024). https://github.com/huggingface/text-generation-inference
8. Kwon, W., et al.: Efficient memory management for large language model serving with pagedattention. In: Proceedings of the ACM SIGOPS 29th Symposium on Operating Systems Principles (2023)
9. Li, S., et al.: Colossal-AI: a unified deep learning system for large-scale parallel training. In: Proceedings of the 52nd International Conference on Parallel Processing, ICPP 2023, pp. 766–775. Association for Computing Machinery, New York (2023). https://doi.org/10.1145/3605573.3605613
10. Li, S., Xue, F., Baranwal, C., Li, Y., You, Y.: Sequence parallelism: long sequence training from system perspective. In: Rogers, A., Boyd-Graber, J., Okazaki, N. (eds.) Proceedings of the 61st Annual Meeting of the Association for Computational Linguistics (Volume 1: Long Papers), Toronto, Canada, pp. 2391–2404. Association for Computational Linguistics (2023). https://doi.org/10.18653/v1/2023.acl-long.134
11. Lin, S., Hilton, J., Evans, O.: TruthfulQA: measuring how models mimic human falsehoods. In: Muresan, S., Nakov, P., Villavicencio, A. (eds.) Proceedings of the 60th Annual Meeting of the Association for Computational Linguistics (Volume 1: Long Papers), Dublin, Ireland, pp. 3214–3252. Association for Computational Linguistics (2022). https://doi.org/10.18653/v1/2022.acl-long.229
12. Liu, Y., Li, S., Fang, J., Shao, Y., Yao, B., You, Y.: Colossal-auto: unified automation of parallelization and activation checkpoint for large-scale models. arXiv preprint arXiv:2302.02599 (2023)
13. Raffel, C., et al.: Exploring the limits of transfer learning with a unified text-to-text transformer. J. Mach. Learn. Res. **21**(1) (2020)
14. vllm-project: vLLM: Easy, fast, and cheap LLM serving for everyone (2024). https://github.com/vllm-project/vllm

15. Walkowiak, B., Walkowiak, T.: Implementation of language models within an infrastructure designed for natural language processing. Int. J. Electron. Telecommun. **70**(1), 153–159 (2024). https://www.ijet.pl/index.php/ijet/article/download/10.24425-ijet.2024.149525/1194

16. Wang, G., Cheng, S., Zhan, X., Li, X., Song, S., Liu, Y.: OpenChat: advancing open-source language models with mixed-quality data. In: The Twelfth International Conference on Learning Representations (2024). https://openreview.net/forum?id=AOJyfhWYHf

17. Wolf, T., et al.: Transformers: state-of-the-art natural language processing. In: Proceedings of the 2020 Conference on Empirical Methods in Natural Language Processing: System Demonstrations, pp. 38–45. Association for Computational Linguistics (2020). https://doi.org/10.18653/v1/2020.emnlp-demos.6

18. Yu, G.I., Jeong, J.S., Kim, G.W., Kim, S., Chun, B.G.: Orca: a distributed serving system for transformer-based generative models. In: 16th USENIX Symposium on Operating Systems Design and Implementation (OSDI 2022), Carlsbad, CA, pp. 521–538. USENIX Association (2022). https://www.usenix.org/conference/osdi22/presentation/yu

19. Zhang, P., Zeng, G., Wang, T., Lu, W.: TinyLlama: an open-source small language model. arXiv preprint arXiv:2401.02385 (2024)

20. Zheng, L., et al.: Judging LLM-as-a-judge with MT-bench and chatbot arena. In: Thirty-Seventh Conference on Neural Information Processing Systems Datasets and Benchmarks Track (2023). https://openreview.net/forum?id=uccHPGDlao

Robustness of Named Entity Recognition Models

Paweł Walkowiak[(✉)] [ID]

Wroclaw University of Science and Technology, Wrocław, Poland
pawel.walkowiak@pwr.edu.pl

Abstract. Transformer based models are gaining popularity on most of the Natural Language Processing (NLP) including sequence labeling task such like Named Entity Recognition (NER). The previous (pre-transformers) state of the art approaches for NER task where models based on Conditional Random Fields (CRF). Adversarial attacks, which involve trying to deceive the model by creating adversarial examples with subtle, almost imperceptible changes, are increasingly posing a threat to all NLP models, including those which solve sequence labeling tasks. When deciding which architecture to use for NER model, robustness of approach could be one of the deciding factors. We are going to examine named entity recognition trained models resilience, focusing on differences between CRF and Transformer based approaches.

Keywords: Adversarial Attacks · NER · Transformer · CRF · Polish

1 Introduction

Adversarial attack is a technique that prepare new spoiled samples, using small hardly noticeable changes, which lead to model incorrect behaviour. They were described in context of usage against neural networks in [20]. The goal of adversarial attacks is to exploit vulnerabilities in machine learning (ML) models and cause them to make mistakes. Adversarial techniques can be used to manipulate the behavior of models in various applications, such as image classification, natural language processing, and autonomous driving systems. Usage of adversarial examples creation methods have raised concerns about the robustness and security of ML models and have led to research efforts to develop more robust and resilient solution. This methods can be separated as white and black box techniques depending on the level of access the attacker has to the targeted model. More detailed taxonomy of adversarial attacks was described in [21]. White-box attacks usually involves full access to the model including architecture, parameters and gradients, allowing to craft the most effective and targeted adversarial examples. Black-box on the other hand involve having limited or no access to the model being attacked. The attacker may access input-output behavior of the model, sometimes including model logits, this type of attacks makes it harder to craft an effective adversarial example but may be closer to real-life scenarios.

© The Author(s), under exclusive license to Springer Nature Switzerland AG 2024
W. Zamojski et al. (Eds.): DepCoS-RELCOMEX 2024, LNNS 1026, pp. 306–315, 2024.
https://doi.org/10.1007/978-3-031-61857-4_30

Named Entity Recognition (NER) is an one of natural language processing (NLP) sequence labeling tasks, given the text input NER model should mark boundaries of each proper name and assign them class labels. Recent survey about actual state of the art for this task was described in [8]. When comes to attacks on NER models classic methods used for attacking classifiers cannot be directly applied, because sequence labeling gives different output that typical classification and attack methods goal function cannot evaluate attack success or failure. Here comes multiple approaches including counting changes in proper names bounds or altering correctly separated entity classification label. Possibility of attacks against NER model comes with high risk, because this models are often part of larger NLP pipeline for example anonymisation. In such scenario, adversary can modify documents in such a way that confidential information will leak in the anonymisation process. This threat makes it necessary to rethink the selection of the right NER model for the given solution. In addition to considering the speed and accuracy of performance on untouched samples from the test collections, it is important to also consider the model's behavior and ability to withstand potential attacks.

Our contributions are to examine which NER model is more resilient against different types of adversarial attacks and whether robustness comes at the cost of reduced accuracy. Specifically we want check differences in behaviour between Conditional Random Fields [9] (CRF) and Transformer based NER models, focusing on models made for Polish namely CRF-based Liner2 [12] and Transformer-based WiNER[1].

2   Related Work

Adversarial samples generation algorithm for classifiers with usage of word-level replacement was described in TextFooler paper [6]. Given a corpus of n sentences $X = X_1, X_2, ...X_n$, a collection of k labels $Y = Y_1, Y_2, ..., Y_k$, and a pre-trained model F that maps X to Y, this method identifies the most significant words by generating a ranking based on the change in prediction before and after removing or masking the word. For word candidates sorted by the decreasing importance, replacement candidates are being generated based in highest cosine similarity between original word and synonym obtained from Glove [16] word embeddings. Final sentences with replaced word for its synonyms are filtered with minimal similarity threshold ϵ using cosine similarity on sentence embeddings from SentenceBERT [17]. The attack is considered successful if example X_{adv}, satisfies the following requirement $F(X_{adv}) \neq F(X)$. While this approach may not be directly applicable for fooling sequence labeling models, certain aspects of it, such as generating synonyms, could be incorporated into a Named Entity Recognition deception technique.

Attack against models based on BERT [4] was previously examined in Breaking BERT [5]. Authors aimed to investigate vulnerabilities in domain specific named entity attack and determine which of BERT variant is the most resilient

---

[1] https://wiki.clarin-pl.eu/pl/nlpservices/list/winer.

to each attack type, focusing on two main spoiling methods: replacing words in context of entities and substitution with other proper names of the same type. For semantic similarity preservation authors used Universal Sentence Encoder (USE) [2], with adequate minimal similarity threshold ϵ set to 0.8, candidate sentences with similarity score lower that ϵ were discarded from attack success verification. Regarding evaluation authors suggest two valuable metrics, percentage of entities mislabelled after attack and partially mislabelled entities, which comes in handy when comes to entity context attack. Those metrics are intuitive conversion of classic single label attack success estimation.

Methods of producing and evaluating adversarial samples for Named Entity Recognition models as well as adversarial training were also described in SeqAttack framework [18]. Framework is based on TextAttack [14] set of adversarial attack methods, mainly targeted on classification models. SeqAttack provides adversarial methods for many strategies including character-level changes, word-level and sentence-level. Authors in experiments focused on attacking BERT based models with untargeted attacks, in which word bugging methods proved to be most effective against models. However word-level replacing produced more fluent adversarial examples. As evaluation metrics, author consider percentage of samples needed to fool the model, percentage of modified tokens, grammar errors and textual similarity. Similarity between original and adversarial samples is obtained by cosine similarity on embeddings from USE.

3   Adversarial Examples

Methodology of adversarial examples creation for NER task was comprehensively described in Breaking BERT [5] paper, for which our experiments inspire, two adversarial attacks for sequence labeling namely Entity Context Attack and Entity Attack. However, the approach used in the previous work cannot be used directly due to examining Polish language models which causes need of proper word synonyms which we managed to provide by usage of pretrained uncontextual word embeddings model fastText [7], precisely its Polish vector model. What is more examined models were accessed through API, which does not provides model logits to address this issue new approach based on masking words measuring only change in proper names were proposed. Apart from using originally proposed Universal Sentence Encoder for candidate sentences similarity preservation we used Sentence-BERT [17] model, with adequate minimal similarity threshold ϵ.

3.1   Entity Context Attack

The method aims to attack neighbourhood of proper names, to cause partially or full change in model sequence labeling. When comes to identifying substitution candidates for the attacks, original methodology uses masking each non named entity word and comparing model logits with original decision. For preparing such samples for Liner2 and WiNER models, such a approach is not possible, because this model usage black-box characteristic. To address this issue, we manage to obtain context candidates by running models with non named entity word masked each time, and comparing model outputs if result entities changed comparing to non masked sample, masked word get score equal to the number of changed entities plus one, one otherwise, components of proper names have zero score. This way sorting by word score method could firstly change words highly important for model output.

3.2 Entity Attack

This method attacks each non zero label, by replacing it with other named entities that share same annotation class. Labels candidates for each class are obtained from other dataset samples. In theory model should treat replaced proper name the same as original one and mark it with accurate class. Example substitution for grand true proper name *John Doe* with class nam_liv_person in sentence: *My name is John Doe and I have two cats.* and substitution candidate with the same class *Marry Goldberg* would be *My name is Marry Goldberg and I have two cats.* This approach is suitable for changing more than one proper name in sample at once, but for sake of this experiments only replacing one proper name at each run is considered.

3.3 Evaluation Metrics

For the purposes of our experiments, we consider the following metrics for evaluating the effectiveness of attacks on models:

Omitted Proper Names. Measure is a ration of omitted proper names counts to all proper names in sample, calculated over each dataset sample, if for given sample i, there are more that one evaluation result (e.g was attacked with multiple adversarial examples), then mean of values is considered.

Partially Omitted Proper Names. Shows what part of grand true proper names were only partially detected, for example for *John Doe*, model marks only word *John* as a named entity, omitting *Doe*.

Changed Entity Type Percent. In case when grand true sequence label is different than predicted keeping a proper name boundaries unchanged, sample is being added to changed entity count. Final metric is mean of each sample changed entity type to all grand true proper names ratio. The three metrics shown above are considered for each data sample, to accumulate result for dataset, we use mean and standard deviation of values. A mean should gather the values to a single number, and a standard deviation is intended to represent the changes in values.

Accuracy and F1-Score. Standard accuracy (ACC) and f1-score (F1) metrics, with difference that are counted over proper names labels including non proper name label. Sequence of label in each sample is concatenated and passed though to metric calculation as flattened list.

4 Results

4.1 Datasets

KPWr [1] is a Polish corpora, with creative commons licensing. Consists of text, manually annotated with part of speech tags, predicate-argument relations, word senses and named entities, which makes it adequate for many sequence labeling task. Corpora includes different genres e.g. Blogs, Science and Law text to provide representatives.

Corpus of Economic News [11] (**CEN**) is a corpora of texts from Polish Wikipedia focused on economy, annotated with 65 categories of proper names. Contains economy specific proper names annotation, provides an alternative domain to KPWr corpora.

For attack evaluation test part of the datasets were used, due to lack of validation split in original Liner2 model training splits. KPWr validation split was part of training for this model, so evaluation on it is pointless (Table 1).

Table 1. Dataset summary, the size of each split and information about test and validation split with NER free samples filtered out.

Dataset	Train	Test	Test Filtered	Validation	Validation Filtered
KPWr	9,210	4,323	2,192	4,748	2,319
CEN	5,800	902	624	875	608

4.2 Models

MorphoDiTa is a part of speech (POS) tagger, originally created for Czech language in paper [19], it's Polish version was adapted for NKJP tagset [15] and deployed in CLARIN-PL infrastructure[2]. The tool operates on a morphological dictionary approach, where words are depicted as trees containing their prefixes, lemmas, and suffixes, forming templates for word representation. MorphoDiTa includes an extensive morphological dictionary which maps possible lemma-tag pairs used for morphological analysis. Using a heuristic algorithm, identifies templates linked to a word's prefix and ending, returning a set of lemma-tag pairs associated with the prefix's lemma and the ending's tag. Obtained lemma-tag pairs are further disambiguated by the part of speech tagger based on rich feature

---

[2] https://services.clarin-pl.eu/dashboard.

averaged perceptron. The tool in its Polish adapted version uses NKJP tagset which is specific for Polish language. Tagset consist of 36 grammatical classes with 13 grammatical categories of possible values. This large set of possible values helps properly describe complex Polish parts of speech. The tool is described as Liner2 needs part of speech tagging at its input so that can properly identify named entities, other NKJP POS taggers can be used for Liner2 but, we chose this as its already part of CLARIN-PL toolkit.

Liner2 [13] is an generic framework, that can various sequence labeling tasks. For purpose of this work, we will focus on its usage in named entity recognition, which was described in paper [12]. Liner processes texts, in prepared form with morphological tagging already done. Author claims that morphological analysis for sequence labeling model can improve the results, but for NER task have small impact. Model processing pipeline consists of three elements namely: statistical model trained on annotated corpus using Conditional Random Fields modeling; set o heuristic to merge, group and filter categories and sent of heuristic for named entity lemmatization. The CRF model includes features of input data which are based on orthography, patterns, morphology, lexicon and WordNet. In this paper we will focus on usage of Liner's fine-grained n82 model version which was trained on KPWr dataset.

WiNER is transformer based named entity recognition model trained on KPWr dataset starting from pretrained Polish RoBERTa [3]. The tool uses transformer encoder namely on RoBERTa [10] for proper names recognition. At its bases model doesn't need part of speech tagged text, but tool benefits from having text tokenization compliant with NKJP tagset and sentence segmentation. Both models gives sequence labeling result using set of 62 classes, compatible with KPWr categories.

Given the prior knowledge about tools processing pipeline for experiments looks consist of two steps: tagging text with **MorphoDiTa** and running NER model **WiNER** or **Liner2** depending on choice using tagger output. For given model, datasets and metrics evaluation of models were conducted, results can be seen in Table 2. Evaluation showed that WiNER model obtained higher accuracy and f1-score on both datasets outperforming Liner2 by 3 to 4% points. Detailed metrics such as Omitted entities percentage seems to confirms that observation as metric for Liner2 have greater by a factor of six to eleven than for WiNER. What is worth to mention is abnormally high partial omit and NTypes percentage for WiNER, this can indicate that model has problems with confusing types and parts of the proper names.

4.3   Entity Attack Results

Results of entity attack conducted on WiNER and Liner2 models on both datasets are presented in Table 3. Experiment setup included replacing one proper name at each adversarial sample and reducing candidates named entity replacement to five. When comes to accuracy and f1-score all model-dataset pairs dropped to level of about 71% to 76% from high above 85%, greatest decrease was obtained for WiNER on both datasets 18 pp. on KPWr and 17 pp. on CEN

Table 2. Results of Liner2 and WiNER models, for two datasets KPWr and CEN on its test split (unattacked). The data cell contains accuracy and f1-score for each model-dataset pair. As well as three proposed detailed metrics percentage of omitted, partially omitted proper names and changed entity type together with standard deviations. Arrows on the end of column headers indicate whether metric should be minimized ↓ or maximized ↑.

Model	Dataset	ACC [%] ↑	F1 [%] ↑	Omit [%] ↓	Partial Omit [%] ↓	NTypes [%] ↓
Liner2	KPWr	86	86	33.61 ± 4.37	6.21 ± 2.66	13.38 ± 2.05
WiNER	KPWr	88	89	5.14 ± 1.00	8.41 ± 3.63	15.53 ± 3.17
Liner2	CEN	88	88	23.76 ± 3.11	6.10 ± 1.11	12.12 ± 2.09
WiNER	CEN	**92**	**92**	**1.69** ± 0.57	**4.52** ± 1.25	**8.09** ± 1.35

Table 3. Entity attack results for Liner2 and WiNER models on KPWr and CEN datasets, with set of metrics defined for unattacked evaluation, accuracy and f1-Score cell contain information about percentage change of metrics relating to unattacked samples values.

Model	Dataset	ACC [%] ↑	F1 [%] ↑	Omit [%] ↓	Partial Omit [%] ↓	NTypes [%] ↓
Liner2	KPWr	74 [−8]	76 [−10]	22.23 ± 22.61	**16.98** ± 17.54	**4.17** ± 4.29
WiNER	KPWr	72 [−16]	71 [−18]	8.67 ± 9.29	22.17 ± 24.05	6.45 ± 7.60
Liner2	CEN	**75** [−13]	**76** [−12]	23.85 ± 19.18	19.49 ± 15.73	5.59 ± 4.54
WiNER	CEN	**75** [−17]	75 [−17]	**6.47** ± 8.01	20.25 ± 24.84	3.45 ± 4.45

f1-score, that may be because of high results on unattacked samples for this model. Result of detailed metrics prove WiNER vulnerability to entity attack as Omit and Partial Omits percentage highly increased. The percentage of omitted proper names for Liner2 only saw a slight change, with a significant shift towards partial omission. However, it is important to note the high standard deviation, which reached values similar to the average. Decrease in changed entity type for both models may be due to most previous mistakes moved were partially or fully missed during inference of adversarial samples.

4.4 Context Attack Results

Context Attack results are presented in Table 4. Obtaining adversarial examples was based on swapping one of proper name neighbors words each time, producing five samples for each named entity, context word selection was based on previously conducted words model importance using multiple model runs with candidates deleted. Basic metrics such as accuracy and f1-score decreased as expected, highest change can be seen on WiNER obtained on KPWr dataset were both accuracy and f1-score dropped by four percentage points. From detailed metrics can be seen that model missed part of proper names or changes its type more often than, observed on unattacked samples results for model. Similar to attacks on entities, metrics evaluating omission and change of types have a large stan-

Table 4. Context attack results for Liner2 and WiNER models on KPWr and CEN datasets, with set of metrics defined for unattacked evaluation. Accuracy and f1-score cell contain information about percentage change of metrics relating to unattacked samples values.

Model	Dataset	ACC [%] ↑	F1 [%] ↑	Omit [%] ↓	Partial Omit [%] ↓	NTypes [%] ↓
Liner2	KPWr	85 [−1]	84 [−2]	28.52 ± 22.92	14.47 ± 12.07	14.97 ± 12.06
WiNER	KPWr	84 [−4]	85 [−4]	4.06 ± 5.50	13.50 ± 18.21	16.59 ± 21.90
Liner2	CEN	87 [−1]	86 [−3]	26.86 ± 19.98	12.81 ± 9.70	19.69 ± 15.08
WiNER	CEN	**90** [−2]	**91** [−1]	**2.21** ± 3.70	**9.87** ± 15.55	**13.64** ± 21.55

dard deviation, especially noticeable for the WiNER model where it exceeds the value of the mean. This can indicate that transformer based model is particularly sensitive on context changes. High standard deviation for WiNER shows that some context attack adversarial samples significantly distort the decisions of the model.

5   Discussion

5.1   Conclusion

In this article we focused on two model and their architectures comparison: Liner2 based on CRF and WiNER that uses RoBERTa. Evaluation using adversarial example produced with entity attack method shown that WiNER model is more vulnerable to such type of attack, than CRF based Liner2, which too recorded a decrease in evaluation metrics, but comparing to unattacked dataset results handle attacks better. During context attack models were affected to a lesser extent, again WiNER model showed highest decrease in accuracy, mainly on KWPr dataset, high standard deviation of detailed metrics for this model suggest significant disort of decisions on some of adversarial examples. What is more altering proper name context exhibits a shift of mistakes towards partially omits and changes of entity types. Summing up, the transformer-based WiNER, despite its performance on unattacked data far superior to Liner2, under attacks obtained results equal to or worse than Liner2, recording a high decrease in accuracy and f1-score. In light of the presented experiments, it can be concluded that for attacking with the described adversarial methods, the CRF-based model is more resilient.

5.2   Future Works

The attacks that are currently being proposed can be viewed as adversarial techniques primarily aimed at BERT-based models. These attacks involve altering either entity context or entity itself, leading to a discrepancy between the proper name and the context in a way that conflicts with the conclusions drawn of BERT-based models. In the future work we would like to address this issue and construct new type of attacks that can advantage from Conditional Random Fields characteristics.

Acknowledgements. The work was financed as part of the investment: "CLARIN ERIC - European Research Infrastructure Consortium: Common Language Resources and Technology Infrastructure" (period: 2024-2026) funded by the Polish Ministry of Science and Higher Education (Programme: "Support for the participation of Polish scientific teams in international research infrastructure projects"), agreement number 2024/WK/01.

References

1. Broda, B., Marcińczuk, M., Maziarz, M., Radziszewski, A., Wardyński, A.: KPWr: towards a free corpus of Polish. In: Calzolari, N., et al. (eds.) Proceedings of the Eighth International Conference on Language Resources and Evaluation (LREC 2012), Istanbul, Turkey, pp. 3218–3222. European Language Resources Association (ELRA) (2012). http://www.lrec-conf.org/proceedings/lrec2012/pdf/965_Paper.pdf
2. Cer, D., et al.: Universal sentence encoder for english. In: Blanco, E., Lu, W. (eds.) Proceedings of the 2018 Conference on Empirical Methods in Natural Language Processing: System Demonstrations, Brussels, Belgium, pp. 169–174. Association for Computational Linguistics (2018). https://doi.org/10.18653/v1/D18-2029
3. Dadas, S., Perełkiewicz, M., Poświata, R.: Pre-training polish transformer-based language models at scale. In: Rutkowski, L., Scherer, R., Korytkowski, M., Pedrycz, W., Tadeusiewicz, R., Zurada, J.M. (eds.) ICAISC 2020. LNCS (LNAI), vol. 12416, pp. 301–314. Springer, Cham (2020). https://doi.org/10.1007/978-3-030-61534-5_27
4. Devlin, J., Chang, M.W., Lee, K., Toutanova, K.: BERT: pre-training of deep bidirectional transformers for language understanding. In: Burstein, J., Doran, C., Solorio, T. (eds.) Proceedings of the 2019 Conference of the North American Chapter of the Association for Computational Linguistics: Human Language Technologies, Volume 1 (Long and Short Papers), Minneapolis, Minnesota, pp. 4171–4186. Association for Computational Linguistics (2019). https://doi.org/10.18653/v1/N19-1423. https://aclanthology.org/N19-1423
5. Dirkson, A., Verberne, S., Kraaij, W.: Breaking BERT: Understanding its Vulnerabilities for Named Entity Recognition through Adversarial Attack. CoRR abs/2109.11308 (2021). https://arxiv.org/abs/2109.11308
6. Jin, D., Jin, Z., Zhou, J.T., Szolovits, P.: Is BERT Really Robust? Natural Language Attack on Text Classification and Entailment. CoRR abs/1907.11932 (2020). http://arxiv.org/abs/1907.11932
7. Joulin, A., Grave, E., Bojanowski, P., Mikolov, T.: Bag of tricks for efficient text classification. In: Proceedings of the 15th Conference of the European Chapter of the Association for Computational Linguistics: Volume 2, Short Papers, pp. 427–431. Association for Computational Linguistics (2017)
8. Keraghel, I., Morbieu, S., Nadif, M.: A survey on recent advances in named entity recognition. CoRR abs/2401.10825 (2024). https://doi.org/10.48550/ARXIV.2401.10825
9. Lafferty, J., Mccallum, A., Pereira, F.: Conditional Random Fields: Probabilistic Models for Segmenting and Labeling Sequence Data, pp. 282–289 (2001)

10. Liu, Y., et al.: RoBERTa: A Robustly Optimized BERT Pretraining Approach. CoRR abs/1907.11692 (2019). http://arxiv.org/abs/1907.11692
11. Marcińczuk, M.: CEN (2007). http://hdl.handle.net/11321/6. CLARIN-PL digital repository
12. Marcińczuk, M., Kocoń, J., Oleksy, M.: Liner2 — a generic framework for named entity recognition. In: Proceedings of the 6th Workshop on Balto-Slavic Natural Language Processing, pp. 86–91. Association for Computational Linguistics (2017). https://doi.org/10.18653/v1/W17-1413
13. Marcińczuk, M., Kocoń, J., Janicki, M.: Liner2 - a customizable framework for proper names recognition for polish. In: Bembenik, R., Skonieczny, L., Rybinski, H., Kryszkiewicz, M., Niezgodka, M. (eds.) Intelligent Tools for Building a Scientific Information Platform. Studies in Computational Intelligence, vol. 467, pp. 231–253. Springer, Heidelberg (2013). https://doi.org/10.1007/978-3-642-35647-6_17
14. Morris, J., Lifland, E., Yoo, J.Y., Grigsby, J., Jin, D., Qi, Y.: TextAttack: a framework for adversarial attacks, data augmentation, and adversarial training in NLP. In: Proceedings of the 2020 Conference on Empirical Methods in Natural Language Processing: System Demonstrations, pp. 119–126 (2020)
15. Patejuk, A., Przepiórkowski, A.: ISOcat Definition of the National Corpus of Polish Tagset (2010)
16. Pennington, J., Socher, R., Manning, C.: GloVe: global vectors for word representation. In: Proceedings of the 2014 Conference on Empirical Methods in Natural Language Processing (EMNLP), Doha, Qatar, pp. 1532–1543. Association for Computational Linguistics (2014). https://doi.org/10.3115/v1/D14-1162
17. Reimers, N., Gurevych, I.: Sentence-BERT: sentence embeddings using siamese BERT-networks. In: Inui, K., Jiang, J., Ng, V., Wan, X. (eds.) Proceedings of the 2019 Conference on Empirical Methods in Natural Language Processing and the 9th International Joint Conference on Natural Language Processing (EMNLP-IJCNLP), Hong Kong, China, pp. 3982–3992. Association for Computational Linguistics (2019). https://doi.org/10.18653/v1/D19-1410
18. Simoncini, W., Spanakis, G.: SeqAttack: on adversarial attacks for named entity recognition. In: Adel, H., Shi, S. (eds.) Proceedings of the 2021 Conference on Empirical Methods in Natural Language Processing: System Demonstrations, Online and Punta Cana, Dominican Republic, pp. 308–318. Association for Computational Linguistics (2021). https://doi.org/10.18653/v1/2021.emnlp-demo.35
19. Straková, J., Straka, M., Hajič, J.: Open-source tools for morphology, lemmatization, POS tagging and named entity recognition. In: Proceedings of 52nd Annual Meeting of the Association for Computational Linguistics: System Demonstrations, Baltimore, Maryland, pp. 13–18. Association for Computational Linguistics (2014). http://www.aclweb.org/anthology/P/P14/P14-5003.pdf
20. Szegedy, C., et al.: Intriguing properties of neural networks. In: Bengio, Y., LeCun, Y. (eds.) 2nd International Conference on Learning Representations, ICLR 2014, Banff, AB, Canada, 14–16 April 2014, Conference Track Proceedings (2014). http://arxiv.org/abs/1312.6199
21. Wang, Y., Sun, T., Li, S., Yuan, X., Ni, W., Hossain, E., Poor, H.V.: Adversarial Attacks and Defenses in Machine Learning-Powered Networks: A Contemporary Survey. CoRR abs/2303.06302 (2023). https://doi.org/10.48550/ARXIV.2303.06302

The Impact of Modes of Locomotion in a Virtual Reality Environment on the Human Being

Marek Woda[(✉)] [iD] and Jakub Michalski

Department of Computer Engineering, Wroclaw University of Technology, Janiszewskiego 11-17, 50-372 Wrocław, Poland
marek.woda@pwr.edu.pl, jakmic100000@gmail.com

Abstract. Virtual reality technology has evolved to the point where users are no longer confined to a single location but can explore virtual worlds created by developers. However, the use of virtual reality often leads to the experience of virtual reality sickness, which poses a limitation to its widespread adoption. A recent study aimed to identify the factors that contribute to virtual reality sickness by examining the different locomotion techniques used in virtual reality and their impact on users. The study employed a simulator sickness questionnaire that was modified to obtain the necessary data. By analyzing these responses, researchers hope to gain insights into the elements of locomotion that are responsible for virtual reality sickness and their effects on human users. Understanding these factors can help improve the design and implementation of virtual reality experiences, making them more enjoyable and reducing the occurrence of virtual reality sickness. The aim of this study is to investigate the impact of locomotion and its elements on humans in virtual reality. This will allow us to determine which elements of locomotion and which types have negative effects. The thesis will present the results of a study to determine which elements of locomotion are most responsible for the occurrence of virtual reality sickness. Once this has been established, it will be possible to determine which elements should be avoided and which can be used to reduce the incidence of virtual reality sickness.

Keywords: virtual reality · simulator sickness questionnaire · locomotion · virtual reality sickness

1 Introduction

With the increasing popularity of VR technology and greater access to devices that simulate virtual environments, the problems associated with using this technology are also increasing. One of the core elements of virtual reality is the simulation of displacement. When moving in a virtual world, there is a conflict between what you see and what you feel. Just like driving a car or sailing a boat, virtual reality can cause a type of motion sickness called virtual reality sickness [8, 10]. There are many factors that can influence the onset of this condition. The experience of movement (locomotion) is not the same for everyone [5, 6]. The way a person is moved can affect their feelings. One tool that can be used to study such a multi-symptom illness is the SSQ questionnaire [4].

W. Zamojski et al. (Eds.): DepCoS-RELCOMEX 2024, LNNS 1026, pp. 316–324, 2024.
https://doi.org/10.1007/978-3-031-61857-4_31

From its inception, virtual reality has been subject to safety and health scrutiny [9], which has slowed the uptake of the technology. Most devices and systems carry warnings about the possibility of epileptic seizures, convulsions, effects on medical equipment and, in children, developmental problems. The use of virtual reality by children is not recommended [2] because it can affect the development of spatial awareness, which can lead to tripping and falling. The most common problem is virtual reality sickness. It is a derivative of motion sickness and has very similar symptoms. The difference between the two is that virtual reality sickness does not require any physical movement, only the observation of movement in the virtual world. The consequences of this, beyond being an illness, are quite undesirable. It can affect the delivery of VR-related training and coaching, while discouraging users from using the technology. Research shows that virtual reality sickness may be one of the main barriers to the development of this technology [1]. The cause of virtual reality sickness is complex.

The main element responsible for this is sensory conflict. This occurs when a person's senses receive conflicting information. In the case of virtual reality, this occurs when the user observes movement in the virtual world and their body does not feel any movement. Then the eyes and inner ear receive stimuli that are sent to the brain. This is the main cause of motion sickness and simulator sickness. In virtual reality, the eyes see that the person is walking and transmit this sensation, but the inner ear, which is responsible for balance, does not feel any movement. At this point, the brain receives conflicting information, which causes a type of motion sickness called virtual reality sickness. Another reason is the low frame rate. When the refresh rate is lower than the brain can work, there is a mismatch between the image and the perceived environment. The result is visible image distortion, which causes virtual reality sickness. The same can happen if animations and position tracking are poorly done.

Failure to correctly track the position of the head or hands can lead to disconnection from the perceived environment and cause virtual reality sickness. In our case, we will be relying on a sensory conflict caused by the different ways of moving in virtual reality. The symptoms of this illness are very complex and can be triggered by many factors. They are not all related to virtual reality and include, for example, age, gender, or background [12]. To diagnose such a condition, it is necessary to develop a suitable system that can collect data about the user and based on this data, determine whether the person is suffering from virtual reality sickness. For this purpose, a questionnaire [4] can be used to investigate the illness of many symptoms that are strongly associated with virtual reality illness. The results of this questionnaire will help to develop a system that will be able to collect data about the user and, based on this data, determine whether a person is experiencing virtual reality sickness, depending on the variable element present, which will be locomotion.

Simulator disease questionnaire is currently one of the most widely used tools for simulator disease. The simulator sickness questionnaire was created by Robert S. Kennedy in 1993 [3]. It was originally developed for the study of military aircraft simulators, but over time it has been adapted for other studies in other fields not only related to the military. It allows to determine which elements of the study influence the components of simulator sickness. Looking at the studies in which it was used for virtual reality, it can be seen that the results for virtual reality are characterised by a high score in the category of disorientation and simulator sickness in oculomotor disorders [11]. From this alone, we can conclude that using this questionnaire in the same way for virtual reality research is not valid and some modifications should have been made to make it suitable for virtual reality research.

The locomotion survey form consists of 18 questions, which the user answers on a 4-point Likert scale [13] from 0 (no effect) to 3 (very intense feeling). It was constructed based on a solution that was developed based on the original [3] and on a modified version proposed for virtual reality testing [4]. The questions address the different symptoms that a user may experience while using virtual reality. They are grouped into relevant categories: (1) *Nausea* (2) *Oculomotor disturbances* (3) *Disorientation*. Scores can be obtained for each category separately and collectively for the entire form.

The values are calculated from the user's responses to the questions. The score is obtained by means of the sum of the scores multiplied by a fixed factor.

Formulae used to calculate scores from questions:

$$N = \sum N_i * 9{,}54 \tag{1}$$

$$O = \sum O_i * 7{,}58 \tag{2}$$

$$D = \sum D_i * 13{,}92 \tag{3}$$

$$T_S = \left(\sum N_i + \sum O_i + \sum D_i \right) * 3{,}74 \tag{4}$$

where:

- N - nausea category score
- O - oculomotor disturbance category score
- D - disorientation category score
- T_S - total questionnaire score

The scores from each category allowed to determine which elements of the simulation are most responsible for the symptoms of virtual reality sickness. Based on the total score, we can assess the quality of the simulation. These values are shown in Table 1.

Table 1. Simulator disease questionnaire results and their meaning

Score	Meaning
0–5	Simulation is very good, causes no symptoms of disease
5–10	Simulation is good, causes minor disease symptoms
10–15	Simulation is average, causes significant disease symptoms
15–20	Simulation is bad, causes very severe disease symptoms

In order to obtain results that are as realistic as possible, the questionnaire had to be modified for use in virtual reality. Modifications were made in accordance with [4]. The modifications made were aimed at making the results more consistent and uniform for people who had no experience with virtual reality.

- *Pre-examination condition survey* - before the examination, the person being examined should complete a questionnaire that will measure their baseline condition. In this way we will be able to determine if the person's condition has deteriorated during the study and by how much.
- *Full reporting of results* - in order to provide a more accurate analysis and better reference to changes during the study, results will be reported in full from all questions and categories.
- *Widening the categorisation of results* - scoring the simulation as very bad at 20 points is too harsh. Especially for people who have no experience with virtual reality.
- *Main assessment* - the quality assessment will mainly be based on calculating the difference between the initial and final state. In this way, we will be able to determine whether the person's condition deteriorated during the study and by how much, and which type of locomotion affected the user the most.

2 Research Methodology

The research was based on introducing subjects [7] to a pre-prepared virtual reality world (in the form of a proprietary application developed in the UNITY environment) and guiding them through a defined movement path. Before the study itself, the age and previous experience of the subject in the virtual reality environment were collected.

During the study, each participant navigated through the VR environment with an overlaid Oculus 2 helmet from META, and data collected related to the user's wellbeing, based on the created virtual reality illness form. The study consisted of five parts - an introductory section (before immersion in VR) and four sections on modes of locomotion. Timing of the passage through the 4 types of locomotion was also done. After each survey, the VR illness form was completed (results in Table 2).

The proprietary VR system supported the following activities:

1. *Learning the principles of navigating the VR world* - the tutorial was intended to familiarise participants with the basic controls and principles of navigating in virtual reality, as well as to introduce people's VR experience to the research environment and alleviate cognitive shock The main place for the user to navigate was a simple tunnel, along with additional elements with which the user could interact. Familiarisation with the mechanics of movement were presented in the form of tasks that the user had to complete in order to move. The task was to use the current mode of locomotion to reach a designated area.
2. *Moving in a continuous motion using a controller* – movement and rotation of the user carried out by means of the appropriate predefined buttons on the controllers
3. *Teleportation using controllers* – movement occurred when the user indicated a destination point on the screen that did not conflict with himself.
4. *Moving along a defined path* – movement along a forced path without the possibility of free movement (only with ability to change movement speed)
5. *Movement directed by hands equipped with controllers* - based on the user's movement in real space. The user can use both hands for locomotion, which gives great freedom

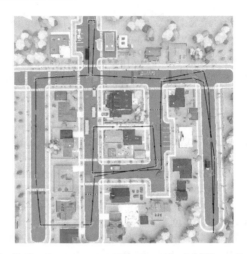

Fig. 1. Defined movement path in the authorial VR application.

of movement. Rotation is based on rotation in real space. To ensure that movement is not active all the time and does not interfere with interaction with objects, the ability to move is activated using a button on the controller.

Twenty-five people, mostly young people, took part in the research. Most of those surveyed had no experience with VR, most were exposed to this type of research environment for the first time (Table 2).

Table 2. VR experience of those taking part in the survey

Experience	Count
I have used virtual reality in the past	7
I use more than once a week	2
I use a couple of times a month	1
I have used it occasionally during the year	3
Never used virtual reality	12

The research took place in a series of meetings with groups of 4 to 8 people. After introducing the survey and setting up the environment, people took it in turns to complete the entire survey scenario (4 types of locomotion). Between changes in the types of locomotion, virtual reality disease surveys were completed. The survey lasted on average about 40 min, with each type of locomotion taking about 10 min (Table 3).

Table 3. Average duration of the study by type of locomotion for the study group

Experience	Avg. Duration [min]
Knob-controlled continuous movement	12,1
Teleportation	8,75
Path-controlled movement	9,15
Hand-controlled continuous movement	8
Average	38

The survey was not completed by 2 people. (One was interrupted by a person with no experience with VR and a person with little experience). The first person started to feel nauseous and strongly dizzy already in the first minutes of the test (the first type of locomotion tested). It may be interesting to note that during the introduction she had no such sensations and there was no indication of this. The second person started to experience nausea and dizziness after moving to the second type of tested locomotion. The symptoms gradually started to manifest themselves in her during the examination on the first stage.

3 Research Findings

Following the surveys in the research environment, indicators were calculated from the survey data using the formulas (1–4) presented earlier. The results were averaged and summarised in Table 4.

Table 4. Aggregate results for the whole test group

Type of locomotion	Total (Avg/STDev)		Nausea (Avg/STDev)		Disorientation (Avg/STDev)		Oculomotor disturbances (Avg/sTDev)	
Knob-controlled continuous movement	15,26	21,93	11,83	18,94	33,41	52,53	13,04	18,51
Teleportation	13,02	22,74	9,16	22,33	29,51	54,31	10,61	17,41
Path-controlled movement	20,79	23,48	20,22	25,50	40,09	51,83	18,80	18,20
Hand-controlled continuous movement	15,11	25,31	11,45	26,71	28,95	53,97	12,73	19,14
Average	16,04		13,17		32,99		13,80	

Analysing the results, it is clear that the movement along the path caused the most problems for the participants. Such poor results may have been influenced by the very characteristics of this movement. The lack of constant control over the direction of movement (unlike the other cases) and the change of control (its different effect on movement) caused increased symptoms of virtual reality sickness.

The best (read least) result was achieved by teleportation. It can be seen that the lack of continuous movement had a positive impact on the user experience. The simple and intuitive controls, as well as the effortless movement, allowed fast movement through the virtual environment without side effects.

Continuous movement and movement with controllers achieved very similar results to each other. Both were based on continuous movement but had completely different forms of control. The first was based on control using the knobs on the controller, and the second was based on moving the hands holding the controllers to change position. The control taking input from the knobs on the controller received slightly worse results.

This may be because users felt they had more control over their bodies when they actually moved and this movement translated directly into what they saw in VR. There was then no sensory conflict associated with seeing movement versus not feeling it.

Analysing the aggregate results, it can be concluded that the type of displacement has a strong influence on the feelings. Leap motion (teleportation), performed significantly better than the other 3 locomotives based on continuous motion. When comparing the other 3 locomotions, continuous motion based on body-related control scores better. Continuous motion based on knob control on the controller had similar results, but slightly worse. The worst results were for a movement that did not require the user to control it and moved it automatically. Depriving the user of free control over movement and displacement through the virtual environment resulted in the greatest sensory conflict with apparent displacement and lack of intention to move.

It is noticeable that the scores for the category 'disorientation' are much higher than for the other categories. This confirms that virtual reality illness is characterised by different symptoms than simulator illness and that the two cannot be equated.

Comparing the results and the standard deviation, it can be seen that the results are very scattered. A large proportion of people reported either very minimal changes or virtually no change in sensations. In addition, there are results from people who experienced very strong symptoms and their results pulled the average upwards strongly. This is particularly noticeable in the case of movement 3, where the standard deviation is the largest.

4 Summary

The main aim of the study was to investigate the effects of four types of virtual reality locomotion on humans. For the study, a virtual reality world was created to show the way the subject moved, and a simulator disease questionnaire was created to collect virtual reality disease symptoms. The form was modified for use with virtual reality. In addition, information was collected about the subjects that could affect the results of the study, such as age and experience with virtual reality. To create the virtual reality world, an application was created in Unity, which allowed for easy implementation of different types of locomotion.

The research was conducted on a study group of 25 people with varying experience with virtual reality. The group was dominated by young people. Before taking part in the study, each participant completed a simulator sickness questionnaire to determine the person's condition prior to the study. Participants then completed the questionnaire again after each stage of the study. The results of the study showed that people respond best to the movement of the virtual world, which does not require constant movement. Locomotion based on teleportation received the best results. Continuous locomotion controlled by controller knobs and locomotion using controller movement achieved very similar results. Both were based on continuous movement. The first locomotion was based on controller movements and the second on hand movements. Both of these movements were found to be worse than teleportation, but both had scores very close to each other and lower than the average. The worst locomotion was found to be movement along a path. A summary of the total results from the study for these locomotives can be seen in Table 5. From these results we can see some correlations. The first is that stepping/intermittent locomotion causes less sensory conflict than continuous locomotion. The second, is the effect of locomotion on control, results in continuous locomotion being better tolerated by humans than locomotion controlled by controllers.

Table 5. Total test result for each type of locomotion

Type of locomotion	Average	STDev
Teleportation	3,02	6, 08
Hand-controlled continuous movement	15, 11	6, 77
Knob-controlled continuous movement	15, 26	5, 86
Path-controlled movement	20, 79	6, 28

Looking at the standard deviation results for the group, it can be seen that there was a wide variation in results. There were people in the group who had no symptoms of the disease and there were also people who had increased symptoms of the disease. Another relationship evident in the results is that by limiting the user's ability to decide the direction of movement, negative feelings increase. Also worth mentioning is that, as expected, the results were greatest in the category of disorientation.

The study also took into account what kind of experience the person studied had. As expected, people with more experience with virtual reality had fewer symptoms of illness and moved more smoothly in the virtual world. Those with no experience moved chaotically, skipped some tasks and had more virtual reality sickness symptoms than the group with high experience. The group with little experience appeared to have the highest symptoms of virtual reality sickness however, compared to the group with no experience, moving around the virtual world was not chaotic and all tasks were completed. From this, it can be seen that experience with virtual reality affects the user's wellbeing - the more often they use it, the better they feel. Additionally, looking at the standard deviation, in the group with a lot of experience the deviation was much smaller than in the group with little and no experience. In the future, it would be useful to test certain patterns of human behaviour and reactions to interactions with the human-computer interface. By extending the research group, it would be necessary to examine the impact of the locomotor elements involved and to determine the direction in which this field should develop in order to make the barrier of virtual reality illness as small as possible.

References

1. Brooks, J.O., et al.: Simulator sickness during driving simulation studies. Accid. Anal. Prev. **42**(3), 788–796 (2010)
2. Araiza-Alba, P., Keane, T., Kaufman, J.: Are we ready for virtual reality in K–12 classrooms? Technol. Pedag. Educ. **31**(4), 471–491 (2022)
3. Biernacki, M.P., Kennedy, R.S., Dziuda, L.: Simulator sickness and its measurement with Simulator Sickness Questionnaire (SSQ). Med. Pr. **67**(4), 545–556 (2016)
4. Bimberg, P., Weissker, T., Kulik, A.: On the usage of the simulator sickness questionnaire for virtual reality research. In: 2020 IEEE Conference on Virtual Reality and 3D User Interfaces Abstracts and Workshops (VRW), pp. 464–467. IEEE, March 2020
5. Bliss, J.P., Tidwell, P.D., Guest, M.A.: The effectiveness of virtual reality for administering spatial navigation training to firefighters. Presence: Teleoper. Virtual Environ. **6**(1), 73–86 (1997)
6. Boletsis, C., Chasanidou, D.: A typology of virtual reality locomotion techniques. Multimodal Technol. Interact. **6**(9), 72 (2022)
7. Bolte, B., Bruder, G., Steinicke, F., Hinrichs, K., Lappe, M.: Augmentation techniques for efficient exploration in head-mounted display environments. In: Proceedings of the 17th ACM Symposium on Virtual Reality Software and Technology, pp. 11–18, November 2010
8. Bonato, F., Bubka, A., Palmisano, S., Phillip, D., Moreno, G.: Vection change exacerbates simulator sickness in virtual environments. Presence: Teleoper. Virtual Environ. **17**(3), 283–292 (2008)
9. Lawson, B.D.: Motion Sickness Symptomatology and Origins (2014)
10. Lo, W.T., So, R.H.: Cybersickness in the presence of scene rotational movements along different axes. Appl. Ergon. **32**(1), 1–14 (2001)
11. Park, G.D., Allen, R.W., Fiorentino, D., Rosenthal, T.J., Cook, M.L.: Simulator sickness scores according to symptom susceptibility, age, and gender for an older driver assessment study. In: Proceedings of the Human Factors and Ergonomics Society Annual Meeting, vol. 50, no. 26, pp. 2702–2706. SAGE Publications, Los Angeles, October 2006
12. Stanney, K.M., Kennedy, R.S., Drexler, J.M.: Cybersickness is not simulator sickness. In: Proceedings of the Human Factors and Ergonomics Society Annual Meeting, vol. 41, no. 2, pp. 1138–1142. SAGE Publications, Los Angeles, October 1997
13. Batterton, K.A., Hale, K.N.: The Likert scale what it is and how to use it. Phalanx **50**(2), 32–39 (2017)

Data Augmentation Techniques to Detect Cervical Cancer Using Deep Learning: A Systematic Review

Betelhem Zewdu Wubineh$^{(\boxtimes)}$ ⬤, Andrzej Rusiecki ⬤, and Krzysztof Halawa ⬤

Faculty of Information and Communication Technology, Wroclaw University of Science and Technology, Wroclaw, Poland

{betelhem.wubineh,andrzej.rusiecki,krzysztof.halawa}@pwr.edu.pl

Abstract. Computer-assisted systems have been widely used as tools to support medical experts in various fields, including the analysis of cervical cytology. However, due to patient privacy and ethical considerations, processing the model is challenging due to insufficient data in medical imaging. Data augmentation has gained popularity as a solution to this problem, especially in sectors where large datasets are unavailable, thereby increasing the size of a training dataset. This study aimed to identify a data augmentation technique to detect cervical cancer. To conduct this analysis, we systematically reviewed secondary studies published between 2017 and 2023 using the search term 'data augmentation', 'cervical cancer' and 'deep learning' from databases including Scopus, Web of Science, PubMed, IEEEXplore, Science Direct, and conducted a manual search on Google Scholar. The results showed that data augmentation techniques are categorized as basic methods and artificial image generation. Among basic data augmentation techniques, rotation and flipping are the most widely used. In the generation of artificial images, DCGAN is used to create high-quality synthetic images. Basic augmentation is used for both segmentation and classification tasks, while artificially generated techniques are used exclusively for classification tasks. Consequently, all these techniques enhance the performance and generalizability of deep learning models by increasing the size of the dataset.

Keywords: Cervical Cancer · Data Augmentation · Deep learning

1 Introduction

With the advancement of artificial intelligence, computer-assisted systems have become essential tools for medical experts in various domains, including cervical cytology analysis [1]. Deep learning (DL) techniques are applied in cytology analysis (Pap smear) to improve the detection of cervical cancer. Among the DL methods, convolutional neural networks (CNN) are one of the most widely used techniques for extracting information from images [2]. CNN is commonly used in classification, which involves distinguishing between benign and malignant tumors; segmentation aims to identify the region of interest for tumor detection and reveal the morphological structure of cells from the pap smear images; and object detection.

© The Author(s), under exclusive license to Springer Nature Switzerland AG 2024
W. Zamojski et al. (Eds.): DepCoS-RELCOMEX 2024, LNNS 1026, pp. 325–336, 2024.
https://doi.org/10.1007/978-3-031-61857-4_32

DL techniques require a substantial amount of data for effective model training [3, 4]. However, the scarcity of sufficient data in medical imaging poses a challenge in processing the model due to concerns about patient privacy [5]. Furthermore, collecting a significant amount of spectral data in clinical practice is expensive, particularly when acquiring a substantial number of cancer cases [6]. It also requires expertise, especially in the field of medical imaging, where labels are the result of a laborious study by one or more human experts [7]. When DL based applications are employed in real-world scenarios, a common issue arises when certain classes contain significantly more examples in the training set than others, known as class imbalance. This imbalance can pose challenges during model training, affecting model performance and introducing bias. Another issue is overfitting, where the model performs well on the training data but poorly on new, unseen data, resulting in poor generalizability. To address these challenges and enhance the learning capacity of DL models, data augmentation serves as a solution by increasing the size of the data set. It has been applied to artificially increase the number of sample images used to train models by augmenting the original dataset [8]. It plays a crucial role in improving model performance, reducing overfitting in limited training datasets, and improving model generalizability [9].

Several research studies have been conducted in this area. In [10], a review of data augmentation techniques for brain tumor segmentation is presented. This review paper concentrates on enhancing and addressing segmentation techniques specifically in the context of brain tumors. Additionally, [11], focuses on augmentation techniques for medical imaging classification tasks. In [12], a review of data pre-processing and data augmentation techniques in also conducted in general. The paper [13], which conducted a review of cervical cancer diagnosis, includes detection and classification techniques. Furthermore, there is a review on generative adversarial networks (GANs) for the segmentation and classification of medical images [14]. All review papers focus on a different perspective; for instance, some focus on segmentation techniques, others on classification techniques, and only on GAN-based augmentation in medical imaging.

Despite the possibility of conducting a review on augmentation techniques in segmentation and classification tasks for detecting cervical cancer, this gap in the literature has not yet been addressed. To fill this void, we propose this review and formulate the following research questions.

RQ1. What types of data augmentation techniques are used to increase the training dataset for detecting and classifying cervical cancer cell types?

RQ2. Which data augmentation techniques are most widely used to increase the size of training set in segmentation and classification tasks in the previous articles?

The purpose of this study is to identify data augmentation techniques for use in the segmentation and classification of Pap smear images to detect cervical cancer.

2   Methods

In this section, the methodology used in the systematic review is described. The analysis is performed by adopting the PRISMA protocol which is used to guide reporting items for systematic reviews and meta-analyses guidelines [15] and is preferred in most review articles [16].

2.1 Search Terms and Information Source

For this review, we used the keywords 'data augmentation', 'cervical cancer', and 'deep learning' to search for research published between January 2017 and December 2023, resulting in a total of 417 articles. These articles were sourced from Scopus, Web of Science, PubMed, IEEEXplore, Science Direct, and manual searches on Google Scholar. The distribution of articles downloaded by database is illustrated in Fig. 1(a).

2.2 Eligible Criteria

The inclusion criteria of the article are: 1) articles published in English, 2) publication year between 2017 and 2023, 3) article type is a journal or proceedings, and 4) articles related to data augmentation techniques in segmentation and classification tasks for diagnosing cervical cancer. On the other hand, the exclusion criteria are 1) papers lacking information on data augmentation techniques and cervical cancer, 2) papers published in a language other than English, 3) articles published before or after the specified date, and 4) papers categorized as review, case study, survey, systematic review, and report types. The selected articles are presented in Fig. 1(b).

2.3 Data Management

We extracted the papers to address the research questions. In responding to the research questions, we categorized the augmentation types as basic and generated artificial data. We determined whether the augmentation technique was used for segmentation or classification tasks and considered the years of publication. Furthermore, we provided details on the augmentation techniques used in the respective categories.

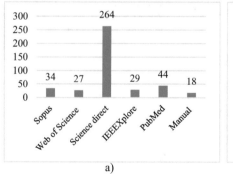
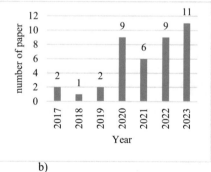

a)

b)

Fig. 1. Number of papers a) downloaded from different databases and b) selected papers based on year proportion

3   Results

In this section, we address the research questions and outline the paper selection process. Initially, a total of 417 articles were sourced from five databases and by manual searches. Subsequently, 53 duplicates were identified and after their removal, 364 unique papers remained. The inclusion and exclusion criteria were then applied, resulting in the exclusion of 301 papers and the inclusion of 63 papers. Among these, 10 articles were found to be not fully available, while 53 articles required further analysis for the final selection. After a thorough review of the full articles, 40 were ultimately selected for the final list. The process to select the studies is shown in Fig. 2.

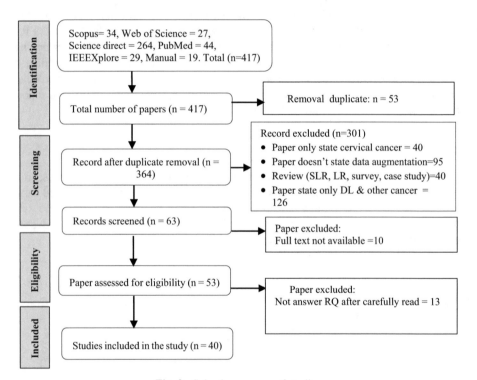

Fig. 2. Selection process of studies

3.1   Data Augmentation Techniques

In this section, we answer the research questions by identifying the augmentation techniques used in segmentation and classification tasks to detect cervical cancer. To answer RQ1, the augmentation techniques are divided into two categories: basic methods and generated artificial image data augmentation techniques.

Basic Data Augmentation Techniques. Is a common technique used in DL to artificially increase the size of a training dataset by applying various transformations to the existing original data. This process helps improve the performance and generalization of DL models. Data augmentation is often employed in computer vision tasks, such as image classification, to generate additional training samples. Here are some basic data augmentation techniques that are used in 34 studies [17–50]. The list of basic augmentation is depicted in Table 1.

Table 1. List of basic augmentation techniques

Basic augmentation	Reference	No of study
Rotation	[17–27, 29–32, 34, 36, 37, 39–47, 49, 50]	29
Flipping	[17, 18, 20–23, 25, 26, 30–35, 37–41, 45–48, 50]	24
Translation	[17, 18, 22, 25–28, 31, 33, 35, 36, 42, 43, 48]	14
Bright & Contrast	[20, 31, 32, 34, 38, 39, 41, 44, 45, 48]	10
Cropping	[17, 27, 29, 33, 36, 37, 42, 46, 47]	9
Zooming	[18, 20, 25, 26, 31, 34, 35]	7
Shearing	[25, 30, 34, 35, 44]	5
Gaussian blur	[21, 29–31, 40]	5
Noise	[24, 30, 40]	3
Others	[20, 22–25, 28, 30, 31, 37, 46]	9

Rotation: rotate images at a specific angle to artificially enhance the data set.

Flipping: creates new images by horizontally or vertically flipping the images.

Translation: involves moving the pixels of an image along the x- and y-axes.

Brightness and contrast: the process of handling variations in lighting conditions.

Cropping: is the variation of an image by randomly or systematically extracting a section.

Zooming: the illusion of a zoomed-in image, wherein the rest of the image is enlarged.

Shearing: a geometric modification of an image's pixels to slope in a specific direction.

Gaussian Blur: This process uses a Gaussian filter to impart a blur effect to an image.

Noise: Addition of noise to an image such as salt-and-pepper and Gaussian noise.

Other: mirroring, mean filter, normalization, full mode, inversion, scaling, and copy paste. Basic data augmentation techniques are depicted in Fig. 3.

Since the test set was intended to assess the generalizability of the model, neither data augmentation nor normalization was performed on the test set [20].

Data Augmentation by Generating Artificial Data. This type of augmentation aims to generate artificial images using the existing original image. As a result, synthetic images that closely resemble the real ones can be produced. The list of the artificial image data augmentation method generated is shown in Table 2.

Generative Adversarial Networks (GANs). GANs use the generator (G) and discriminator (D) to train the network. In the generator, a random noise matrix is used as

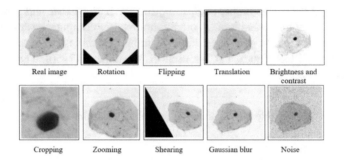

Fig. 3. Sample basic augmentation images

Table 2. List of generative artificial data augmentation techniques

Generating artificial data augmentation technique	Reference	No of study
Generative Adversarial Networks (GANs)	[51, 52]	2
Deep Convolutional Generative Adversarial Network (DCGAN)	[53]	1
Residual Condition Generative Adversarial Networks (RCGAN)	[54]	1
Self-Attention Generative Adversarial Network (SAGAN)	[55]	1
Vector Quantized Generative Adversarial Network (VQGAN)	[56]	1

input to create a synthetic image. It consists of a convolutional layer, batch normalization, an activation function, and a pooling layer. The aim is to fool the discriminator, making it difficult for the discriminator to determine whether the data are real or fake [51]. Meanwhile, the discriminator evaluates whether the image is real or synthetic. The discriminator must be optimized when the generator produces data [52].

Deep Convolutional Generative Adversarial Network (DCGAN). It is a type of GAN network designed to produce high-quality images. In the generator, transpose convolutional blocks are used for up-sampling instead of pooling layers. Additionally, each block is made up of a batch normalization, ReLU activation function, except for the last layer, which uses a tanh activation function [53]. However, in the discriminator, stride convolutions are used and inputs include both the real image and the image generated from the generator. Each block has batch normalization and a LeakyReLU activation function, except for the last layer, which uses a Softmax activation function.

Residual Condition Generative Adversarial Networks (RCGAN). Residual connections are frequently employed to address the problem of vanishing gradients and facilitate the training of extremely deep networks. The generator uses the least squares loss function and is made up of several residual networks as input, while the discriminator uses the loss of penalty for gradient [54]. In the generator, labels and random Gaussian noise are utilized to create fake samples. The discriminator uses a convolutional network,

LeakyReLU activation function, and batch normalization during training. The discriminator takes the generated image, the real image, and the sample label as input, with the final input being the real image.

Self-Attention Generative Adversarial Network (SAGAN). Self-attention is the primary key that enables the model to produce images by focusing on various areas of the input space. During the generation process, the network can assign different weights to different input components at various points, improving the overall quality of the images produced [55]. SAGAN facilitates long-range dependency modelling for attention-driven data augmentation. Therefore, data augmentation was used to improve the generalizability of the data by redistributing them among training classes.

Vector Quantized Generative Adversarial Network (VQGAN). This type of neural network architecture combines transformers and convolutional neural networks to produce artificial images. To maintain an acceptable perceptual quality even when increasing the compression rate, the use of a discriminator and perceptual loss in VQGAN is emphasized. A Squeeze-and-Excitation (SE) block is added to the encoder due to its high capacity to generalize across many data sets and significantly improve the performance of convolutional neural networks [56].

3.2 Review Papers Proportion

To address RQ2, we present the proportion of the most widely used augmentation techniques. Basic data augmentation techniques are predominant in most of the papers. The percentage distribution of the number of studies is shown in Table 3.

Table 3. The proportion of the review study

Augmentation method	Number of study	% value	Segmentation		Classification	
			Number of study	% value	Number of study	% value
Basic augmentation	34	82.9%	11	32.3%	23	67.7%
Artificial generated image	6	17.1%	0	0%	6	100%

4 Discussion and Limitations

Basic data augmentation techniques are straightforward to implement and involve a quick computational process. These techniques can be applied to individual images to modify the geometric structure of the image, with the aim of increasing the training data set's size to enhance model performance. However, the test set is not augmented, as it is used to evaluate the model and assess its generalizability. Of a total of 40 reviewed articles, 34 of them employed basic data augmentation techniques, while 6 papers used artificially generated image augmentation techniques with 82.9% and 17.1% respectively. In the case of basic augmentation techniques, out of a total of 34 articles, 11 studies (32.35%) focused on segmentation tasks and 23 studies (67.65%) were dedicated to classification tasks. On the other hand, for artificially generated images, all augmentation techniques were applied in the context of classification tasks.

This systematic review identifies the state-of-the-art and provides insights into data augmentation techniques in cervical cancer detection. The results indicate that basic augmentation is employed for both segmentation and classification tasks, while artificially generated techniques are used exclusively for classification. Synthesizing findings from existing articles aims to contribute to a more informed understanding of data augmentation and when to implement it in cervical cancer detection. Future work will involve a comparative analysis of the performance outcomes of these techniques.

Our study has some limitations. We downloaded the papers using three keywords; however, adding the keywords 'segmentation' and 'classification' in the search term could have allowed us to review more papers. Additionally, future work will include a comparison of DL algorithms with and without using data augmentation techniques to identify which augmentation technique affects the performance of the result. Furthermore, it is important to identify which data augmentation techniques have not been addressed in detecting cervical cancer.

5   Conclusions

In this systematic review, we analyzed 40 articles describing data augmentation techniques in segmentation and classification tasks for cervical cancer detection. When faced with a scarcity of images in the medical field, data augmentation emerges as a valuable solution. Moreover, it effectively addresses two common challenges: class imbalance and overfitting. Basic data augmentation techniques are prominently featured in most reviewed papers, with a notable emphasis on classification tasks. Consequently, data enhancement has been helpful in improving the performance and generalizability of the models. In the generation of artificial data, new high-quality images are produced, particularly in the context of DCGAN.

References

1. Win, K.P., Kitjaidure, Y., Hamamoto, K., Aung, T.M.: Computer-assisted screening for cervical cancer using digital image processing of pap smear images. Appl. Sci. **10**(5) (2020). https://doi.org/10.3390/app10051800
2. Chlap, P., Min, H., Vandenberg, N., Dowling, J., Holloway, L., Haworth, A.: A review of medical image data augmentation techniques for deep learning applications. J. Med. Imaging Radiat. Oncol. **65**(5), 545–563 (2021). https://doi.org/10.1111/1754-9485.13261
3. Muchlinski, D.: Machine learning and deep learning. In: Elgar Encyclopedia of Technology and Politics, pp. 114–118 (2022). https://doi.org/10.4337/9781800374263.machine.learning.deep.learning.muchlinski.xml

4. Wan, T., Xu, S., Sang, C., Jin, Y., Qin, Z.: Accurate segmentation of overlapping cells in cervical cytology with deep convolutional neural networks. Neurocomputing **365**, 157–170 (2019). https://doi.org/10.1016/j.neucom.2019.06.086

5. Zhou, S.K., et al.: A review of deep learning in medical imaging: imaging traits, technology trends, case studies with progress highlights, and future promises. Proc. IEEE **109**(5), 820–838 (2021). https://doi.org/10.1109/JPROC.2021.3054390

6. Zhang, X., et al.: Application of spectral small-sample data combined with a method of spectral data augmentation fusion (SDA-Fusion) in cancer diagnosis. Chemom. Intell. Lab. Syst. **231**(September), 104681 (2022). https://doi.org/10.1016/j.chemolab.2022.104681

7. Yadav, S.S., Jadhav, S.M.: Deep convolutional neural network based medical image classification for disease diagnosis. J. Big Data **6**(1) (2019). https://doi.org/10.1186/s40537-019-0276-2

8. Shorten, C., Khoshgoftaar, T.M.: A survey on image data augmentation for deep learning. J. Big Data **6**(1) (2019). https://doi.org/10.1186/s40537-019-0197-0

9. Sandfort, V., Yan, K., Pickhardt, P.J., Summers, R.M.: Data augmentation using generative adversarial networks (CycleGAN) to improve generalizability in CT segmentation tasks. Sci. Rep. **9**(1), 1–9 (2019). https://doi.org/10.1038/s41598-019-52737-x

10. Nalepa, J., Marcinkiewicz, M., Kawulok, M.: Data augmentation for brain-tumor segmentation: a review. Front. Comput. Neurosci. **13**(December), 1–18 (2019). https://doi.org/10.3389/fncom.2019.00083

11. Hussain, Z., Gimenez, F., Yi, D., Rubin, D.: Differential data augmentation techniques for medical imaging classification tasks. In: AMIA Annual Symposium Proceedings. AMIA Symposium, vol. 2017, no. May, pp. 979–984 (2017)

12. Maharana, K., Mondal, S., Nemade, B.: A review: data pre-processing and data augmentation techniques. Glob. Transit. Proc. **3**(1), 91–99 (2022). https://doi.org/10.1016/j.gltp.2022.04.020

13. Mustafa, W.A., Alias, N.A., Jamlos, M.A., Ismail, S., Alquran, H.: A recent systematic review of cervical cancer diagnosis: detection and classification. J. Adv. Res. Appl. Sci. Eng. Technol. **28**(1), 81–96 (2022). https://doi.org/10.37934/araset.28.1.8196

14. Jeong, J.J., Tariq, A., Adejumo, T., Trivedi, H., Gichoya, J.W., Banerjee, I.: Systematic review of Generative Adversarial Networks (GANs) for medical image classification and segmentation. J. Digit. Imaging **35**(2), 137–152 (2022). https://doi.org/10.1007/s10278-021-00556-w

15. Kamioka, H.: Preferred reporting items for systematic review and meta-analysis protocols (PRISMA-p) 2015 statement. Jpn. Pharmacol. Ther. **47**(8), 1177–1185 (2019)

16. Sarkis-Onofre, R., Catalá-López, F., Aromataris, E., Lockwood, C.: How to properly use the PRISMA statement. Syst. Rev. **10**(1), 13–15 (2021). https://doi.org/10.1186/s13643-021-01671-z

17. Alyafeai, Z., Ghouti, L.: A fully-automated deep learning pipeline for cervical cancer classification. Expert Syst. Appl. **141**, 112951 (2020). https://doi.org/10.1016/j.eswa.2019.112951

18. Pramanik, R., Biswas, M., Sen, S., de Souza Júnior, L.A., Papa, J.P., Sarkar, R.: A fuzzy distance-based ensemble of deep models for cervical cancer detection. Comput. Methods Programs Biomed. **219**, 106776 (2022). https://doi.org/10.1016/j.cmpb.2022.106776

19. Sabeena, K., Gopakumar, C.: A hybrid model for efficient cervical cell classification. Biomed. Sig. Process. Control **72**(PA), 103288 (2022). https://doi.org/10.1016/j.bspc.2021.103288

20. Atkinson Amorim, J.G., et al.: A novel approach on segmentation of AgNOR-stained cytology images using deep learning. In: Proceedings - IEEE 33rd International Symposium on Computer-Based Medical Systems, vol. 2020-July, pp. 552–557 (2020). https://doi.org/10.1109/CBMS49503.2020.00110

21. Cao, L., et al.: A novel attention-guided convolutional network for the detection of abnormal cervical cells in cervical cancer screening. Med. Image Anal. **73**, 102197 (2021). https://doi.org/10.1016/j.media.2021.102197
22. Kurita, Y., et al.: Accurate deep learning model using semi-supervised learning and Noisy Student for cervical cancer screening in low magnification images. PLoS ONE **18**(5 May), 1–17 (2023). https://doi.org/10.1371/journal.pone.0285996
23. Xue, D., et al.: An application of transfer learning and ensemble learning techniques for cervical histopathology image classification. IEEE Access **8**, 104603–104618 (2020). https://doi.org/10.1109/ACCESS.2020.2999816
24. Benhari, M., Hossseini, R.: An Improved Fuzzy Deep Learning (IFDL) model for managing uncertainty in classification of pap-smear cell images. Intell. Syst. Appl. **16**(July), 200133 (2022). https://doi.org/10.1016/j.iswa.2022.200133
25. Chitra, B., Kumar, S.S.: An optimized deep learning model using mutation-based atom search optimization algorithm for cervical cancer detection. Soft. Comput. **25**(24), 15363–15376 (2021). https://doi.org/10.1007/s00500-021-06138-w
26. Sato, M., et al.: Application of deep learning to the classification of images from colposcopy. Oncol. Lett. **15**(3), 3518–3523 (2018). https://doi.org/10.3892/ol.2018.7762
27. Sompawong, N., et al.: Automated pap smear cervical cancer screening using deep learning. In: Proceedings of the Annual International Conference of the IEEE Engineering in Medicine and Biology Society EMBS, pp. 7044–7048 (2019). https://doi.org/10.1109/EMBC.2019.8856369
28. Xia, M., Zhang, G., Mu, C., Guan, B., Wang, M.: Cervical cancer cell detection based on deep convolutional neural network. In: Chinese Control Conference, CCC, vol. 2020-July, pp. 6527–6532 (2020). https://doi.org/10.23919/CCC50068.2020.9188454
29. Zhang, T., et al.: Cervical precancerous lesions classification using pre-trained densely connected convolutional networks with colposcopy images. Biomed. Sig. Process. Control **55**, 101566 (2020). https://doi.org/10.1016/j.bspc.2019.101566
30. Kalnhor, M., Shinde, S., Wajire, P., Jude, H.: CerviCell-detector: an object detection approach for identifying the cancerous cells in pap smear images of cervical cancer. Heliyon **9**(11), e22324 (2023). https://doi.org/10.1016/j.heliyon.2023.e22324
31. Aina, O.E., Adeshina, S.A., Adedigba, A.P., Aibinu, A.M.: Classification of Cervical Intraepithelial Neoplasia (CIN) using fine-tuned convolutional neural networks. Intell. Med. **5**(September), 100031 (2021). https://doi.org/10.1016/j.ibmed.2021.100031
32. Miranda Ruiz, F., et al.: CNN stability training improves robustness to scanner and IHC-based image variability for epithelium segmentation in cervical histology. Front. Med. **10**(July), 1–14 (2023). https://doi.org/10.3389/fmed.2023.1173616
33. Zhao, J., He, Y.J., Zhou, S.H., Qin, J., Xie, Y.N.: CNSeg: a dataset for cervical nuclear segmentation. Comput. Methods Programs Biomed. **241**(May), 107732 (2023). https://doi.org/10.1016/j.cmpb.2023.107732
34. Wong, L., Ccopa, A., Diaz, E., Valcarcel, S., Mauricio, D., Villoslada, V.: Deep learning and transfer learning methods to effectively diagnose cervical cancer from liquid-based cytology pap smear images. Int. J. Online Biomed. Eng. **19**(4), 77–93 (2023). https://doi.org/10.3991/ijoe.v19i04.37437
35. Asawara, C., Homma, Y., Stuart, S.: Deep Learning Approaches for Determining Optimal Cervical Cancer Treatment, pp. 1–8 (2017)
36. Lin, Y.C., et al.: Deep learning for fully automated tumor segmentation and extraction of magnetic resonance radiomics features in cervical cancer. Eur. Radiol. **30**(3), 1297–1305 (2020). https://doi.org/10.1007/s00330-019-06467-3
37. Pacal, I., Kılıcarslan, S.: Deep learning-based approaches for robust classification of cervical cancer. Neural Comput. Appl. **35**(25), 18813–18828 (2023). https://doi.org/10.1007/s00521-023-08757-w

38. Pal, A., et al.: Deep multiple-instance learning for abnormal cell detection in cervical histopathology images. Comput. Biol. Med. **138**(May), 104890 (2021). https://doi.org/10.1016/j.compbiomed.2021.104890

39. Rahaman, M.M., et al.: DeepCervix: a deep learning-based framework for the classification of cervical cells using hybrid deep feature fusion techniques. Comput. Biol. Med. **136**(May), 104649 (2021). https://doi.org/10.1016/j.compbiomed.2021.104649

40. Sornapud, S., et al.: DeepCIN: attention-based cervical histology image classification with sequential feature modeling for pathologist-level accuracy. J. Pathol. Inform. **11**(40), 1–9 (2020). https://doi.org/10.4103/jpi.jpi

41. Tian, W., et al.: Development and validation of a deep learning algorithm for pattern-based classification system of cervical cancer from pathological sections. Heliyon **9**(8), e19229 (2023). https://doi.org/10.1016/j.heliyon.2023.e19229

42. Chandran, V., et al.: Diagnosis of cervical cancer based on ensemble deep learning network using colposcopy images. Biomed Res. Int. **2021** (2021). https://doi.org/10.1155/2021/5584004

43. Guo, Y., Wang, Y., Yang, H., Zhang, J., Sun, Q.: Dual-attention EfficientNet based on multi-view feature fusion for cervical squamous intraepithelial lesions diagnosis. Biocybern. Biomed. Eng. **42**(2), 529–542 (2022). https://doi.org/10.1016/j.bbe.2022.02.009

44. Zhang, Y., Zall, Y., Nissim, R., Satyam, Zimmermann, R.: Evaluation of a new dataset for visual detection of cervical precancerous lesions. Expert Syst. Appl. **190**(October), 116048 (2022). https://doi.org/10.1016/j.eswa.2021.116048

45. Chauhan, N.K., Singh, K., Kumar, A., Kolambakar, S.B.: HDFCN: a robust hybrid deep network based on feature concatenation for cervical cancer diagnosis on WSI pap smear slides. Biomed. Res. Int. **2023** (2023). https://doi.org/10.1155/2023/4214817

46. Zhao, Y., Fu, C., Xu, S., Cao, L., Ma, H.F.: LFANet: lightweight feature attention network for abnormal cell segmentation in cervical cytology images. Comput. Biol. Med. **145**(November), 105500 (2022). https://doi.org/10.1016/j.compbiomed.2022.105500

47. Brenes, D., et al.: Multi-task network for automated analysis of high-resolution endomicroscopy images to detect cervical precancer and cancer. Comput. Med. Imaging Graph. **97**(February) (2022). https://doi.org/10.1016/j.compmedimag.2022.102052

48. Alemi Koohbanani, N., Jahanifar, M., Zamani Tajadin, N., Rajpoot, N.: NuClick: a deep learning framework for interactive segmentation of microscopic images. Med. Image Anal. **65** (2020). https://doi.org/10.1016/j.media.2020.101771

49. Kurnianingsih, et al.: Segmentation and classification of cervical cells using deep learning. IEEE Access **7**, 116925–116941 (2019). https://doi.org/10.1109/ACCESS.2019.2936017

50. Maurya, R., Pandey, N.N., Dutta, M.K.: VisionCervix: Papanicolaou cervical smears classification using novel CNN-Vision ensemble approach. Biomed. Sig. Process. Control **79**(P2), 104156 (2023). https://doi.org/10.1016/j.bspc.2022.104156

51. Yu, S., et al.: Generative adversarial network based data augmentation to improve cervical cell classification model. Math. Biosci. Eng. **18**(2), 1740–1752 (2021). https://doi.org/10.3934/MBE.2021090

52. Xu, L., Cai, F., Fu, Y., Liu, Q.: Cervical cell classification with deep-learning algorithms. Med. Biol. Eng. Comput. **61**(3), 821–833 (2023). https://doi.org/10.1007/s11517-022-02745-3

53. Xue, Y., Zhou, Q., Ye, J., Long, L.R., Antani, S., Cornwell, C., Xue, Z., Huang, X.: Synthetic augmentation and feature-based filtering for improved cervical histopathology image classification. In: Shen, D., Liu, T., Peters, T.M., Staib, L.H., Essert, C., Zhou, S., Yap, P.-T., Khan, A. (eds.) MICCAI 2019. LNCS, vol. 11764, pp. 387–396. Springer, Cham (2019). https://doi.org/10.1007/978-3-030-32239-7_43

54. Chen, S., Gao, D., Wang, L., Zhang, Y.: Cervical cancer single cell image data augmentation using residual condition generative adversarial networks. In: 2020 3rd International Conference on Artificial Intelligence and Big Data, ICAIBD 2020, no. x, pp. 237–241 (2020). https://doi.org/10.1109/ICAIBD49809.2020.9137494
55. Khan, A., Han, S., Ilyas, N., Lee, Y.M., Lee, B.: CervixFormer: a multi-scale swin transformer-based cervical pap-smear WSI classification framework. Comput. Methods Programs Biomed. **240** (2023). https://doi.org/10.1016/j.cmpb.2023.107718
56. Zhao, C., Shuai, R., Ma, L., Liu, W., Wu, M.: Improving cervical cancer classification with imbalanced datasets combining taming transformers with T2T-ViT, vol. 81, no. 17 (2022). https://doi.org/10.1007/s11042-022-12670-0

Orthogonal Transforms in Neural Networks Amount to Effective Regularization

Krzysztof Zając[✉] , Wojciech Sopot , and Paweł Wachel

Faculty of Information and Communication Technology, Wrocław University of
Science and Technology, Wrocław, Poland
{krzysztof.zajac,wojciech.sopot,pawel.wachel}@pwr.edu.pl

Abstract. We consider applications of neural networks in nonlinear system identification and formulate a hypothesis that adjusting general network structure by incorporating frequency information or other known orthogonal transform, should result in an efficient neural network retaining its universal properties. We show that such a structure is a universal approximator and that using any orthogonal transform in a proposed way implies regularization during training by adjusting the learning rate of each parameter individually. We empirically show in particular, that such a structure, using the Fourier transform, outperforms equivalent models without orthogonality support.

Keywords: Neural Networks · Nonlinear Dynamics · Orthogonal Transform · System Identification

1 Introduction

Neural networks are a very general type of parametric model, capable of learning various relationships between variables in various modalities. This generality is one of their greatest strengths, but special cases designed for specific applications exist, usually adjusting the general structure by incorporating useful *a priori* knowledge about the task. Among different tasks of modern system modelling and identification, one can consider the problem of estimating the system's output, conditioned on its past values and excitation. In this context, two classes of problems are distinguished: *simulation modelling* and *predictive modelling* [21]. *Simulation modelling* is a task in which outputs are predicted based only on the input signal, while *prediction modelling* requires measurements of the system trajectory to predict future states (hybrid approaches are also possible). This paper considers simulation modelling, which can be framed as an optimization problem of n-step ahead prediction based on m samples of the input sequence. The training dataset contains measurements of input and output signals from the system, which are grouped into windows of fixed length.

We hypothesise that adding frequency information to the network's structure will be useful inductive knowledge for dynamical system identification in a simulation setting. A similar approach was used to derive Fourier neural operators,

© The Author(s), under exclusive license to Springer Nature Switzerland AG 2024
W. Zamojski et al. (Eds.): DepCoS-RELCOMEX 2024, LNNS 1026, pp. 337–348, 2024.
https://doi.org/10.1007/978-3-031-61857-4_33

which achieve good performance in fluid dynamics using predictive modelling [9,15], while still being very general and applicable to various kinds of physical or engineering systems. Different orthogonal transforms, such as wavelet transforms are also considered in literature [18,28]. The number of methods developed for the identification of nonlinear systems is very large. Some of them include domain knowledge [10,16,23], while others are very general and require only very mild assumptions about the nature of modelled systems [7,17,27,31,32]. One of the most successful classes of models applied to the identification of nonlinear dynamics are neural networks [1,8,19]. Nevertheless, even given the generality of the network structure, many specialized networks have been developed specifically for nonlinear dynamics and usually achieve better results than the less specialized structures [2,6]. Often those specialized networks are built using general *a priori* knowledge about the nature of the system; sometimes, they have knowledge about physical equations baked into the architecture [13]. Our hypothesis leads to the formulation of a network structure resembling that of Fourier Neural Operator [15]. Such a model processes the system input signal in parallel in the time and frequency domain. We theoretically analyse this structure, showing that both branches are universal approximators, so such a structure also retains this property originally proved for a feed-forward network. We also show how adding such transforms impacts the learning process, namely that it effectively scales the gradient, for each parameter separately. Both those results hold for any orthogonal transform. We implement such a dual model and empirically apply it for simulation modelling, where it outperforms models without added frequency information[1].

2   Dual-Orthogonal Neural Network

The investigated structure of a dual neural network with orthogonal transform is an extension of standard feed-forward network. It is designed to incorporate useful information about the input signals by transforming it into a different basis. The input of the block is a sequence of measurements of a dynamical system, its output is also a sequence, possibly with different length. The input can be multi-dimensional (as well as the output). In experiments, we are using the Fourier transform, following the assumption that frequencies of the input signal will be useful for identification.

2.1   Dual-Orthogonal Block

Each dual block we consider consists of two parallel branches. One operates in the time domain and is a linear block processing the input sequence, whereas the other one is designed to focus on the space in a different basis, depending on the orthogonal transform used. Block combining more than two representations

---

[1] Model implementation and datasets used for experiments are open-source, available at https://github.com/cyber-physical-systems-group/kernel-supported-neural-networks.

are also conceivable, but we leave them for future research. Outputs of both branches are added together and passed through a non-linear activation function σ. Each block has four hyper-parameters: the length of the *input* sequence, S_i, the length of the *output* sequence it produces, S_o, and the input and output dimensionalities, D_i and D_o, respectively. The output of each time-branch can be expressed by equation

$$\hat{h}_l = xW_l^{\mathsf{T}} + b_l, \tag{1}$$

where $x \in \mathbb{R}^{S_i}$ is a row vector with length S_i, and W_l is a matrix of learnable parameters with shape $[S_o \times S_i]$, and b_l is a bias vector with length S_o. The output $\hat{h}_l \in \mathbb{R}^{S_o}$ is a row vector with length S_o. Elements of the time branch are denoted with subscript l to distinguish them from the elements of the transformed branch, denoted with subscript t. The orthogonal block uses an orthogonal transform to convert the signal, applies learned linear transformation in this space and then converts back to the original domain

$$\hat{h}_t = \mathcal{T}^{-1}\left(\mathcal{T}(x)W_t^{\mathsf{T}} + b_t\right), \tag{2}$$

where $\mathcal{T}$ denotes the orthogonal transform applied to the signal, and $\mathcal{T}^{-1}$ is its inverse. The matrix of parameters W_t can be real or complex-valued. In general, the shape of matrices W_l and W_t is the same, but in some cases, such as Fourier transform certain parameters might be simplified. The output of each block is computed as the sum of representations produced by both branches with a nonlinear activation function σ, *i.e.*,

$$\hat{y} = \sigma(\hat{h}_l + \hat{h}_t). \tag{3}$$

The structure of the dual-orthogonal neural network can be extended to the system with multiple-input multiple-output (MIMO) systems by extending the parameter matrices and using reshape operation (Fig. 1).

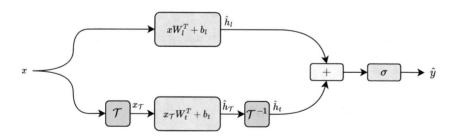

Fig. 1. Schematic representation of a dual-orthogonal block structure

2.2   Frequency-Supported Block

In this study we are particularly focused on the dual block using Fourier transform, which will be further called *frequency-supported block* and multiple such blocks, arranged sequentially will be called *frequency-supported neural network*. The usage of the Fourier transform is dictated by the fact, that many practical dynamical systems can be modelled in frequency space effectively, so this transformation should be useful inductive bias. We use real-valued Fourier transform [25], so only the positive side of the spectrum is processed due to the assumption that such a model will process only real-valued signals. Due to the FFT algorithm used in implementation, it is most efficient on sequence lengths being a power of 2, *cf.* [3]. In such case, the learned parameters are complex-valued in this branch. Using RFFT as $\mathcal{T}$ in formulation given by (2) allows us to reduce the number of parameters (weight matrix has shape $1/2S_o \times 1/2S_i$ and bias vector has length $1/2S_o$), due to our assumption about the real-valued signal.

2.3 N-Step Ahead Prediction

Dataset is composed of input and output measurements of the system of interest aligned in time with constant sampling between measurements. Moreover, the model structure requires the input and output to be of the known and constant length (input and output length can be different). To simulate a dynamical system using our approach, the input signal needs to be split into windows, for which outputs will be predicted, where each window of inputs corresponds to a single window of outputs. In practice, short output windows and long input windows are most efficient, with shortest possible output length of 1 often resulting in best performance. In the experiments, time windows were created using two parameters, m for input length and n for output length and time index s (we denote index as sequence step s, while t is used for transform), which was moved along the sequence of sampling times S. During training, the overlap is possible and sometimes useful so the same measurement can appear in different parts of the input or target sequence

$$\{(U_{(s-m):s}, Y_{(s-n):s}) \mid s \in S\}. \tag{4}$$

For multi-dimensional systems, multiple vectors for excitation or output measurements can be included in the training dataset, all aligned using the same time index.

3 Theoretical Properties

Feed-forward neural networks are known to have universal approximation property, which was originally proven in [4] and extended in particular in [11,12,26]. Informally, the universal approximation property guarantees that for any n-dimensional function f from a given space, there exists a feed-forward neural network, $G(x)$, of the form given in (6), such that $|G(x) - f(x)| < \epsilon$ for arbitrarily small $\epsilon > 0$. In the case of simulation modelling of dynamical systems, the input to the network, $x \in \mathbb{R}^N$, is an input signal with finite time steps. Nevertheless, to guarantee this property, *cf.* [4], the network is required to have an infinite number[2] of neurons (also called units) in the hidden layer with sigmoidal activation. Furthermore, it is required that

$$\sigma(x) = \begin{cases} 1 & \text{when} \quad x \to +\infty \\ 0 & \text{when} \quad x \to -\infty. \end{cases} \tag{5}$$

The representation presented by [4] allows showing that both time and orthogonal branches of dual-orthogonal network have universal approximation properties. For the time branch (*i.e.* orthogonal network with $\hat{h}_t \equiv 0$), this is straightforward, as the general form given in Eq. (6) expresses the same computation as the time branch

$$G(x) = \sigma(xW_l^\mathsf{T} + b_l)W_s + b_s. \tag{6}$$

The learned parameters are in the hidden layer of the network, while W_s and b_s are additional readout parameters, which are also present in the original formulation [4].

3.1   Orthogonal Branch as Universal Approximator

The orthogonal branch (*i.e.* dual-orthogonal network with $\hat{h}_l \equiv 0$), is a universal approximator, which can be shown based on Cybenko's proof since it is possible to write it in a way equivalent to $G(x)$ network structure, utilizing the properties of the transformation matrix, given above. The general form of such branch in matrix notation is given by equation

$$G_T(x) = \sigma\left((xTW_t^\mathsf{T} + b_t)T^{-1}\right)W_s^\mathsf{T} + b_s, \tag{7}$$

where $T = [\phi_0, \phi_1, \phi_2, \cdots]$ denotes the transition matrix of orthogonal transform, with ϕ_i being i-th column vector of T's base. Such a matrix is infinitely wide, similarly to the hidden layer of a universal approximation network. When T is Fourier transform, the matrix T is the DFT matrix [30]. T^{-1} corresponds to the inverse orthogonal transform, it is guaranteed to exist since T is unitary. Given that parameters can take any value, it is possible to find matrix W_t and complementing bias vector b_t, such that general form (7) is equivalent to universal feed-forward network (6)

$$W_f = TW_t^\mathsf{T}T^{-1}$$
$$b_f = b_tT^{-1}. \tag{8}$$

Those values can be computed analytically and they are presented in Eq. (8). This property only holds for square matrices, since T is always square by definition. After plugging in those values, the frequency branch has all the properties of a feed-forward network.

---

[2] For certain sophisticated activations, the number of units does not need to be infinite [24]. These models are however of lower practical potential.

3.2   Gradients in Orthogonal Branch

To show the influence of orthogonal transform on the learning process, we analyze the orthogonal branch at n-th layer (we set $W_l = 0, b_l = 0$), which is given by

$$B^{(n)}(x) = \sigma\left(\left(xTW_t^\mathsf{T} + b_t\right)T^{-1}\right). \tag{9}$$

We calculate the derivative of the output $B^{(n)}(x)$ with respect to one of the matrix's W_t parameters, which we denote $W_{t_{ab}}$,

$$\frac{\partial B^{(n)}(x)}{\partial W_{t_{ab}}} = xH_{ab}^\mathsf{T} \circ \sigma'\left(\left(xTW_t^\mathsf{T} + b_t\right)T^{-1}\right), \tag{10}$$

where elements of matrix H_{ab} are defined as $H_{ab_{ij}} = \phi_{ia}\phi_{bj}$. In special case of Fourier basis ($T \equiv F^\mathsf{T}$, where F is DFT matrix) $H_{ab_{ij}} = \frac{\omega^{ia+bj}}{N}$ and $\omega = e^{\frac{-2\pi i}{N}}$. For any orthogonal transform, elements of H_{ab} can be derived by writing the transformation as matrix multiplication. Derivative for bias vector will have an analogous form. For the learning process we assume that one of the variations of the stochastic gradient algorithm is used [20] to compute updates of parameters θ (weights and biases of the network) in batches. Updates are computed according to the general rule

$$\theta_{e+1} = \theta_e - \alpha g_e, \tag{11}$$

where g_e denotes the gradient of the parameters with respect to the loss function at a certain step e and α is a constant learning rate, typically set upfront. Values of H_{ab} are determined per-parameter for every batch of training data, therefore such an approach can also be interpreted as adaptive per-parameter learning rate changing. In practice, adaptive algorithms [5] or learning rate scheduling is used [14]. Learning rate schedule can be interpreted as using learning rate, which is a function of step e or depends on past values of the loss L. In both cases, the learning rate is the same for all parameters, so our interpretation of the impact of orthogonality on the learning process is the same as for the constant learning rate. For adaptive algorithms, such as Adam, the learning rate is set per parameter based on the history of its gradients

$$\theta_{e+1} = \theta_e - \frac{\alpha}{\sqrt{\hat{v}_e} + \phi}\hat{m}_e, \tag{12}$$

where $\hat{m}_e$ and $\hat{v}_e$ are computed using decaying averages of past gradients, with decay speed controlled by additional hyper-parameters, which are similar to learning rate: β_1 and β_2. Values $\hat{m}_e$ and $\hat{v}_e$ are first and second moment respectively, i.e.,

$$\hat{m}_{e+1} = \beta_1\hat{m}_e + (1 - \beta_1)g_e,$$
$$\hat{v}_{e+1} = \beta_2\hat{v}_e + (1 - \beta_2)g_e^2. \tag{13}$$

Therefore, when each derivative is scaled by the influence of orthogonal transform, this can be interpreted as adaptive scaling of β_1 and β_2 learning rates, which are famously difficult to set. Every first-order optimization algorithm used in the neural network should behave similarly with respect to the orthogonal transform.

4   Numerical Experiments

Three numerical experiments were run to verify the hypothesis that adding frequency information to neural networks will be empirically beneficial. One consists of a toy problem with a static system, while the other two were benchmarks selected from system identification literature. We have trained three models, mostly focusing on the improvement between models with and without frequency information. The frequency-supported neural network, (i), abbreviated FSNN, described in the previous section and consisting of a number of frequency-supported blocks stacked together. The FNN model, (ii) (*i.e.* frequency neural network) consisting of only frequency blocks, which is a subset of the FSNN architecture, where $\hat{h}_l \equiv 0$. The final model (iii) was a regular feed-forward network processing the signal using delayed input measurements (we refer to it as MLP), which also is a subset of FSNN with $\hat{h}_t \equiv 0$. Additionally, selected models were re-implemented and run on the same benchmark problems, or when available, the results were transferred from original papers. For our three core architectures, we performed a hyper-parameter search and reported the best results for each of them.

4.1   Evaluation

All the above algorithms were evaluated using root mean squared error (RMSE), with physical units added, where possible. RMSE was computed as

$$RMSE(y, \hat{y}) = \sqrt{MSE(y, \hat{y})} = \sqrt{\frac{1}{N} \sum_{i=1}^{N} (y - \hat{y})^2}. \tag{14}$$

Additionally, a normalized mean squared error was also evaluated, *i.e.* the ratio of RMSE and standard deviation of the predicted value (reported as a percentage)

$$NRMSE(y, \hat{y}) = \frac{RMSE(y, \hat{y})}{\sigma_y}. \tag{15}$$

4.2   Static System with Frequency Input

A simple static affine system with an input signal consisting of pure-tone sine waves was created to test FSNN structure, under conditions most suitable for it. The input signal was generated using a sum of frequencies drawn from uniform distribution: $\alpha \sim \mathcal{U}(-5, 5)$, which is given by Eq. (16)

$$u(t) = \sum_{i=1}^{5} sin(\alpha_i t). \tag{16}$$

The output of the system was generated using two additional parameters, randomly drawn from the uniform distribution $\mathcal{U}(-5, 5)$, similar to the excitation frequencies. Those are denoted as β_1, β_2 in

$$f(u) = \beta_1 u(t) + \pi \beta_2. \tag{17}$$

Note, that the best-performing model, *DynoNet* [6], needs information about the structure of the system to perform well, as it is composed of static nonlinear and learnable linear blocks, while other models do not (Table 1 and Fig. 2).

Table 1. Evaluation results for selected models on test dataset for a static affine system with frequency input, where $\#P$ is the number of parameters

Model	$\#P$	RMSE	NRMSE
FNN	1610	$10.3 \cdot 10^{-3}$	0.30%
MLP	2157	$5.4 \cdot 10^{-3}$	0.16%
FSNN	887	$1.4 \cdot 10^{-3}$	0.04%
DynoNet	49	$0.3 \cdot 10^{-3}$	0.02%

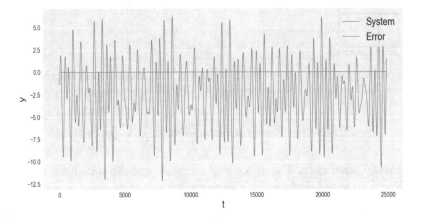

Fig. 2. Simulation error computed for best FSNN model on the test dataset for a static affine system with frequency input

4.3 Wiener-Hammerstein Benchmark

The Wiener-Hammerstein benchmark [22] is a well-known benchmark for the identification of nonlinear dynamics. It consists of two linear blocks and a static non-linearity, which were implemented using an electronic RLC circuit with a diode. The measurements of this system are used to create training and test datasets for the model. In this benchmark, we do not observe the performance change, as the number of parameters increases. We report results for two sizes of MLP model, the best in terms of RMSE and orders of magnitude smaller model, with marginally worse performance. The same behaviour was observed for FNN and FSNN (Table 2 and Fig. 3).

Table 2. Evaluation results for selected models on test dataset for Wiener-Hammerstein benchmark compared to selected results reported in literature

Model	#P	RMSE	NRMSE
FNN	7856	1.9 mV	0.78%
DynoNet	63	1.2 mV	0.50%
MLP	1193	1.1 mV	0.46%
Large MLP	1379841	0.9 mV	0.38%
FSNN	1591	0.5 mV	0.22%
State-Space Encoder	21410	0.2 mV	0.10%

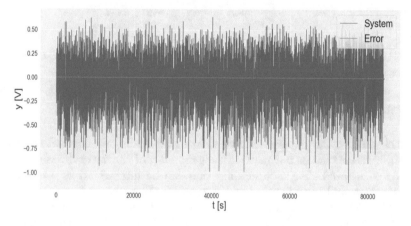

Fig. 3. Simulation error computed for best FSNN model on the test dataset for Wiener-Hammerstein benchmark

4.4 Silverbox Benchmark

Silverbox benchmark is an electronic implementation of the Duffing oscillator, which can be modelled using a second-order LTI system with a polynomial non-linearity in the feedback [29] (Table 3 and Fig. 4).

Table 3. Evaluation results for selected models on test dataset for Silverbox benchmark compared to selected results reported in literature

Model	#P	RMSE	NRMSE
FNN	14192	4.1 mV	7.69%
MLP	37313	3.9 mV	7.32%
DynoNet	81	2.9 mV	5.39%
FSNN	69719	2.3 mV	4.31%
State-Space Encoder	19930	1.4 mV	2.60%

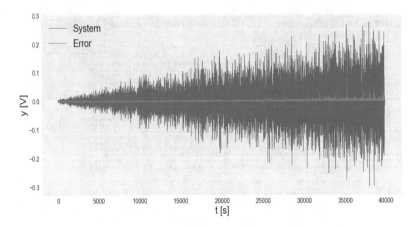

Fig. 4. Simulation error computed for best FSNN model on the test dataset for Silverbox benchmark

5   Discussion

In conclusion, in the paper, we have shown that adding orthogonal transforms to neural networks does not restrict the generality of the network. Moreover, we have demonstrated that such modification can be interpreted as a method for regularization and initialization of the network. Experimental results, through the application of the Fourier transform, support the hypothesis that effective bases for system identification exist. Generally, it is likely to derive more useful bases for identification from physical insight or a priori knowledge about system structure. Therefore, the dual-orthogonal structure is a way of making neural models specialized for specific identification tasks and is worth further investigation.

References

1. Andersson, C., Ribeiro, A., Tiels, K., Wahlström, N., Schön, T.: Deep convolutional networks in system identification. In: 2019 IEEE 58th Conference on Decision and Control (CDC), pp. 3670–3676 (2019). https://doi.org/10.1109/CDC40024.2019. 9030219
2. Beintema, G., Tóth, R., Schoukens, M.: Nonlinear state-space identification using deep encoder networks. In: Proceedings of Learning for Dynamics and Control, vol. 144, pp. 241–250 (2021)
3. Brigham, E.O., Morrow, R.E.: The fast fourier transform. IEEE Spectr. **4**(12), 63–70 (1967). https://doi.org/10.1109/MSPEC.1967.5217220
4. Cybenko, G.: Approximation by superpositions of a sigmoidal function. Math. Control Signal Syst. **2**, 303–313 (1989). https://doi.org/10.1007/BF02551274
5. Diederik, P.K., Ba, J.: Adam: a method for stochastic optimization (2017)
6. Forgione, M., Piga, D.: dynoNet: a neural network architecture for learning dynamical systems. Int. J. Adapt. Control Signal Process. **35**(4), 612–626 (2021)

7. Geneva, N., Zabaras, N.: Modeling the dynamics of PDE systems with physics-constrained deep auto-regressive networks. J. Comput. Phys. **403**, 109056 (2020). https://doi.org/10.1016/j.jcp.2019.109056. https://www.sciencedirect.com/science/article/pii/S0021999119307612
8. Geneva, N., Zabaras, N.: Transformers for modeling physical systems. Neural Netw. **146**, 272–289 (2022). https://doi.org/10.1016/j.neunet.2021.11.022
9. Guibas, J., Mardani, M., Li, Z., Tao, A., Anandkumar, A., Catanzaro, B.: Adaptive fourier neural operators: efficient token mixers for transformers (2022)
10. Hjalmarsson, H., Schoukens, J.: On direct identification of physical parameters in non-linear models. IFAC Proc. Vol. **37**(13), 375–380 (2004). https://doi.org/10.1016/S1474-6670(17)31252-1. https://www.sciencedirect.com/science/article/pii/S1474667017312521
11. Hornik, K., Stinchcombe, M., White, H.: Multilayer feedforward networks are universal approximators. Neural Netw. **2**(5), 359–366 (1989). https://doi.org/10.1016/0893-6080(89)90020-8. https://www.sciencedirect.com/science/article/pii/0893608089900208
12. Hornik, K., Stinchcombe, M., White, H.: Universal approximation of an unknown mapping and its derivatives using multilayer feedforward networks. Neural Netw. **3**(5), 551–560 (1990). https://doi.org/10.1016/0893-6080(90)90005-6. https://www.sciencedirect.com/science/article/pii/0893608090900056
13. Karniadakis, G., Kevrekidis, I.G., Lu, L., Perdikaris, P., Wang, S., Yang, L.: Physics-informed machine learning. Nat. Rev. Phys. **3**, 422–440 (2021)
14. Kim, C., Kim, S., Kim, J., Lee, D., Kim, S.: Automated learning rate scheduler for large-batch training (2021)
15. Li, Z., et al.: Fourier neural operator for parametric partial differential equations. In: International Conference on Learning Representations (2021). https://openreview.net/forum?id=c8P9NQVtmnO
16. Ljung, L., Zhang, Q., Lindskog, P., Juditski, A.: Estimation of grey box and black box models for non-linear circuit data. IFAC Proc. Vol. **37**(13), 399–404 (2004). https://doi.org/10.1016/S1474-6670(17)31256-9. https://www.sciencedirect.com/science/article/pii/S1474667017312569
17. Mzyk, G., Wachel, P.: Wiener system identification by input injection method. Int. J. Adapt. Control Signal Process. **34**(8), 1105–1119 (2020). https://doi.org/10.1002/acs.3124. https://onlinelibrary.wiley.com/doi/abs/10.1002/acs.3124
18. Navaneeth, N., Tripura, T., Chakraborty, S.: Physics informed WNO (2023)
19. Ribeiro, A., Tiels, K., Aguirre, L., Schön, T.: Beyond exploding and vanishing gradients: analysing RNN training using attractors and smoothness. In: Proceedings of the Twenty Third International Conference on Artificial Intelligence and Statistics, vol. 108, pp. 2370–2380 (2020)
20. Ruder, S.: An overview of gradient descent optimization algorithms (2017)
21. Schoukens, J., Ljung, L.: Nonlinear system identification: a user-oriented road map. IEEE Control Syst. Mag. **39**(6), 28–99 (2019)
22. Schoukens, J., Suykens, J., Ljung, L.: Wiener-hammerstein benchmark. In: 15th IFAC Symposium on System Identification (SYSID 2009) (2009). https://www.nonlinearbenchmark.org/benchmarks/wiener-hammerstein#h.2wdiw8u9jr39
23. Schoukens, M., Pintelon, R., Rolain, Y.: Identification of wiener-hammerstein systems by a nonparametric separation of the best linear approximation. Automatica **50**(2), 628–634 (2014). https://doi.org/10.1016/j.automatica.2013.12.027. https://www.sciencedirect.com/science/article/pii/S0005109813005864
24. Shen, Z., Yang, H., Zhan, S.: Deep network approximation: achieving arbitrary accuracy with fixed number of neurons. J. Mach. Learn. Res. 1–60 (2022)

25. Sorensen, H., Jones, D., Heideman, M., Burrus, C.: Real-valued fast fourier transform algorithms. IEEE Trans. Acoust. Speech Signal Process. **35**(6), 849–863 (1987). https://doi.org/10.1109/TASSP.1987.1165220
26. Stinchcombe, M., White, H.: Universal approximation using feedforward networks with non-sigmoid hidden layer activation functions. In: International 1989 Joint Conference on Neural Networks, vol. 1, pp. 613–617 (1989). https://doi.org/10.1109/IJCNN.1989.118640
27. Tanaka, G., et al.: Recent advances in physical reservoir computing: a review. Neural Netw. **115**, 100–123 (2019). https://doi.org/10.1016/j.neunet.2019.03.005. https://www.sciencedirect.com/science/article/pii/S0893608019300784
28. Tripura, T., Chakraborty, S.: Wavelet neural operator for solving parametric partial differential equations in computational mechanics problems. Comput. Methods Appl. Mech. Eng. **404**, 115783 (2023). https://doi.org/10.1016/j.cma.2022.115783. https://www.sciencedirect.com/science/article/pii/S0045782522007393
29. Wigren, T., Schoukens, J.: Three free data sets for development and benchmarking in nonlinear system identification. In: European Control Conference (ECC), pp. 2933–2938 (2013). https://doi.org/10.1007/BF02551274. https://www.nonlinearbenchmark.org/benchmarks/silverbox
30. Winograd, S.: On computing the discrete fourier transform. Math. Comput. **32**(141), 175–199 (1978). http://www.jstor.org/stable/2006266
31. Łagosz, S., Wachel, P., Śliwiński, P.: A dual averaging algorithm for online modeling of infinite memory nonlinear systems. IEEE Trans. Autom. Control **68**(9), 5677–5684 (2023). https://doi.org/10.1109/TAC.2022.3225506
32. Śliwiński, P., Marconato, A., Wachel, P., Birpoutsoukis, G.: Non-linear system modelling based on constrained volterra series estimates. IET Control Theory Appl. **11**(15), 2623–2629 (2017). https://doi.org/10.1049/iet-cta.2016.1360. https://ietresearch.onlinelibrary.wiley.com/doi/abs/10.1049/iet-cta.2016.1360

Classification of European Residential Tall Buildings – Application Assumptions

Tomasz Zamojski[(✉)] [iD]

Wrocław University of Science and Technology, 27 Wybrzeże Stanisława Wyspiańskiego, 50-370 Wrocław, Poland
`tomasz.zamojski@pwr.edu.pl`

Abstract. Residential high-rise buildings are the domain of large cities, urban agglomerations and metropolitan areas. The dynamic urbanization of the residential environment is resulting in increasingly dense population and development, which, unfortunately, is associated with limited access to housing resources for the average citizen. In this article, special interest is focused on the evaluation and categorization of the composition, functions and structures of residential high-rise buildings and their close surroundings (neighborhoods). The proposed application, based on the collected data about the customer, his needs and preferences arising from his personal life, professional work, educational needs, financial situation, etc., will give a preliminary answer as to whether the "indicated" apartment in a tall building meets his requirements and possibilities. This article is an attempt develop an application to classify European residential tall buildings and their living environment with a use of machine learning techniques and models. The presented assumptions of the application are a preliminary approach to the problem of supporting the process of finding housing (for rent or purchase) that meets the requirements of habitat.

Keywords: European residential high-rise buildings · Classification · Machine Learning · Artificial Intelligence · Vertical Habitat · Architecture · Real Estate

1 Introduction

Residential high-rise buildings are the domain of large cities, urban agglomerations and metropolitan areas. The dynamic urbanization of the residential environment is resulting in increasingly dense population and development, which, unfortunately, is associated with limited access to housing resources for the average citizen. The search for a suitable apartment in modern metropolises is time-consuming, as it is necessary to determine the location of the building, the apartment in the structure of the building, assess the environment in which the proper-ty is located, learn about the financial conditions (price, credit and its terms), etc. Added to this is the problem of whether the apartment in question meets my and my family's housing needs and preferences.

The premise of the article and the presented research results is to demonstrate that the use of **artificial intelligence** tools and methods can be useful in selecting an apartment

© The Author(s), under exclusive license to Springer Nature Switzerland AG 2024
W. Zamojski et al. (Eds.): DepCoS-RELCOMEX 2024, LNNS 1026, pp. 349–359, 2024.
https://doi.org/10.1007/978-3-031-61857-4_34

in European high-rise buildings, taking into account information about the availability of related functions in the neighborhood, public areas, amenities and services, nearby transportation, etc., i.e. factors that affect the quality of life of residents and the price of the apartment. The proposed application should be supplemented with a module for identifying the needs and preferences of the apartment buyer.

The application (system), let's tentatively call it *My Nest*, is intended to support a customer looking for a suitable apartment for himself in a metropolitan development, which should be a habitat for him [1, 2]. The user (customer) will collect with his smartphone, photos of buildings, their surroundings and, most importantly, photos of those parts of the building where his future apartment is potentially located. The *My Nest* application, based on the collected data about the customer, his needs and preferences arising from his personal life, family life, professional work, educational needs, financial situation, etc., will give a preliminary answer as to whether the "indicated" apartment meets his requirements and possibilities.

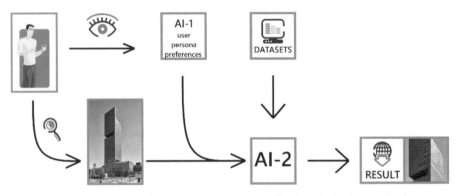

Fig. 1. Diagram of the functionality of the *My Nest* application. Author's own development.

The basic functionality of the *My Nest* application is shown in Fig. 1:

- AI-1 - a web-based module for tracking and collecting data on customer needs and preferences
- Construction of the image processing and classification module,
- DATASETS - developers' database system - technical parameters of apartments, buildings, districts, prices and terms of transactions (e.g., credit)
- AI-2 - housing selection system.

The task of the AI-1 module, which tracks a customer's online behavior, is to generate a list of needs, priorities and preferences used by the AI-2 system to select an apartment. AI-1's inputs are collected against the backdrop of daily life, some of which are quantitative (number of children, their age, distance to work, type of school, etc.) and some qualitative, such as a love of walks in the park or shopping malls, with qualitative assessments in many cases determining the final choice of a particular apartment.

The data provided by developers is quantitative in nature, but is often reluctant to be disclosed, as so-called trade secrets apply. The processing of these large data sets (Big

Data) is hindered in many cases by an excessive desire for benefits, and this on the part of both the developer and the customer.

The "tall building" module, based on photos provided by the customer, for example, classifies the buildings in question and the apartments indicated in them in terms of height and/or location, which is the basis for the AI-2 system to verify that they meet the customer's needs and preferences (the result of the AI-1 system) and his physical needs and financial capabilities (DATASETS).

Implementing AI-1, AI-2 and the "high building" system requires sophisticated artificial intelligence tools using **cognitive systems** and the correspondingly sophisticated databases needed to learn them - an additional problem is profiling this data.

Further considerations are limited to a preliminary analysis of the possibility of classifying tall buildings in European urban development.

2   Research Status

The most developed applications of artificial intelligence in the real estate and housing industry are concerning [3]:

- Data analytics platforms based on artificial intelligence solutions help property managers gain insights into tenant preferences, enable personalization of amenities, services and marketing campaigns. Targeted property development increases tenant satisfaction, improves retention rates and attracts new residents,
- Forecasting demand for real estate and construction materials,
- "Chat bots" and virtual assistants, enabling them to respond instantly to tenant inquiries, schedule maintenance and offer personalized recommendations - saving time and increasing service efficiency,
- Optimization of property management processes - customer service and property maintenance,
- Anticipate maintenance needs and optimize energy consumption (heating, cooling, lighting),
- Ensuring the security of residents and their property through real-time analysis and reporting of potential threats, and enabling biometric identification and virtual keys.

Among the numerous applications related to the real estate market, it is worth mentioning:

- **Skyline AI Virtual Neighborhoods** (New York, NY) [4] enables computerized analysis of the real estate market and improves the accuracy of predicting current and future property values and rents. Prediction of current and future property values and rents is done using supervised learning[1] algorithms. Once the model is trained, the algorithm is able to correctly identify class labels for unseen cases. This is because the new clusters generated can be used as additional data points to enrich the data set used in the supervised models. This process is commonly referred to as feature engineering [5, 6].

---

[1] Supervised learning is a type of machine learning that involves learning a function that maps input data to output data based on labeled training data. The supervised learning algorithm analyzes the training data and creates an inferred function that can be used to map new examples.

- **Zillow** (Seattle, Washington) [7] Uses artificial intelligence to process digital photos to estimate property values. The technology is implemented in a neural network trained on millions of photos and home values (prices). Based on the photo of a new listing, an estimate of the property's value is obtained. According to Zillow, the resulting estimates - which are also based on additional indicators such as area - predict property values with a national median error of 2.4% [8]
- **Trulia** (San Francisco, California) [9]. In the Trulia real estate application (portal), artificial intelligence "learns" the user's (customer's) tastes and preferred, search terms. Based on this, recommendations are made based on sites frequently visited by users with similar tastes [8]. The Trulia system personalizes real estate (or more specifically, home) seekers by identifying the customer and learning and remembering their unique preferences and search criteria. On this basis, machine learning technology is developed, including computer vision, a recommendation system and engagement models. Using computer vision, the type of room in a photo and its features (such as wood floors, granite countertops, etc.) can be identified and extracted. The recommendation system uses a similarity graph to see how consumers who have expressed interest differ. The system then checks all the other properties in Trulia's inventory and determines which ones might be of interest to the customer [10].

It's worth noting that the aforementioned systems use real estate databases and datasets in the United States, where they are much more extensively recorded, collected and made available for acquisition. The aforementioned systems are dedicated to North American residents, and the Zillow and Trulia platforms largely deal with the valuation of single-family homes in the US.

The solutions proposed in the article focus on the area of Europe, high multi-family housing and an attempt to assess their residential environment and close proximity.

3 Assumptions of a Housing Environment Assessment System Based on Artificial Intelligence

The task of the *My Nest* app is to integrate and process data such as photos, comments, ratings, locations, neighborhood data, etc., to enable categorization of the high-rise housing environment based on a photo taken with a smartphone. A user, using a smartphone, would like to determine which height category a building belongs to, how apartment prices vary depending on their location in a high-rise structure, what problems residents report, the distance to neighborhood amenities and their type. The system filters the user's previously gathered personal preferences and financial capabilities to then suggest the most favorable housing solutions.

Today's personal devices - smartphones, equipped with the latest operating systems, high-speed processors, graphics cards, GPS, IMUs, cameras, using pre-installed software and Internet access ("the cloud") have become an inseparable companion. Smartphones are able to record our life functions, choices, preferences, needs, goals, frustrations, lifestyles - they are able to co-create a mental map of the user (AI-1 system in Fig. 1) with appropriate artificial intelligence tools. Based on it, it is possible to suggest specific solutions for individual users or groups of users, such as multi-generational families (cf. Trulia system). Figure 2 shows the results of the author's research on defining housing needs for three types of tenants, with these needs going beyond metric needs and including broadly defined well-being as defined by Zbigniew Bać and the Scientific School of Habitat [1].

Fig. 2. Tables depicting the various personality types and preferences of app users, their ages, locations, occupations, family status, their lifestyles, personal and professional goals, needs, frustrations; (a) multi-generational family; (b) "singles" and young couples; (c) young couples with children; (from author's own archive, developed as part of Bertelsmann & Udacity - AI Manager Scholarship, September 2021)

The concept of building a classification system for European residential high-rise buildings is based on acquiring input datasets from the Google Maps platform:

- images of tall residential buildings (photos taken and shared by platform users),
- parameters of locations,
- user ratings, feedback on the satisfaction of residents and users of tall buildings,
- neighborhood, data about the area (e.g., related features, amenities located in and around the high-rise building, their ratings, details about nearby places, prices, ETA (estimated time of arrival) and directions).

On the Google Maps platform, tall buildings are defined by geographic locations and described as points of interest (POIs), with access to each group of data within the Google Maps API/Google Place API:

- Place Search - list of places based on users' location or search term
- Place Details - detailed information about a specific place, including user reviews
- Place Photos - access to millions of place-related photos stored in the Google Place database [11–13].

The extracted data from the Google Maps platform (Places API) will be subjected to machine learning based in the next steps:

- image recognition
- natural language processing (NLP)
- geographic location data
- Sentiment data (opinions, comments, impressions).

The operation of the system is based on the transfer of acquired data sets to the Google Cloud Platform (GCP) and its engines: VERTEX AI and AutoML [14–16] allowing the use of machine learning models that do not require users to have advanced programming skills, but process images (photographs), natural language in the form of comments and opinions, and emotional sentiments.

4 Data Acquisition

An important stage of the preparatory work for the construction of the application is the collection of data, such as photos of different categories of European residential tall buildings. In his dissertation, the author collected examples of tall residential buildings in Europe and conducted comparative research on them. On this basis, a basic dataset was collected for training and evaluation of the model. The dataset was divided into four height classes of buildings: low (11–21 stories), medium-high (22–34 stories), high (35–60 stories) and super-high (60+ stories). The list and tables contained in the author's doctoral dissertation describe also the number of apartments in the building, because the number of apartments and their density translates into the comfort of life and the identity of users in the residential environment. It distinguishes accompanying functions present in the building or its immediate vicinity, constituting amenities for residents, which affects the quality of their life as well as the price of the apartment. The height classes and other characteristic features by which European tall residential buildings were grouped were described in more detail in the author's earlier works [2, 17–19]. Due to the extensive format and content of the tables, which significantly exceeds the guidelines of this publication and thus significantly affects the readability of the content, the author decided not to publish them. However, they are available for viewing in the doctoral thesis [2].

5 Experiment - Verification of the Usefulness of Machine Learning Models for Image Classification of European Residential Tall Buildings

As part of the research, an experiment was conducted with the intention of verifying the suitability of selected machine learning models for the implementation of a small part of the described application, responsible for the classification of European residential high-rise buildings based on images (Fig. 3).

The experiment compared the usability of two classification models developed on Google Cloud Platform (GCP) - VERTEX AI API: AutoML (VERTEX AI) and AutoML (the default). VERTEX API allows commercially (for a fee) to train high-quality custom machine learning models.

Documentation of Google Cloud Platform's AutoML and Vertex AI models [14] suggests using at least 100 images for each category for machine learning, although optimal collections should have about 1,000 examples each. This is a difficult task to accomplish without incurring additional costs. For the experiment, images of tall buildings completed between 2000 and 2024 in Europe, were obtained from the Council of Tall Buildings and Urban Habitat [20], which were grouped by height class.

The experiment was conducted based on a basic collection of acquired data with the number of 580 photos, which were assigned labels according to the building's affiliation

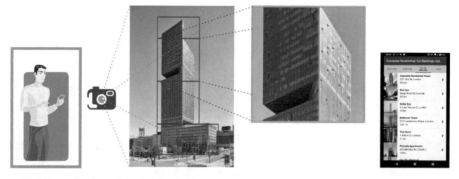

Fig. 3. Evaluation of the height and location of the apartment. Author's own development

to a height class. The collection consists of photographs of buildings of four height classes: 83 low, 198 medium, 247 high and 52 super-high residential buildings. The collections of medium and high buildings are the groups with the largest number of examples, the collections of low and super-high buildings are much less numerous, which means that the collection of European high-rise residential buildings collected for the study is characterized by an unbalanced number of examples (unbalanced dataset), which affects the results of the machine learning process. After assigning labels to the collected images, the dataset was divided 8/1/1 into training data/validation data/test data.

The labeled dataset (images) was subjected to machine learning based on two models available within the GCP platform - VERTEX AI (AutoML Image Classification): AutoML (default - highlighted in red in the diagrams) and AutoML (Vertex AI - highlighted in blue). Time of machine learning model: 8 node hours (Figs. 4, 5 and 6).

Fig. 4. Experiment - division of data into: training set (80%), validation set (10%), test set (10%)

Fig. 5. Experiment - abundance of labeled images according to classes expressed by the number of floors.

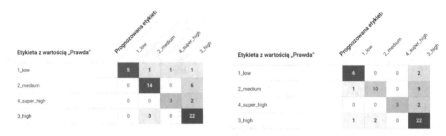

Fig. 6. Experiment - confusion matrices (Confusion matrix: A confusion matrix shows how often a model has correctly predicted an outcome. For incorrectly predicted outcomes, the matrix shows what the model predicted instead. The confusion matrix helps understand where the model "confuses" the results [21, 22].) of both models for test data: a). AutoML model (VERTEX AI); b). AutoML model (default); confidence threshold (Confidence threshold: The confidence score determines which forecasts to return. The model returns forecasts with this or a higher value. A higher confidence threshold increases precision but decreases sensitivity. Vertex AI returns confidence scores at different threshold values to show how the threshold affects precision and sensitivity [21].): 0.5

Tables 1 and 2 summarize the numerical results to compare the computational accuracy of the two models.

The results of the experiment show that it is possible to classify European high-rise buildings based on images and using machine learning models on the Google Cloud Platform. It is worth noting that the AutoML (Vertex AI) model is more efficient, i.e. more

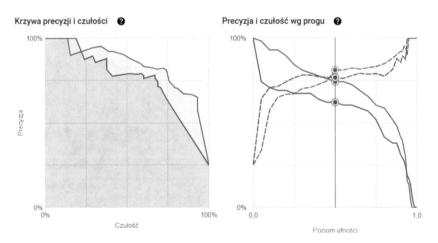

Fig. 7. Experiment - comparison of results of two models; blue: AutoML model (VERTEX AI); red: AutoML model (default); a). precision (Precision: the fraction of classification predictions generated by the model that were correct [21, 22].) and recall (Recall (Sensitivity): the fraction of predictions for a given class that the model correctly predicted. Also referred to as the true positive rate [21, 22].) curves; b). Curve of precision (red dashed line - AutoML model (default); blue dashed line, - AutoML model (VERTEX AI)) and recall (red solid line - AutoML model (default); blue solid line - AutoML model (VERTEX AI)) at set *threshold*: 0.5

accurate, more precise and more sensitive, but it is a commercial model and therefore often beyond the reach of researchers with limited financial resources (Fig. 7).

Table 1. Results of AutoML model (Vertex AI) for confidence threshold: 0.5

Class	Accuracy	Precision	Recall	F1-Score	PR AUC	Log loss	Macro-average F1Score
Low	0,942	100%	62,5%	0,769	0,834	0,291	0,779
Medium	0,835	82,4%	70%	0,757			
High	0,853	75%	84%	0,792			
Super-high	0,927	100%	60%	0,75			
All	0,834	81,1%	74,1%	–			

PR AUC: The area under the precision-recall (PR) curve, also referred to as the average precision. This value ranges from zero to one, where a higher value indicates higher model quality [21].
Log loss function: The cross entropy between model predictions and target values. This value ranges from zero to infinity, where a lower value indicates higher model quality [21].
F1 Score: The harmonic average of precision and sensitivity. F1 is a useful indicator if you are looking for a balance between precision and sensitivity and the class distribution is uneven [21, 22].

Table 2. Results of AutoML model (default) for confidence threshold: 0.5

Class	Accuracy	Precision	Recall	F1-Score	PR AUC	Log loss	Macro-average F1 Score
Low	0,757	85,7%	75%	0,799	0,759	0,556	0,698
Medium	0,701	88,9%	40%	0,552			
High	0,821	69%	80%	0,741			
Super-high	0,891	100%	40%	0,571			
All	0,759	76,6%	62,1%	-			

As a result of the scarcity of labeled examples in the low and super-high building classes, we observe a disorder - a very high model accuracy score on the test set, while the medium and high classes, consisting of more than 100 collected examples, get a lower score. In the situation of unbalanced datasets, in order to assess the goodness of the machine learning model, we should rely on the calculation of the Macro-average F1 Score.

To improve the performance of the machine learning model using the GCP - VERTEX AI API, it would be necessary to balance the number of labeled examples in each class collection. It is suggested to add more images visually similar to false negatives and false positives in each class to the collection. It is also necessary to re-examine whether the labels are correctly assigned to the images. As well, try to create new iterations of the machine learning model (Fig. 8).

Fig. 8. AutoML model (VERTEX AI) results for a class of low-rise residential buildings.

6 Summary

The presented assumptions of the application are a preliminary approach to the problem of supporting the process of finding housing (for rent or purchase) that meets the requirements of habitat.

The results of the experiment show the usefulness of the tools and models developed by Google Cloud Platform. There is certainly a need for a cheaper, more accessible application dedicated to European and Polish solutions. Unfortunately, an almost insurmountable problem for a modest researcher is the creation of (or access to) databases of buildings, locations, prices and, above all, the opinions of users and their neighbors.

Further work on building the application should focus on improving the classification of buildings and, above all, those parts of the building where the designated apartment is located, which is directly related to the price, square footage and standard of equipment, as well as the layout of apartments and the building. Another area of research is the introduction of a natural language comment and opinion module.

References

1. Bać, Zb.: Theory of Habitat: Contemporary Context, Oficyna Wydawnicza Politechniki Wrocławskiej, Wrocław, Poland (2019). ISBN 978-83-7493-114-4
2. Zamojski, T.: Residential skyscrapers in Europe in 2000–2019. Faculty of Architecture and Urban Planning, Wrocław University of Science and Technology. Doctoral dissertation (2020)

3. How Artificial Intelligence Is Impacting Multifamily Real Estate. https://www.forbes.com/sites/forbesbusinesscouncil/2021/03/25/how-artificial-intelligence-is-impacting-multifamily-real-estate/. Accessed 25 Jan 2024

4. Skyline AI Virtual Neighborhoods, New York, USA. https://www.skyline.ai/. Accessed 04 Mar 2024

5. Sequencing the DNA of Real Estate: An AI-Driven Approach for Comparing Assets. https://medium.com/skyline-ai/sequencing-the-dna-of-real-estate-an-ai-driven-approach-for-comparing-assets-a867dbfd6eb3. Accessed 04 Mar 2024

6. Gaining Insights about Real Estate Properties From Online Reviews Using NLP. https://medium.com/@_orcaman/gaining-insights-about-real-estate-properties-from-online-reviews-using-nlp-211c7754d1c. Accessed 04 Mar 2024

7. Zillow, Seattle, USA. https://www.zillow.com/. Accessed 04 Mar 2024

8. AI in Real Estate: 21 Companies Defining the Industry. https://builtin.com/artificial-intelligence/ai-real-estate. Accessed 04 Mar 2024

9. Trulia, San Francisco, USA. https://www.trulia.com/. Accessed 04 Mar 2024

10. Definition of artificial intelligence and how Trulia uses it. https://www.trulia.com/blog/tech/artificial-intelligence-defined/. Accessed 04 Mar 2024

11. Google Maps Platform, Web Services, Places API. https://developers.google.com/places/web-service/overview. Accessed 10 Mar 2024

12. Google Maps Platform. https://mapsplatform.google.com/. Accessed 10 Mar 2024

13. Google Maps Platform, Web, Maps Static API. https://developers.google.com/maps/documentation/maps-static. Accessed 10 Mar 2024

14. Google Cloud, VETREX AI, AutoML Beginner's Guide. https://cloud.google.com/vertex-ai/docs/beginner/beginners-guide. Accessed 10 Mar 2024

15. Google Cloud, Natural Language AI. https://cloud.google.com/natural-language?hl=pl. Accessed 10 Mar 2024

16. Google Cloud, VERTEX AI, Innovate faster with enterprise-ready artificial intelligence enhanced by Gemini models. https://cloud.google.com/vertex-ai?hl=en#innovate-faster-with-enterprise-ready-ai-enhanced-by-gemini-models. Accessed 10 Mar 2024

17. Zamojski, T.: European high-rise residential buildings - an attempt at typology. In: Bać, Zb. (eds.) Theory of Habitat: Contemporary Context, pp. 245–269. Oficyna Wydawnicza Politechniki Wrocławskiej, Wrocław, Poland (2019)

18. Zamojski, T.: European high rise residential buildings. Vertical habitat. Space Form (49), 113–132 (2022). https://doi.org/10.21005/pif.2022.49.C-03

19. Zamojski, T.: Humanization of European residential high-rise buildings by designing common zones. Space Form (54), 31–54 (2023). https://doi.org/10.21005/pif.2023.54.B-02

20. Council of Tall Buildings and Urban Habitat. https://www.ctbuh.org/. Accessed 04 Mar 2024

21. Google Cloud, VERTEX AI, Evaluate AutoML classification and regression models. https://cloud.google.com/vertex-ai/docs/tabular-data/classification-regression/evaluate-model. Accessed 10 Mar 2024

22. Precision, Recall, Accuracy, and F1 Score for Multi-Label Classification. https://medium.com/synthesio-engineering/precision-accuracy-and-f1-score-for-multi-label-classification-34ac6bdfb404. Accessed 14 Mar 2024

Author Index

W. Zamojski et al. (Eds.): DepCoS-RELCOMEX 2024, LNNS 1026, pp. 361–362, 2024.
https://doi.org/10.1007/978-3-031-61857-4

Printed in the United States
by Baker & Taylor Publisher Services